WARREN F. WALKER, Jr., Ph.D.

PROFESSOR OF BIOLOGY, OBERLIN COLLEGE

THIRD EDITION

VERTEBRATE

DISSECTION

W. B. SAUNDERS COMPANY
Philadelphia and London

Reprinted October, 1965

Vertebrate Dissection

PREFACE TO THIRD EDITION

The purpose of this book, as with previous editions, has been to fulfill the need for a manual on the systemic plan that covers the anatomy of a few animals with reasonable thoroughness, but is not encyclopedic in scope. Studying comparative anatomy in the laboratory by tracing the changes that occur in one organ system through a series of animals, and then returning to consider another system has certain advantages over the study of one animal completely and then another. The laboratory work correlates better with the text and lectures, which usually follow the systemic approach. But more than this, the student has a better opportunity to visualize the transformations that occurred in a given organ system during the evolution of animals, and this is one of the main objectives of comparative anatomy. One valid criticism of the systemic approach, however, is that the student does not gain as thorough a concept of any vertebrate as a whole as he would through the systematic approach, but it seems to me that this is outweighed by the advantages gained. The systemic approach poses certain difficulties in the storage of specimens, but I have not found these critical for a one semester course. Dogfish and mudpuppies can be tagged and stored in the containers in which they are shipped, and mammals keep well wrapped in a cloth soaked in embalming fluid and put in the plastic bags in which they are usually shipped. If for practical or other reasons, the instructor prefers to use the systematic approach, this manual can be adapted by selecting the appropriate sections from the table of contents.

The central theme of this manual is a study of the major anatomical transformations that have occurred in the vertebrates during their evolution from the fish to the mammalian stage, with the view of making the anatomy of the mammal meaningful.

There is much about mammalian anatomy, including that of man, that is unintelligible unless one knows something of the anatomy of lower forms. One cannot, of course, examine the actual evolutionary sequence to mammals in living vertebrates, but one can simulate this sequential study to some extent by examining certain of the living lower vertebrates. For most organ systems, the more important stages can be seen by dissecting examples from three levels of organization — the fish, the primitive tetrapod, and the mammal. The dogfish (*Squalus*) is used as an example of a primitive, jawed fish; the mudpuppy (*Necturus*), as an example of a primitive terrestrial vertebrate; and both the cat (*Felis*) and rabbit (*Lepus*), as examples of the mammal.

The directions for the mammal are written so that either a cat or a rabbit, or both, may be dissected. At Oberlin College, one half of the class dissects a cat, and one half a rabbit. Thus a student who is dissecting a cat has a neighbor dissecting a rabbit. At the end of an exercise, the students briefly compare dissections. But even if this comparing were not done, natural inquisitiveness would make it nearly impossible for the student not to imbibe a certain amount of information about the species his neighbor is dissecting. Although the dissection by the class of two mammals is not common practice, I believe that it has many advantages. It serves to emphasize the basic similarity of animals at one level of organization. The student sees that the cat and rabbit have much in common, and can better realize that even man would share many of these features. At the same time, the student has the opportunity to make comparisons between animals at the same level and to see the diverse features that are superimposed upon a common pattern. While most of the differences between the cat and rabbit are the result of adaptation to different modes of life, some doubtless result from chance divergence. A few of the differences also illustrate primitive and advanced stages in the evolution of a structure within the mammalian level of organization. The duplex uterus of the rabbit and the bipartite uterus of the cat are a case in point. Finally, the student becomes acquainted with two animals widely used in experimental work.

In addition to the jawed fish, primitive tetrapod, and mammal, directions are included for the study of representative lower chordates and an agnathous fish (the lamprey), for those who have time to supplement the major sequence by examining these more primitive types.

Flexibility has been an aim throughout, for the length and emphasis of comparative anatomy courses are subject to much variation. Enough material is included for an intensive course.

On the other hand, the directions are written so that much can be omitted. For example, the lower chordates and lamprey could be omitted, leaving the jawed fish, primitive tetrapod, and mammal. A still shorter course could also omit the primitive tetrapod. An alternate way of shortening the course would be the omission of some parts of certain organ systems — the muscles of the hind limb, certain of the sense organs, the lymphatic system, etc. Certain dissections could also be replaced by demonstrations. Regardless of what, if anything, is omitted, the sections that would be left are rather thorough, for in my opinion it is a better educational experience for the student to do a limited amount of material well than to cover a lot of ground superficially.

Some textual material has been incorporated with the laboratory directions. This is not intended to replace a text, but represents essentially summaries intended to make the laboratory work more meaningful, and statements pointing out the ways in which the species being studied resembles or departs from a generalized type of the group.

I have continued to take the opportunity in this edition of making numerous minor changes and corrections throughout the text. In particular I have added some short sections explaining the functional and structural interrelationships of various organs, e.g., the trunk skeleton of mammals, the jaws of cats and rabbits, myomeres of fish, respiratory organs of a fish, etc. This is an aspect of anatomy receiving increasing attention nowadays, and it is also one that can be introduced particularly well in the laboratory.

I have also given considerable attention to terminology. Unfortunately many names have often been used for the same structure. Latin codes of terminology have been agreed upon for human anatomy, the most recent one being the Nomina Anatomica Paresiensia (1955). This is a terminology with which prospective medical students and anatomists should begin to become aquainted, but it is not binding upon other mammals, and it is sometimes quite inappropriate for lower vertebrates. However, in describing mammalian anatomy in this edition I have made a more extensive use of the Nomina Anatomica. When a single term is used for a mammalian structure, it is usually the Nomina Anatomica term or its anglicized version. But sometimes the term has been altered in such a way as to make it more appropriate for a quadruped; thus superior vena cava becomes anterior vena cava. When two terms are mentioned, I have cited the Nomina Anatomica term first if it is the one that is preferred, e.g., **tympanic cavity**, or **middle ear cavity**. If the Nomina Anatomica term is less commonly used, I have cited it in parentheses after the more familiar one, e.g., **dermis (corium)**. Mammalian terms have been applied to lower vertebrate structures where deemed appropriate.

Many new figures have been added to this edition so that the manual can also serve as a partial atlas of the anatomy of the animals considered.

As the emphasis in biological teaching shifts more and more toward the experimental side of the field, less and less time seems to be available for certain of the essential classic courses, including comparative anatomy. Figures can save a considerable amount of the student's time, both in finding the structures and in serving as a record of his observations. But the student should be cautioned that figures should not serve as a substitute for careful dissection and observation. Most instructors give several examinations on the specimens, which helps to insure that careful work is done. Many of the figures show not only the organs being considered but the relationship of these organs to surrounding parts. This should facilitate finding and remembering the structures concerning.

Fifteen figures in the previous edition have been replaced by ones more appropriate for a laboratory manual, and 31 additional figures have been added, bringing the new total to 136.

I am indebted to many for help in preparing the third edition of this book. Students and colleagues have sent comments and useful suggestions to me or to the publishers. I wish to thank all of them, known and unknown to me, and I earnestly hope that they will continue to call my attention to any errors or deficiencies that they may find in this edition. Their past support and encouragement have greatly increased the merit of this manual. Most of the new drawings have been prepared by my student. Mr. H. Jon Janosik, and I am much indebted to his skill in portraying my dissections upon paper. I am also greatly indebted to Mr. William A. Osburn, Medical Art Director of the W. B. Saunders Co., for preparing many other drawings and for the work entailed in adding color to most of the drawings of blood vessels. My wife has again been most helpful in lending encouragement and in proofreading the materials in their various stages of writing. Again I should like to pay tribute to my former teacher and mentor, Professor Alfred S. Romer who has indirectly influenced this book in many ways. I also wish to thank the staff of the W. B. Saunders Company who have been most helpful and encouraging in the preparation of all editions of this book.

Oberlin, Ohio   WARREN F. WALKER, JR.

CONTENTS

Chapter **1**

The Lower Chordates 1

Subphylum Hemichordata.. 1
Subphylum Urochordata 3
    External Features 3
    Dissection... 4
Subphylum Cephalochordata 6
    External Features 7
    Whole Mount Slide 8
    Cross Sections 11
        (A) Common Features in the Sections..................... 11
        (B) Section Through Oral Hood 12
        (C) Section Through Pharynx............................. 12
        (D) Section Through Intestine 13
        (E) Section Through Anus................................ 13

Chapter **2**

The Lamprey — A Primitive Vertebrate 14

External Features... 14
    Sagittal and Cross Sections....................................... 17
        The Skeletal and Muscular Systems 18
        The Nervous System and Sense Organs 18
        The Coelom and the Digestive and Respiratory Systems 20
        The Circulatory System 21
        The Excretory and Genital Systems 22
The Ammocoetes Larva 22

Chapter **3**

The Evolution and External Anatomy of Vertebrates 26

Vertebrate Evolution.. 26
External Anatomy 30

Fishes 30
    (A) General External Features· 31
    (B) Integumentary Derivatives 33
Primitive Tetrapods 34
    (A) General External Features 34
    (B) Integumentary Derivatives 36
Mammals 36
    (A) General External Features 36
    (B) Integumentary Derivatives 38

Chapter **4**

The Axial and Visceral Skeleton 40

Fishes 41
  Postcranial Axial Skeleton 41
    (A) Relationships of the Vertebral Column 41
    (B) Vertebral Regions and Caudal Vertebrae 42
    (C) Trunk Vertebrae 43
    (D) Ribs 44
    (E) Median Fin Supports 44
  Head Skeleton 45
    (A) Composition and Structure of the Chondrocranium 45
    (B) Sagittal Section of the Chondrocranium 47
    (C) Visceral Skeleton 48
    (D) Dermal Bones 50
Primitive Tetrapods 51
  Postcranial Axial Skeleton 51
    (A) Vertebral Regions 52
    (B) Trunk Vertebrae 53
    (C) Ribs 53
    (D) Sternum 53
  Head Skeleton of Primitive Tetrapods and Necturus 56
    (A) Entire Skull 58
    (B) Chondrocranium 60
    (C) Lower Jaw 60
    (D) Teeth 61
    (E) Hyoid Apparatus 61
  Head Skeleton of the Reptile 62
    (A) General Features of the Skull 62
    (B) Composition of the Skull 63
    (C) Lower Jaw 64
    (D) Hyoid Apparatus 66
Mammals 66
  Postcranial Axial Skeleton 66
    (A) Vertebral Groups 66
    (B) Thoracic Vertebrae 67
    (C) Lumbar Vertebrae 68
    (D) Sacral Vertebrae 69
    (E) Caudal Vertebrae 69
    (F) Cervical Vertebrae 69
    (G) Ribs 70
    (H) Sternum 70
  Head Skeleton 72
    (A) General Features of the Skull 74
    (B) Composition of the Skull 76

(C) Sagittal Section of the Skull 79
(D) Foramina of the Skull 80
(E) Lower Jaw 82
(F) Teeth 82
(G) Hyoid Apparatus 83

Chapter **5**

The Appendicular Skeleton 84

Fishes 84
   Pectoral Girdle and Fin 85
     (A) Squalus 85
     (B) Amia 87
   Pelvic Girdle and Fin 87
Primitive Tetrapods 88
   Appendicular Skeleton of Necturus 90
     (A) Pectoral Girdle and Appendage 90
     (B) Pelvic Girdle and Appendage 90
   Appendicular Skeleton of the Turtle 91
     (A) Pectoral Girdle and Appendage 91
     (B) Pelvic Girdle and Appendage 94
Mammals 96
   Pectoral Girdle and Appendage 98
   Pelvic Girdle and Appendage 100

Chapter **6**

The Muscular System 103

Fishes 113
   Axial Muscles 113
     (A) Typical Myomeres of the Trunk and Tail 113
     (B) Epibranchial Muscles 115
     (C) Hypobranchial Muscles 115
     (D) Extrinsic Muscles of the Eye 116
   Appendicular Muscles 116
     (A) Muscles of the Pectoral Fin 116
     (B) Muscles of the Pelvic Fin 117
   Branchiomeric Muscles 117
Primitive Tetrapods 119
   Axial Muscles 121
     (A) Muscles of the Trunk 121
     (B) Hypobranchial Muscles 122
   Appendicular Muscles 123
     (A) Pectoral Muscles 123
     (B) Pelvic Muscles 125
   Branchiomeric Muscles 127
Mammals 128
   Skinning, and Integumentary Muscles 129
   Posterior Trunk Muscles 130
     (A) Hypaxial Muscles 130
     (B) Epaxial Muscles 132
   Pectoral Muscles 132
     (A) Pectoralis Group 132

(B) Trapezius and Sternocleidomastoid Group...............135
(C) Remaining Superficial Muscles of the Shoulder.........137
(D) Deeper Muscles of the Shoulder.......................138
(E) Muscles of the Brachium 140
(F) Muscles of the Forearm 141

Pelvic Muscles of the Cat................................141
(A) Lateral Thigh and Adjacent Muscles..................142
(B) Superficial Muscles on the Lateral
        Surface of the Pelvic Girdle.....................143
(C) Quadriceps Femoris Complex 144
(D) Posteromedial Thigh Muscles 144
(E) Deeper Muscles of the Pelvic Girdle 144
(F) Iliopsoas Complex and Adjacent Muscles..............146
(G) Muscles of the Shank 146

Pelvic Muscles of the Rabbit............................146
(A) Lateral Thigh Muscles 147
(B) Superficial Muscles on the Lateral Surface
        of the Pelvic Girdle.............................148
(C) Quadriceps Femoris Complex 148
(D) Posteromedial Thigh Muscles 149
(E) Deeper Muscles of the Pelvic Girdle 150
(F) Iliopsoas Complex and Adjacent Muscles
        (Cat and Rabbit) 151
(G) Muscles of the Shank (Cat and Rabbit) 152

Anterior Trunk Muscles..................................152
(A) Hypaxial Muscles....................................152
(B) Epaxial Muscles 155

Hypobranchial Muscles...................................156
(A) Posthyoid Muscles 156
(B) Prehyoid Muscles 157

Branchiomeric Muscles...................................158
(A) Mandibular Muscles 158
(B) Hyoid Muscles.......................................160
(C) Posterior Branchiomeric Muscles.....................160

Chapter **7**

The Sense Organs 161

The Lateral Line System...................................162
  Pit Organs..162
  Ampullae of Lorenzini 162
  Lateral Line Canals...................................163
The Eye and Associated Structures.........................164
  Fishes 165
  Mammals...168
The Nose..171
  Fishes 171
  Primitive Tetrapods...................................172
  Mammals...172
The Ear...174
  Fishes 175
  Primitive Tetrapods...................................177
  Mammals...178

Chapter **8**

The Nervous System 180

Fishes...184
   Dorsal Surface of the Brain....................................184
   Cranial and Occipital Nerves..................................187
      (A) Nervus Terminalis......................................187
      (B) Olfactory Nerve 188
      (C) Optic Nerve 188
      (D) Oculomotor Nerve 188
      (E) Trochlear Nerve 190
      (F) Abducens Nerve 190
      (G) Trigeminal Nerve 190
      (H) Facial Nerve 191
      (I) Statoacoustic Nerve 192
      (J) Glossopharyngeal Nerve 192
      (K) Vagus Nerve 192
      (L) Occipital Nerves 193
   Ventral Surface of the Brain...................................200
   Ventricles of the Brain..201
   The Spinal Cord and Spinal Nerves.............................203
Primitive Tetrapods 203
   Dorsal Surface of the Brain....................................203
   Cranial and Occipital Nerves..................................205
   Ventral Surface of the Brain...................................206
Mammals..206
   Meninges 207
   External Features of the Brain and the Stumps
     of the Cranial Nerves 209
      (A) Telencephalon 209
      (B) Diencephalon 210
      (C) Mesencephalon 212
      (D) Metencephalon 212
      (E) Myelencephalon 214
   Sagittal Section of the Brain 215
   Dissection of the Cerebrum 217
   Spinal Cord and Spinal Nerves 219
      (A) The Cord and Roots of the Spinal Nerves 219
      (B) The Spinal Nerves and Brachial Plexus................221

Chapter **9**

The Coelom and the Digestive and Respiratory Systems 224

Fishes...229
   Pleuroperitoneal Cavity and Its Contents 229
      (A) Body Wall and Pleuroperitoneal Cavity.................229
      (B) Visceral Organs 230
      (C) Mesenteries 232
      (D) Further Structure of the Digestive Organs.............232
   Pericardial Cavity...235
   Oral Cavity, Pharynx, and Respiratory Organs 235
Primitive Tetrapods 239
   Pleuroperitoneal Cavity and Its Contents 240
      (A) Body Wall and Pleuroperitoneal Cavity 240
      (B) Visceral Organs 240

(C) Mesenteries...242
(D) Further Structure of the Digestive Organs.............242
Pericardial Cavity..243
Oral Cavity, Pharynx, and Respiratory Organs 244
Mammals..246
Digestive and Respiratory Organs of the Head and Neck 247
(A) Salivary Glands...247
(B) Oral Cavity 249
(C) Pharynx..250
(D) Larynx, Trachea, and Esophagus 250
Thorax and Its Contents..252
(A) Pleural Cavities 252
(B) Pericardial Cavity 254
Peritoneal Cavity and Its Contents..............................254
(A) Body Wall and Peritoneal Cavity.......................254
(B) Abdominal Viscera and Mesenteries....................254
(C) Further Structure of the Digestive Organs.............258

Chapter **10**

The Circulatory System 262

Fishes 265
External Structure of the Heart..................................266
Venous System..266
(A) Hepatic Portal System and Hepatic Veins 266
(B) Renal Portal System 267
(C) Systemic Veins..268
Arterial System..270
(A) Branchial Arteries 270
(B) Dorsal Aorta and Its Branches 276
Internal Structure of the Heart 278
Pericardioperitoneal Canal 279
Primitive Tetrapods...280
Heart and Associated Vessels.....................................281
Venous System..282
(A) Hepatic Portal System......................................282
(B) Ventral Abdominal Vein 282
(C) Renal Portal System 283
(D) Posterior Systemic Veins 284
(E) Anterior Systemic Veins 284
Arterial System..285
(A) Aortic Arches and Their Branches.......................285
(B) Dorsal Aorta and Its Branches 289
Mammals 290
Heart and Associated Vessels.....................................291
Venous System..292
(A) Hepatic Portal System......................................292
(B) Posterior Systemic Veins 292
(C) Anterior Systemic Veins 295
Arterial System..302
(A) Aortic Arches and Their Branches 302
(B) Dorsal Aorta and Its Branches 306
Bronchi and Internal Structure of the Heart....................308
Lymphatic System...310

Chapter **11**

The Excretory and Reproductive Systems......312

Fishes.......315
    Kidneys and Their Ducts 316
    Male Urogenital System 316
    Female Urogenital System......319
    Reproduction and Embryos......320
Primitive Tetrapods 322
    Kidneys and Their Ducts......322
    Male Urogenital System 323
    Female Urogenital System......324
Mammals......325
    Excretory System......326
    Male Reproductive System......328
    Female Reproductive System......331
    Reproduction and Embryos......336

Appendix **I**

Terms for Directions, Planes, and Sections 339

Appendix **II**

The Preparation of Specimens 341

Appendix **III**

References 345

General......345
Fishes 346
Amphibians and Reptiles 347
Mammals 348

INDEX......351

1 / The Lower Chordates

THE PHYLUM Chordata is a large and diversified group containing about 35,000 living species, ranging from small wormlike forms to man. But all tend to have three characters in common at some stage of their life cycle: a **notochord**, a **single**, **dorsal**, **tubular nerve cord**, and **pharyngeal gill slits** or **pouches**. Variations in these diagnostic characteristics are the chief criteria for dividing the phylum into its four subphyla: **Hemichordata**, **Urochordata**, **Cephalochordata**, and **Vertebrata** or **Craniata**.

The relationship of the first three subphyla (the lower chordates or protochordates) to the vertebrates is uncertain. Most authorities believe that they occupy a position somewhere between the vertebrates and some invertebrate group, a possibly early echinoderms, or some unknown type ancestral to both echinoderms and chordates. This is not to say, however, that living lower chordates are themselves the direct ancestors of vertebrates.

Courses in comparative anatomy are concerned primarily with the vertebrates, but it is desirable for the student to have some idea of the nature of the lower chordates. In particular, the student should become acquainted with their general external features, and understand how the three fundamental chordate characters are represented. The cephalochordates, which are closer in structure to the vertebrates, are often studied in more detail. If it is desirable to survey the lower chordates quickly, much of what follows could be studied on demonstration preparations.

SUBPHYLUM HEMICHORDATA

Present-day hemichordates include a few rare, colonial, deep-sea forms, the pterobranchs, which resemble the graptolites of ancient Cambrian seas, and wormlike types known as enteropneusts, or acorn worms, The latter are found burrowing in the sand and mud of tidal flats and shallow coastal waters, where they are sometimes abundant. *Balanoglossus* and *Saccoglossus* (*Dolichoglossus*) are common genera along North American coasts.

Examine the external features of one of the enteropneusts (Fig. 1-1). You may have to place the specimen in a pan of water and use a hand lens to see certain structures. The body is divided into three distinct

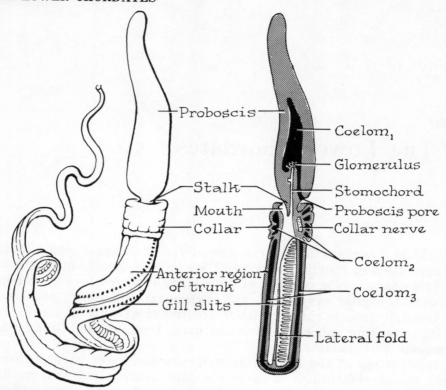

Figure 1–1. Left, external view of *Saccoglossus;* right, sagittal section of the anterior end. (From Villee, Walker, and Smith, General Zoology; external view after Bateson.)

regions: an anterior **proboscis,** a **collar,** and a long **trunk.** The proboscis attaches by a narrow stalk to the encircling collar located just posterior to it. The proboscis nesting in the collar often gives the appearance of an acorn in its cup — a fact that gives the common name to the group. The proboscis and collar assist the ciliated epidermis in burrowing. The collar coelom fills, thus inflating the collar and anchoring the worm in its burrow. Then the deflated proboscis is pushed forward by the action of muscles within it; its coelom is filled and it anchors the worm. Finally the collar coelom is emptied, and the worm is pulled forward to the proboscis. The **mouth** is situated inside the front of the collar, ventral to the proboscis stalk. As the worm moves forward, sand and organic debris enter. The **anus** is at the very posterior end of the trunk.

Numerous, small, paired, **external gill slits** can be seen dorsolaterally on the anterior portion of the trunk. As many as 150 pairs have been counted on a 16-inch specimen. Each external gill slit leads to a gill pouch which connects with the pharynx by way of an unusual U-shaped internal gill slit. It is of interest that at one stage of development the gill slits of *Amphioxus* (a cephalochordate) have an identical appearance. The gill slits enter the dorsal half of the pharynx, which is separated by a longitudinal fold from the ventral part of the pharynx through which sand and food pass (Fig. 1–1).

In mature individuals, prominent **genital ridges** will be found just ventral to the posterior external gill slits, and extending a short distance caudad. The sexes are separate, but cannot be distinguished externally. In some, but not all species, conspicuous **hepatic ridges** will be seen posterior to the genital ridges. When present, they are the outward manifestation of hepatic cecae that bud off the simple, straight intestine.

A longitudinal middorsal, and a similar midventral, ridge can also be seen on the trunk. Each contains a superficial and solid **nerve strand** that extends forward into the collar. Within the collar, the dorsal strand rolls up upon itself to form a hollow neural tube. Neurons from an echinoderm-like subepidermal nerve plexus connect with the nerve strands.

As to the diagnostic chordate characteristics, pharyngeal gill slits are well represented. Nerve strands are present, but they are not exactly in the form of a single, dorsal, tubular nerve cord. A diverticulum from the anterior end of the mouth cavity extends into and helps to stiffen the proboscis (Fig. 1-1). It consists partly of a chitinous plate and partly of vacuolated cells resembling those of a notochord. Some investigators consider it to represent a rudimentary notochord (the hemichordates get their name from this "half notochord"), but others question this interpretation and prefer to avoid any implications of homology by calling it a **stomochord**. Since all of the chordate characteristics are not well represented at any stage of development, some authors prefer to consider the Hemichordata as a distinct phylum related to, but not a part of, the phylum Chordata.

SUBPHYLUM UROCHORDATA

Members of the subphylum Urochordata are odd marine animals most of which are encased in a leathery membrane, called the tunic; hence the animals are commonly referred to as tunicates. Although some are pelagic, the most familiar are the sessile sea squirts belonging to the class **Ascidiacea**. They are found attached to submerged objects in coastal waters, or occasionally partly buried in the sand or mud. Many sea squirts are colonial but some are solitary. Their anatomy can be studied conveniently on one of the latter types, such as **Molgula**, which is one of the most abundant along the Atlantic coast.

External Features

Examine a specimen in a pan of water and notice its saclike appearance. Two spoutlike openings, or siphons, will be seen near the top of the animal (Fig. 1-2). When a living specimen is touched, water is expelled with considerable force through both openings. The name sea squirt is derived from this habit. Normally, however, water enters the organism through the topmost aperture (**incurrent siphon**) and is discharged through the opening that is set off on one edge (**excurrent siphon**). The margin of the incurrent siphon bears six small tentacles. Also notice the external covering or **tunic**. It contains a large amount of tunicin, a com-

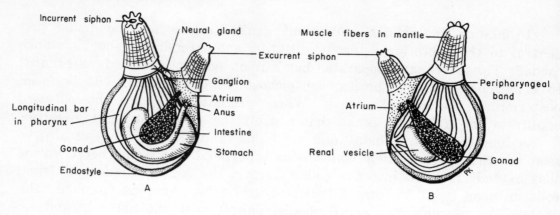

Molgula

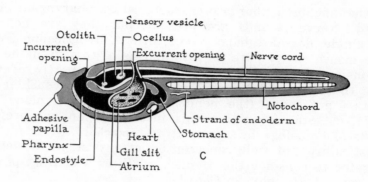

Figure 1-2. *A* and *B*, left and right side views, respectively, of a dissection of *Molgula*. All of the tunic and most of the mantle have been removed. *C,* diagrammatic lateral view of a larval ascidian. (*C* from Villee, Walker, and Smith, General Zoology.)

plex polysaccharide similar to cellulose found in plant cell walls. The tunic is secreted by cells derived from the underlying bodywall. Minute, hairlike processes extend out from the tunic and help to anchor the animal in or on its substrate. The lower part of the tunic may have sand grains adhering to the hairs, for *Molgula* often lies partly buried in the sand.

Although the shape of the animal appears rather asymmetrical, the sea squirt really has a modified bilateral symmetry. If one compares a larva with an adult (Fig. 1-2), one will notice that the region between the two shiphons represents the dorsal surface, and the rest of the edge of the adult, the ventral surface. The incurrent siphon lies anteriorly and the excurrent siphon posteriorly.

Dissection

To get at the inside of the animal, cut through the tunic beneath the excurrent siphon and extend the cut around the edge and base of the sac to a point near the incurrent siphon. Reflect the tunic, observing that it is attached to the rest of the body only at the siphons. Detach the tunic at these points. A number of bundles of longitudinal and circular muscle fibers lie within the thin body wall, or **mantle**, and aid in expelling water. The mantle is nearly transparent, and many of the internal organs can be seen beneath it. To see them more clearly, carefully peel off the mantle without unduly injuring organs that may adhere to it. This part of the dissection should be done beneath water.

Study the dissection and compare it with Figure 1-2. You may need a hand lens to see certain structures. The incurrent siphon leads into a large, thin-walled, vascular **pharynx**, which occupies most of the inside of the body. In many sea squirts, the wall of the pharynx is divided into a number of rectangular areas by longitudinal and transverse bars, and within each area the wall is perforated by rows of gill slits. In *Molgula*, the **longitudinal bars** are grouped into conspicuous folds, transverse bars are not apparent, and the **gill slits** are microscopic. To see them clearly, it is necessary to remove a piece of the pharynx wall, and prepare a wet mount of it. The gill slits will appear as arclike slits arranged in spirals. Although it will not be seen in this type of preparation, the bars between the slits contain blood vessels in which gas exchange with the environment occurs, and are covered with cilia which create the current of water that passes through the animal. The gill slits do not lead directly to the outside but into a delicate chamber, the **atrium**, located on each side of the pharynx. The lateral portions of the atrium may not be seen, but they converge posteriorly to form a more conspicuous median atrial chamber which opens to the surface through the excurrent siphon.

The fold along the ventral suface of the pharynx is the **endostyle**, or **hypopharyngeal groove**. It is generally more conspicuous, and more deeply grooved on the inside, than other longitudinal folds. Certain cells of the endostyle are ciliated; some are glandular and secrete a mucus which entraps minute food particles in the water; and some produce iodinated proteins as do the cells of the vertebrate thyroid gland. The food-containing mucous band is moved toward the incurrent siphon by the cilia. Near the anterior end of the pharynx, it is carried to the dorsal side by lateral **peripharyngeal bands**. These may be hard to see. Then the mucous string moves posteriorly to the esophagus along a middorsal fold called the **dorsal lamina**. In some tunicates, other lateral folds connect the endostyle with the dorsal lamina. Thus the pharynx of the sea squirt is primarily a food gathering device, although some gas exchange between the blood and environment does occur in its walls.

The rest of the alimentary canal of *Molgula* lies on the left side of the pharynx. A short **esophagus** leads from the posterior end of the pharynx to a slightly expanded and elongate **stomach**, and this is followed by an **intestine**. The esophagus, stomach, and first part of the intestine form a C-shaped loop. Then the intestine doubles on itself, goes back beside the stomach and esophagus, and opens at the anus into the median portion of the atrial chamber. The stomach differs from that of vertebrates because it, and minute glandular folds evaginated from it, secrete enzymes that act upon carbohydrates and fats as well as upon proteins. The intestine appears to be primarily absorptive.

In mature individuals, large **gonads** will be seen. In *Molgula* there is one on each side of the pharynx. Inconspicuous genital ducts lead from them to the median portion of the atrium. Tunicates are hermaphroditic, but generally not self-fertilizing. However, *Molgula* can fertilize itself.

An oval, **renal vesicle** lies ventral to the right gonad. Many waste products, including concretions of uric acid, accumulate in the vesicle and stay there until the death of the animal, for there are no excretory ducts.

A small oval-shaped structure will be seen on the dorsal edge of the pharynx between the two siphons. This is the **neural gland**, which has been homologized with the vertebrate hypophysis. It is connected with the pharynx by a minute duct visible only in special preparations. Extracts of the gland injected into vertebrates have many of the effects of pituitary extracts, and there is some evidence that it has gonadotropic effects in tunicates. It may also act as a chemoreceptor, testing the water current entering the pharynx. Carefully pull off the gland, and you will see beneath it an elongated nerve **ganglion**. Other internal organs are not usually seen in this type of dissection.

Aside from the abundant pharyngeal gill slits, there is little about an adult sea squirt that would suggest a chordate. However, the other diagnostic features of the phylum are represented in the free-swimming, tadpole-shaped, larval stage that most sea squirts pass through (Fig. 1–2, C). The tail of the larva is supported by a notochord and above this is a single, dorsal, tubular nerve cord that extends forward and expands into a sensory vesicle containing organs of equilibrium and light sense. The name of the subphylum is derived from the position of the larval notochord. At metamorphosis, the larva attaches to the substratum by its anterior end, and its tail atrophies. The notochord and nerve cord are lost, and the sensory vesicle is reduced to the nerve ganglion.

SUBPHYLUM CEPHALOCHORDATA

The subphylum Cephalochordata includes the lancelet or *Amphioxus* (*Branchiostoma*[1]) and several very similar types. All are superficially fishlike animals that have an extremely long notochord extending beyond the nerve cord to the very front of the animal. This extreme extension of the notochord is probably correlated with the burrowing habits of the animal. It also gives the name Cephalochordata to the subphylum. These animals are found in coastal waters, usually lying partly buried in the sand with only their front end protruding, for like the sea squirts, *Amphioxus* is a filter-feeder. At times they actively

---

[1] Although *Branchiostoma* (Costa 1834) has priority over *Amphioxus* (Yarrel 1836) as the technical name, the term *Amphioxus* is so familiar that it is customary to retain it, at least as a common name.

swim to new feeding sites, but their locomotion is not very efficient because stabilizing fins are poorly developed. In the United States, they are found south from Chesapeake and Monterey bays. Since many of the features of *Amphioxus* are believed to be very primitive, and hence may throw light on the origin of vertebrate structure, it will be considered in more detail than other lower chordates.

External Features

Examine a preserved specimen of *Amphioxus* in a pan of water. You will need a hand lens to see certain structures. Its shape is highly streamlined, or fusiform, being elongate, flattened from side to side (compressed), and pointed at each end (Fig. 1-3). Segmental, V-shaped muscle blocks, the **myomeres**[2], can be seen through the transparent epidermis. The apex of each V points anteriorly. Note the **myomeres** extend nearly the length of the body, those at either end of the animal being slightly smaller. Their number is a specific character ranging, in American species, from about 55 to 75. The lines of separation between the myomeres are connective tissue partitions called **myosepta** or **myocommata**.

A **dorsal fin** extends along the top of the body, and a **ventral fin** will be seen beneath the posterior quarter of the animal. The dorsal and ventral fins are continuous around the tail and expand slightly in this region to form a **caudal fin**. A pair of ventrolateral fins, or **metapleural folds**, continue foward from the anterior end of the ventral fins (Fig. 1-4).

The metapleural folds end a short distance from the front of the body. In front of them, and ventral to the anterior few myomeres, you will see a transparent chamber called the **oral hood**. The mouth is located

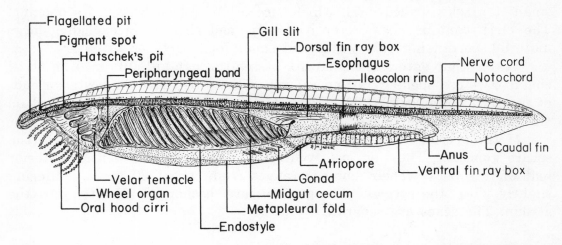

Figure 1-3. A lateral view of a whole mount slide of a young specimen of *Amphioxus*.

[2] **Myotome** and **myomere** are terms for the primitive muscle segments. They are often used synonymously, but I follow the usage of **myotome** for an embryonic muscle segment and **myomere** for an adult segment.

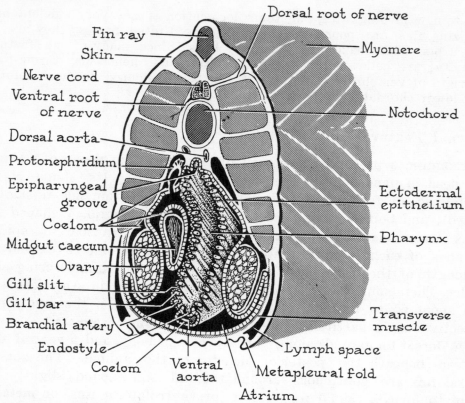

Figure 1–4. A diagrammatic cross section through the posterior part of the pharynx of *Amphioxus*. (From Villee, Walker, and Smith, General Zoology.)

deep within this chamber and will not be seen at this time, but the opening of the oral hood on the ventral surface can be seen. It is fringed with small tentacles called **cirri**, which are often folded across its opening. The cirri contain chemoreceptive cells, and also aid in excluding large material, permitting only water and small food particles to enter.

Water that enters the pharynx passes through gill slits into an atrial chamber whose opening (**atriopore**) you will see between the posterior ends of the two metapleural folds. The intestine opens by an **anus** located on the left side of the caudal fin.

If the specimen is mature, you will see on each side a row of whitish, square **gonads**. They lie just ventral to the myomeres in the anterior half of the body. Their number ranges from about 20 to 35, differing slightly with the species. The gametes are discharged directly into the atrium. The sexes are separate.

Whole Mount Slide

Study a stained microscope slide of a small specimen of *Amphioxus* under the low power of the microscope. Note its fusiform shape, and find the structures described earlier: myomeres; dorsal, ventral, and caudal fins; metapleural folds; oral hood and cirri; atriopore; anus; gonads.

The **myomeres** have been cleared to render them somewhat transparent, so they will not be seen as plainly as in the preserved specimen. But you should see indications of them, at least just ventral to the dorsal fin.

Observe that the **dorsal** and **ventral fins** are supported by small, transparent blocks, called **fin ray boxes**. In order to see the **metapleural folds**, you will have to focus sharply on the surface. Each will appear as a horizontal line, parallel and slightly dorsal to the ventral edge of the body.

Since small specimens from which slides are made are generally not sexually mature, the **gonads** will not be fully developed, and may be absent. If present, they will appear as a row of lightly staining, oval structures close to the ventral edge of the body. The largest ones are just anterior to the atriopore.

Notice the **notochord** located in the back dorsal to the dark-staining alimentary canal. It extends nearly the entire length of the animal in the general position of the vertebral column of higher chordates, and it has a comparable function—namely, to provide support and prevent the body from shortening when the myomeres contract.

The single, dorsal, tubular **nerve cord** will appear as a dark-staining band lying dorsal to the notochord. Its position may be recognized by the dark pigment granules along its ventral edge. Each granule represents parts of a simple photoreceptor. Notice that they are particularly numerous near the front of the animal. Why? The nerve cord ends in a blunt point anteriorly. There is no expanded brain. A prominent **pigment spot** of unknown function will be seen in front of the nerve cord. Sharp focusing will also reveal a clear, saclike structure just dorsal to the front of the nerve cord. It is called the **flagellated pit**, and is believed to be a chemoreceptor. It occurs only on the left side of the snout. Embryologically, it connects with the nerve cord.

Examine the region of the **oral hood** in detail. The **cirri** have little processes along their edges, and each cirrus is supported by a skeletal rod. All the rods connect with a common basal rod. Ciliated grooves, or bands, are located on the inside of the lateral walls of the oral hood. In a lateral view, they appear as large, dark-staining, fingerlike lobes extending forward from a common basal band. This complex is called the **wheel organ**, and it functions to draw a current of water into the organism. The dorsalmost lobe, which is called **Hatschek's groove**, is longer than the others. Slightly anterior to the middle of Hatschek's groove, you will see a region where the groove is deeper and forms a pit that extends dorsally to overlap the right side of the notochord. This is **Hatschek's pit**. Besides aiding in the ciliary current, Hatschek's groove and pit secrete mucus that helps to entrap minute food particles in the water.

Posterior to the wheel organ you will notice a dark-staining line that is approximately in the transverse plane. This is the **velum** — a transverse partition that forms the posterior wall of the oral hood. The mouth, which cannot be seen in this view, is located in its center, and is fringed with **tentacles**. The tentacles can be seen extending either anteriorly or posteriorly from the velum. They, too, act as strainers and probably contain chemoreceptive cells.

The mouth leads into a large **pharynx**, most of whose lateral walls are perforated by numerous elongate **gill slits** with ciliated **gill bars** between them. In favorable specimens, supporting rods may be seen within the gill bars. In mature specimens there are over 200 gill bars. These provide a very large ciliated surface that plays the major role in moving water and food particles through the pharynx. Food is entrapped in mucus as will be described, and the water escapes through the gill slits. The pharynx is primarily a food concentrating mechanism, and its role in respiration evolved later. Although blood vessels pass through the gill bars, no gills are present. Because of the great activity of the ciliated cells in this region, it is even possible that the blood leaving the bars contains less oxygen than that entering them. The major site of blood aeration appears to be the general body surface. As in the sea squirts, water does not pass directly to the outside, but through the gill slits into an **atrium**. The only part of the atrium to be seen in this view is the clear space ventral to the pharynx and continuing beneath the gut to the **atriopore**. The atrium is formed by the down growth of folds of the body wall around the gill slits, and it serves to protect the delicate gill bars.

The longitudinal band that extends along the entire floor of the pharynx is the **endostyle** or **hypopharyngeal groove**, which was referred to in the section on urochordates. Its function is the same, for it secretes mucus in which minute food particles become entrapped, and its cilia carry the mucus anteriorly. The string of mucus passes to the dorsal side of the pharynx along the gill bars and peripharyngeal bands. A **peripharyngeal band** is located on each side of the front of the pharynx, and appears as a dark-staining line extending from the ventral edge of the velum diagonally dorsad and caudad just above the anterior gill slits. The mucus sheet is carried posteriorly along a middorsal **epipharyngeal groove**. Experiments have shown that certain endostylar cells also concentrate radioactive iodine, and Barrington has proposed that these cells are homologous to those of the vertebrate thyroid gland. In this case the iodinated proteins are discharged into the gut and absorbed further caudad.

The posterior end of the pharynx floor extends diagonally dorsad, and just behind the last gill slit the pharynx leads into a short, narrow **esophagus**. The outlines of the esophagus are often obscured by a large midgut cecum, but can be seen if you look carefully. The top of the esophagus lies just ventral to the notochord, and its floor will appear as a longitudinal line extending posteriorly a short distance from the bottom of the last gill slit.

The diameter of the alimentary tract increases two- or three-fold just posterior to the esophagus, for the large **midgut cecum** has evaginated at this point. The midgut cecum extends toward the front of the animal, lying along the ventral side of the esophagus and right side of the pharynx. It is located within the atrium. The midgut cecum secretes digestive enzymes, and in this respect it differs from the vertebrate liver.

Posterior to the cecum, the **midgut** narrows and is followed by a very deeply staining segment of the alimentary canal called the **ileocolic ring**. At this point, cilia impart a rotary action to the cord of mucus and food in the entire midgut. This presumably aids in the discharge of enzymes by the midgut cecum and in mixing them with the food mass. A still narrow **hindgut**, or **intestine**, follows the ileocolic ring and opens on the body surface at the **anus**. Remains of microorganisms, including the shells of diatoms, are often seen in the intestine.

Other internal structures are not seen in this type of preparation.

Cross Sections

(A) *Common Features in the Sections*

The anatomy of *Amphioxus* will be understood better if slides of representative cross sections can be examined. While studying such sections, compare them with the generalized diagram shown in Figure 1–4, and correlate the appearance of organs in this view with their appearance in the whole mount.

Many things will look much the same in any section. The **dorsal fin** will be recognized along with the hollow-appearing **fin ray box**. Actually the fin ray box contains a gelatinous connective tissue. The skin consists of an **epidermis** of simple columnar epithelium supported by a thin layer of connective tissue (the **dermis**). **Myomeres** will appear as several oval chunks of tissue beneath the skin. They are separated from each other by the **myosepta**.

The **nerve cord** is the large tubular structure slightly ventral to the dorsal fin. Its cavity, the **neurocoel** is very narrow. In favorable sections, you will see lateral nerves arising from the cord. As in vertebrates, there are **dorsal** and **ventral roots** to the nerves; however, they do not unite in *Amphioxus*, but run directly to the tissues. The ventral roots carry motor fibers to the myomeres; the dorsal roots, sensory fibers to the integument and motor fibers to the ventral, nonmyotomal muscles.

You will see the **notochord** just beneath the nerve cord. It consists of vacuolated cells that are distended with fluid and held tightly together by a firm connective tissue sheath. You should be able to see the sheath, but the details of the cells will not be apparent.

(B) *Section Through Oral Hood*

In a section taken near the front of the animal, you will see an open space beneath the notochord and myomeres. This is the **oral hood**. The ciliated grooves of the wheel organ will appear as thicker patches of epithelium on the inside of the wall of the hood. Hatschek's groove is the most dorsal of these, and is located a bit to the right of the median plane. The oral hood opens ventrally, but some sections may be taken just posterior to the opening. Pieces of **cirri** probably will be seen in the section.

(C) *Section Through Pharynx*

In a section through the pharynx, the **metapleural folds** will be seen projecting from the ventrolateral portion of the body. There is a prominent **lymph space** in each. The wrinkled body wall between them contains a **transverse sheet** of muscle that extends from the myomeres on one side to those on the other. This layer serves to compress the atrial cavity dorsal to it, and thus aids in expelling water. The **pharynx** occupies most of the center of the section, and is surrounded laterally and ventrally by the **atrium**. Note the numerous **gill bars** that form its walls and the **gill slits** between them. The deeply grooved **endostyle** will be seen in the floor of the pharynx, and a similar **epipharyngeal groove** in its roof.

In certain sections, pieces of the **gonads** push into the atrium from the body wall, carring the lining of the atrium before them. An ovary consists of many large nucleated cells; testis tissue appears as small dark dots or fine tubules.

If the section is taken near the posterior end of the pharynx, the hollow, oval-shaped **midgut cecum** will be observed lying on the right side of the pharynx. It first appears to be completely within the atrium, but actually is covered with a layer of atrial epithelium.

A coelom is present in *Amphioxus*, but in a highly modified form. Close examination of the section will reveal certain of its subdivisions. A pair of **dorsal coelomic canals** are located slightly lateral to several of the most dorsal gill bars. The atrium in this region is a narrow space between the gill bars and the dorsal coelomic canals. Another coelomic canal will be found ventral to the epithelium of the endostyle. The **subendostylar coelom** connects with the dorsal coelomic canals by small coelomic passages within every other gill bar. Portions of these may be seen. Another portion of the coelom will be seen lateral to the gonads. It, too, connects with the dorsal coelomic canals, but this connection often disappears in the adult. Finally, a very narrow coelomic space may be seen between the cells of the midgut cecum and the surrounding cells of the atrial epithelium.

Excretion is by way of clusters of flame cells or **protonephridia** that lead from the dorsal coelomic canals to the atrium. Portions of them may be found. Certain blood vessels may also be found, but the circulatory system will not be considered in detail. It is of interest, however, that the general course of blood flow is the same as in a vertebrate, i.e., from the tissues anteriorly to the ventral side of the pharynx, dorsally through the gill bars, and posteriorly in a dorsal aorta to the body. There is no well developed heart, and many of the vessels are contractile. The tissues are supplied by open lacunae, for true capillaries are absent.

(D) *Section Through Intestine*

In a section through the intestine posterior to the atriopore, the metapleural folds are absent, and the **ventral fin** is present instead. Such a section will also show the **intestine** lying within a large coelomic space, for the separate coelomic passages of the pharyngeal region have coalesced. A posterior extension of the atrium lies on the right side of the intestine and coelom.

(E) *Section Through Anus*

A section through the anus passes through the **caudal fin**. Note that this fin is narrower and higher than either the dorsal or ventral fins which it replaces. The intestine opens at the **anus** on the left side of the fin.

All of the three diagnostic characteristics of chordates are well represented in the cephalochordates. There can be no doubt of the affinities of the group. In addition, *Amphioxus* has many features that are found in the vertebrates, but not in the other lower chordates. Notable among these are the myomeres, a ventral, glandular diverticulum of the alimentary canal, and a postanal tail. *Amphioxus* is certainly closer to vertebrates than any of the other lower chordates, but it has certain peculiarities, including the atrial chamber, nephridia-like excretory organs, and an extreme anterior extension of the notochord, that remove it from direct ancestry to the vertebrates. It may represent a somewhat parallel, active line of chordate evolution, or possibly a specialized and degenerate side branch of a stock that gave rise to vertebrates.

2 / The Lamprey–A Primitive Vertebrate

THE VERTEBRATES, or craniates, constitute the largest of the chordate subphyla. They differ from the lower chordates in many particulars, but notably in having at least traces of a vertebral column, and a skeletal encasement for the brain (brain case or cranium). Generally, the vertebral column is well developed in the adult, and replaces the notochord as the axial support for the body.

Although the anatomy of vertebrates will be approached primarily by studying each organ system in a representative series, it seems appropriate to begin by examining the over-all structure of a primitive member of the group. This will acquaint the student with the basic organization of the vertebrate body, and give him some idea of the initial structure of vertebrates. The most primitive vertebrates are jawless fishes of the class Agnatha. The most primitive agnathans, in turn, were the ostracoderms. These fishes became extinct about 300,000,000 years ago, but the modern cyclostomes (class Agnatha, order Cyclostomata) appear to be fairly direct descendants of this ancient vertebrate stock. Cyclostomes are represented today by the hagfishes and the lampreys. A favorable species for study is the large sea lamprey, *Petromyzon marinus*. This species enters fresh water to breed, and has recently become permanently established in the Great Lakes.

The lamprey shows a mixture of primitive and specialized characters. In studying the animal, the student should try to separate one from the other, for the aim is to learn the anatomy of the lamprey not for its own sake, but for the information it may provide concerning the structure of ancestral vertebrates. Among the major specializations of the lamprey are its eel-like shape, the absence of an armor of dermal bone, and its mode of feeding. Lampreys attach to other fishes by means of a suctorial buccal funnel, rasp away their prey's flesh, and suck its blood. A number of modifications of the digestive and respiratory systems are correlated with this habit.

EXTERNAL FEATURES

Examine the external features of a lamprey, noting its eel-like shape and the scaleless, slimy skin. The body can be divided into **head**, **trunk**,

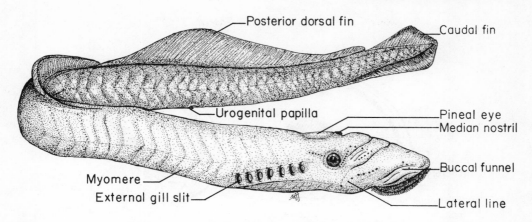

Figure 2–1. Lateral view of the sea lamprey, *Petromyzon marinus.*

and **caudal** regions. The head extends posteriorly through the gill or branchial area; the caudal region, or tail, posteriorly from the cloacal aperture. The body is rounded in cross section anteriorly, but progressing posteriorly along the trunk and tail, the body becomes compressed, or flattened from side to side. This provides a good surface for the undulations of the body during locomotion.

Observe that the only fins present are in the median plane — two **dorsal**, and a symmetrical **caudal fin** (Fig. 2–1). Lateral (paired) fins, usually found in other groups of fishes, are absent. The median fins are supported by slender, cartilaginous **fin rays**, which can be seen best if you cut across the fin in the frontal plane.

Notice the **buccal funnel** at the front of the head. It is fringed with **papillae**, and lined with **horny teeth**. The papillae are sensory, and also enable the lamprey to get a tight seal when it attaches to another fish. The "teeth" are composed of cornified cells, and hence differ from the true teeth of higher vertebrates. A protrusible **tongue** is situated near the center of the funnel. It is outlined by a ring of dark tissue, and is provided with small, rasplike, horny teeth. The **mouth opening** is just dorsal to the tongue. No jaws are present. This is a primitive characteristic, but the horny teeth and the tongue are specializations for the animal's blood sucking mode of feeding. A single, **median nostril** is located far back on the top of the head.

A median nostril is an unusual, but an ancient feature, for it is also found in certain of the ostracoderms. The manner in which the nostril is displaced during embryonic development from the more usual ventral position is shown in Figure 2–2. This condition derives from the tremendous enlargement of the upper lip of the embryo to form the buccal funnel.

Just posterior to the nostril, you will see an oval area that is often slightly depressed, and generally a lighter color than the rest of the skin. The **pineal eye**, a primitive feature also found in the ancient ostracoderms, lies beneath this depigmented skin. Experiments have shown

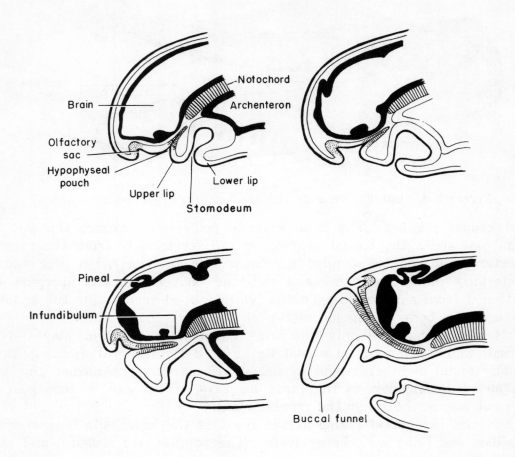

Figure 2–2. Diagrams in the sagittal plane of four stages in the development of the head region of *Petromyzon*. Note in particular how the originally independent hypophysis and olfactory sac become crowded together, and are pushed onto the top of the head by the enlargement of the upper lip. This region also differs from gnathostomes in that the hypophysis invaginates anterior to the stomodeum rather than from within the stomodeum. (From Parker and Haswell, A Text-Book of Zoology, Macmillan Company. After Dohrn.)

that the pineal eye detects changes in light and initiates diurnal color changes in larval lampreys, probably by its effect upon the hypothalamus and pituitary gland. It is possible that other physiologic activities are adjusted to the diurnal cycle in a similar way. A pair of conventional but lidless **eyes** are on the sides of the head. Seven oval **external gill slits** are located behind the eyes. A large number of gill slits characterizes primitive fishes.

If you let the head dry a bit, and examine it with a hand lens, you may be able to detect groups of pores, or little bumps, arranged in short lines. One group is found posterior to the top of the lateral eye, another extends from the underside of the eye anteriorly and dorsally. A third group is located on the ventral side of the head, posterior to the buccal funnel. These, together with other less conspicuous pores, are parts of the **lateral line system**, a group of sense organs associated with detecting vibrations and movements in the water (p. 162).

The body musculature consists chiefly of segmented **myomeres**,[3] whose outlines can be seen through the skin of the trunk and tail, especially in smaller specimens. Each myomere is roughly W-shaped, the top of the W being anterior. Furthermore, each is continuous from its dorsal to ventral end, there being no interruption near its middle as in higher vertebrates.

On the underside of the posterior end of the trunk, you will see a shallow pit called the **cloaca**. It receives the excretory and genital products, which leave through the tip of a small **urogenital papilla**, and also the opening of the **intestine** located anterior to the papilla. The opening of the cloaca to the surface is called the **cloacal aperture**, or, sometimes, anus.[4]

Sagittal and Cross Sections

Study the internal structure of the lamprey by examining a midsagittal section (Fig. 2-3) and a series of cross sections (Fig. 2-4). It is desirable to work in groups for this portion of the work. Certain students should prepare cross, and the rest sagittal sections. A good series of cross sections is as follows: (1) through the pineal and lateral eyes, (2) through a pair of external gill slits near the middle of the branchial region, (3) about one inch behind the branchial region, (4) near the middle of the trunk, (5) about an inch anterior to the cloaca, (6) through the tail, In preparing the sagittal sections use a large knife, if one is available, and be particularly careful to cut the head and branchial region

---

[3] See footnote 2, p. 7.
[4] The term anus, unfortunately, has been variously used — for the cloacal aperture, for the opening of the intestine into the cloaca, and for the opening of the intestine onto the body surface in mammals. The last is the original usage.

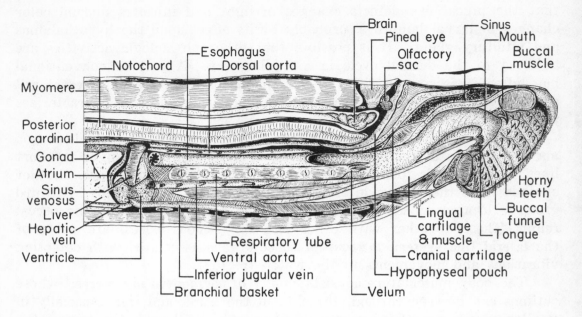

Figure 2–3. Sagittal section through the anterior portion of a lamprey.

as close to the midsagittal plane as possible. It will probably be necessary to dissect the sagittal section a bit more as you proceed.

The Skeletal and Muscular Systems

Study the better, or larger, sagittal section correlating the appearance of structures in this view with their appearance in the cross sections. Notice that the main skeletal axis is a long **notochord** extending from the posterior end of the body to a point beneath the middle of the **brain**. It has a gelatinous texture, but is enclosed in a strong fibrous sheath. It is firm yet flexible, and probably serves primarily to prevent the body from telescoping when the myomeres contract. In favorable cross sections, cartilaginous blocks (**arcualia**) will be seen above the notochord on either side of the **spinal cord**. They constitute the rudiments of vertebral neural arches. Other cartilages surround parts of the brain, and extend into the roof of the buccal funnel. These are parts of the primary brain case, or **chondrocranium**. A long median **lingual cartilage** extends into the tongue. Still other cartilages will be found beneath and lateral to the gill region, and posterior to the heart. They form a **branchial basket** that supports the gill region. The branchial basket occupies a more superficial position than the visceral arches of other fishes, but its relationship to the various cranial nerves is the same, and it may be homologous to the visceral skeleton. Examine a special preparation of the skeletal system, if available, to better appreciate this system.

The muscular system consists primarily of the segmented myomeres already observed. Note their appearance in the sections. Each myomere consists of bundles of longitudinal muscle fibers that attach onto connective tissue septa, **myocommata**, between the myomeres. Waves of contraction passing alternately down the two sides of the body cause the lateral undulations of the trunk and tail. Jets of water expelled from the gill slits may also aid locomotion. The movements of the buccal funnel and tongue during feeding are caused by the intricate musculature you see associated with these structures.

The Nervous System and Sense Organs

The brain and the spinal cord have been seen lying above the notochord. Did any of the lower chordates have an enlargement of the nerve cord comparable to a brain? The lamprey's brain has poorly developed acousticolateral centers and the cerebellum is small, but in other major respects it is similar to the dogfish's brain (p. 184).

Notice the connections of the nostril in the sagittal section. It first leads into a dark **olfactory sac** located anterior to the brain. The internal surface of the sac is greatly increased by numerous folds. Then a **hypophyseal pouch** continues from the entrance of the olfactory sac and passes ventral to the brain and the anterior end of the notochord. Respiratory movements of the pharynx squeeze the end of the hypophyseal pouch as one squeezes the bulb of a medicine dropper, thereby moving water in and out of the olfactory sac. Much of the inconspicuous pituitary gland is derived from an embryonic hypophysis, but in the higher vertebrates it invaginates from the roof of the mouth (stomodeum, Fig. 2-2).

Details of the pineal eye are not apparent in these sections, but the well developed lateral eyes show in one of the cross sections. If you make another cross section just posterior to the lateral eye, you will see part of the inner ear. It will appear as a bit of tissue imbedded in a cavity of the chondrocranium lateral to the brain.

The ear of fishes will be studied at a later date, but only the inner part is present. As you will learn (Fig. 7-7), the inner ear normally has three semicircular canals. The lamprey, however, has only two. This condition was also found in certain of the ostracoderms.

The Coelom and the Digestive and Respiratory Systems

In the sagittal section, it will be seen that the mouth opening leads into a **buccal cavity**, which extends caudad to the level of the anterior end of the notochord. A pair of **buccal glands**, which can be seen in the first cross section (Fig. 2-4, A), lie ventral and lateral to the buccal cavity. Inconspicuous ducts lead from them to the underside of the tongue. Their secretion acts as an anticoagulant, and is also hemolytic and cytolytic. The posterior end of the buccal cavity leads into two tubes — an **esophagus** dorsally and a **respiratory tube** ventrally. The esophagus can be recog-

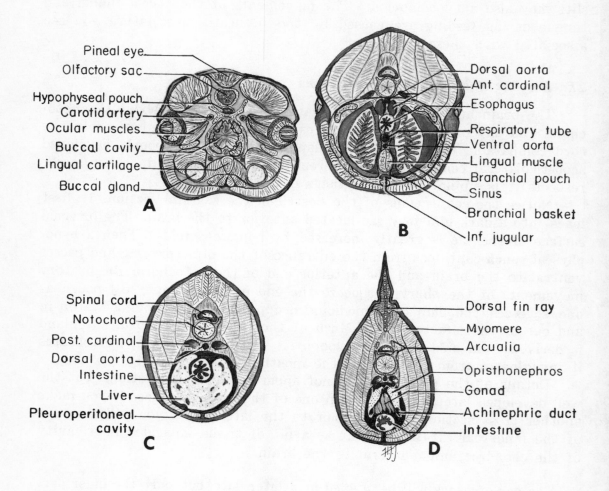

Figure 2–4. Representative cross sections of a lamprey. *A*, through the pineal and lateral eyes; *B*, through a branchial pouch; *C*, through the liver; *D*, through the posterior part of the trunk.

nized by the numerous oblique folds in its lining. Slightly posterior to the gill region, the esophagus leads into a long, straight **intestine** that continues to the cloaca. The internal surface of the intestine is increased by longitudinal folds, one of which is particularly prominent, has a somewhat spiral course, and is called the **spiral valve**. Otherwise there is little differentiation to the intestine. A true stomach is absent, and this is believed to be a primitive feature.

A large, and in the sagittal section, triangular-shaped **liver** is located beneath the anterior portion of the intestine. It grows out from the intestine embryologically, and remains connected to it by an inconspicuous bile duct.

The **pancreas** of higher vertebrates is not present as a gross organ. But Barrington (1945) has found cells comparable to the enzyme-secreting cells of the pancreas scattered in the epithelium lining that part of the intestine adjacent to the liver, and groups of cells comparable to the endocrine portion of the pancreas (islets of Langerhans) imbedded in the wall of the intestine and liver. Experimental destruction of these cells causes a rise in blood sugar.

The intestine and liver lie within a division of the **coelom** known as the **pleuroperitoneal cavity**. The intestine of vertebrates is usually supported by a long dorsal mesentery, but this is reduced in the lamprey to a few strands surrounding blood vessels that pass to the intestine. The other division of the coelom, which is located anterior to the liver and around the heart, is the **pericardial cavity**. Pleuroperitoneal and pericardial cavities are separated from each other by a **transverse septum**, which, in the lamprey, contains a cartilage of the branchial basket.

Returning to the respiratory tube, it will be seen that its entrance is guarded by a series of tentacles that constitute the **velum**, and its wall is perforated by seven **internal gill slits**. By looking at a cross section, and using a probe, you will note that each internal gill slit leads into an enlarged gill, or **branchial pouch** (Fig. 2-4, *B*), lined with gill lamellae. The pouches open to the surface through the **external gill slits**.

Both the respiratory tube and esophagus of the adult develop from a longitudinal division of the larval pharynx. The resulting separation of digestive and respiratory tracts is correlated with the lamprey's mode of feeding. When attached to its prey, the lamprey respires by pumping water in and out of the external gill slits. Some water may seep into the respiratory tube, but this would not interfere with feeding. The most active phase of respiration is expiration, for then the branchial muscles constrict the pouches, and water is forcibly expelled. Inspiration results primarily from the elastic recoil of the branchial basket. When not feeding, some water may enter the gill pouches through the mouth and internal gill slits—the usual situation in fishes.

The Circulatory System

Study the **heart** in the sagittal section. It consists of a thin-walled sinus venosus located in the center of the pericardial cavity between the atrium and ventricle. The sinus venosus receives all of the venous drainage of the body and leads into the large atrium. The **atrium** is generally filled with hardened blood, and is hence dark in color. It is located lateral to the sinus venosus and fills most of the left side of the pericardial cavity. It leads into a muscular **ventricle** located in the right ventral portion of the pericardial cavity. A conus arteriosus, found in most fishes, is not present, so the ventral aorta leaves directly from the ventricle.

The **ventral aorta** extends forward beneath the respiratory tube giving off eight paired afferent branchial arteries, which may not be seen, that pass through capillaries in the gills to enter the dorsal aorta. The first of these leads to a vestigial branchial pouch located anterior to the first well developed pouch. The **dorsal aorta** can be found in the cross sections just ventral to the notochord. It receives aerated blood from the gills by way of efferent branchial arteries, and continues posteriorly, supplying the body musculature and viscera. In the tail, it is called the **caudal artery**. The head is supplied chiefly by a pair of **carotid arteries** (Fig. 2-4, A) that leave from the front of the dorsal aorta.

A **caudal vein** is located ventral to the artery. The characteristic renal portal system of fishes is absent, for the caudal vein does not go to the kidneys, but directly to the paired **posterior cardinals**. These veins can be seen in cross sections of the trunk on either side of the dorsal aorta (Fig. 2-4, C and D). At the level of the heart, the posterior cardinals normally unite with **anterior cardinals** and turn ventrally as the **common cardinals**, or **ducts of Cuvier**, to enter the sinus venosus. In the adult lamprey, however, the left duct of Cuvier disappears. An inconspicuous hepatic portal system runs from the alimentary canal to the liver. From here blood passes to the sinus venosus through an **hepatic vein**.

The head and the branchial region and their numerous venus sinuses are drained by a pair of **anterior cardinals** (Fig. 2-4, B), which lead posteriorly to the right duct of Cuvier, and by a median, **inferior jugular vein** that enters the sinus venosus independently. The anterior cardinals are located lateral to the notochord; the inferior jugular vein is located ventral to the prominent tongue musculature in the floor of the branchial region.

A spleen is absent in lampreys.

The Excretory and Genital Sytems

The excretory organs consist of paired **opisthonephric kidneys** that appear as flaps suspended from the dorsal wall of the posterior half of the pleuroperitoneal cavity (Fig. 2-4, D). The gonad may have to be pushed aside to see them. Each is drained by an **archinephric duct**, which runs along its free border. Cut away the lateral body wall in the cross section segment that contains the cloaca, and observe that the two archinephric ducts unite at the posterior end of the pleuroperitoneal cavity to form a **urogenital sinus** which opens at the tip of the urogenital papilla. You may be able to pass a bristle into an archinephric duct and out the urogenital papilla.

Although the gonads (**ovary** or **testis**) develop from paired primordia, that of the adult is a large median organ which fills most of the pleuroperitoneal cavity. It is supported by a **mesentery**. Genital ducts are absent in both sexes, the gametes being discharged directly into the coelom. They leave the coelom through paired **genital pores** located on either side of the urogenital sinus at the extreme posterior end of the pleuroperitoneal cavity.

THE AMMOCOETES LARVA

After ascending rivers in the spring months to spawn, the marine lamprey dies. The eggs hatch into a larva that once was thought to be a distinct animal, named *Ammocoetes*. The larval stage lasts five to seven years, during which time the larva attains a length of 4 to 6 inches. It then undergoes a metamorphosis to the adult, and descends the streams. The adult lives in the ocean or in larger inland lakes for a year or two.

Examine a whole mount slide of a small *Ammocoetes* larva through the microscope (Fig. 2-5). Note its fusiform, or streamlined, shape, and the **dorsal** and **caudal fin** which are continuous with each other. The fine dark lines, or specks, on the body surface are pigment cells (**chromatophores**). If the pigment is dispersed, it will be seen that each chromatophore consists of a central area from which branching processes radiate. Although the **myotomes** have often been rendered transparent, their outlines may show as faint lines on the surface.

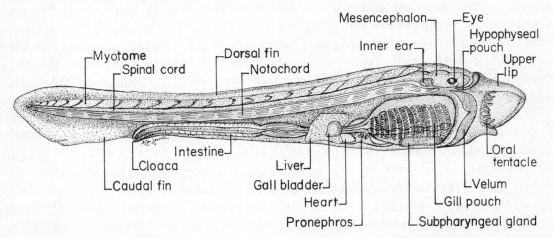

Figure 2-5. Lateral view of a whole mount slide of the *Ammocoetes* larva.

The **spinal cord** appears as a dorsal, dark-staining band that is enlarged anteriorly to form the **brain**. In favorable specimens, you will be able to see that the brain is composed of several lobes separated by constrictions, the largest and most posterior of which is the hindbrain (**rhombencephalon**), the next the midbrain (**mesencephalon**), and the most anterior (sometimes subdivided) the forebrain (**prosencephalon**). The **notochord** appears as a light-staining, longitudinal band ventral to the spinal cord and the posterior two divisions of the brain.

Turning to the sense organs, you will see a small surface protuberance anterior to the brain. This is the **median nostril** which leads into the **hypophyseal pouch**, ventral to the brain. If apparent, each **lateral eye** is represented by a round, dark spot lying between the mesencephalon and prosencephalon. An evagination from the posterior portion of the roof of the prosencephalon is the primordium of the **pineal eye**; the large, clear vesicle overlapping the front of the rhombencephalon is the primordium of the **inner ear**.

Many of the specializations of the adult digestive and respiratory systems are absent in the larva for it feeds by sifting minute food particles from the water. Observe that the **upper lip** has already enlarged to form the primordium of the buccal funnel. The **lower lip** appears as a transverse shelf. The **mouth opening** is surrounded by a series of **oral tentacles**, which function as strainers and as sensory organs. Behind the mouth, there is a clear chamber, the **buccal cavity**, bounded posteriorly by a pair of large, muscular flaps, the **velum**. Movements of the velum aid in bringing a current of water and food into the **pharynx** behind it. It is of interest that the feeding current is caused by muscular action of the velum and pharynx rather than by ciliary action as it is in lower chordates. This is more efficient, and is undoubtedly a factor that permits the *Ammocoetes* larva to attain a considerably larger size than any of the lower chordates. Note the seven large **gill pouches** in the pharyngeal region. They are lined with **gill lamellae**, and open through small, round, **external gill slits** which may be seen by sharp focusing on the surface. Ventral to the pharynx you will find a large, dark-staining, elongate body called the **subpharyngeal gland**. It secretes mucus, which is discharged through a pore into the pharynx. Details of the feeding currents are not completely known, but the mucus probably rotates and entraps food particles as it passes posteriorly into the narrow **esophagus**. The subpharyngeal gland has many of the relationships of the lower chordate endostyle, and may be its homologue. Certain of its cells also produce iodoproteins, and these cells transform into the thyroid gland of the adult. Posterior to the esophagus, the alimentary canal widens to form the **intestine**, which continues to the **cloaca**.

The **liver** is located adjacent to the posterior portion of the larval esophagus, and contains a large, clear vesicle, the **gall bladder**. Notice the **heart** lying ventral to the esophagus in front of the liver. Between the heart and esophagus you will see a few bell-shaped or finger-like processes, or tubules, which are parts of the larval **pronephric kidney**.

As you have noticed, the *Ammocoetes* larva lacks many of the specializations of the adult lamprey. It also resembles *Amphioxus* in many fundamental characteristics, but again lacks such specializations of *Amphioxus* as the atrium. For these reasons, the *Ammocoetes* larva is of great phylogenetic interest, and may give some idea of the nature of the missing link between the vertebrates and lower chordates.

3 / The Evolution and External Anatomy of Vertebrates

WITH THE ANATOMY of a primitive vertebrate as a starting point, one can proceed to a consideration of the evolution of the higher vertebrates. But before examining the details of vertebrate anatomy, it is desirable to survey briefly the general course of vertebrate evolution. If possible, this should be illustrated by conducting the class through a museum, or through a synoptic collection.

VERTEBRATE EVOLUTION

The subphylum **Vertebrata** is divided into eight classes, four of which are essentially terrestrial (superclass **Tetrapoda**). Their relationships are depicted in Figure 3–1.

Class Agnatha. As can be seen from Figure 3–1, the ancestral class was the **Agnatha**. This class is an ancient one, the earliest fossils being found in Ordovician deposits over 400,000,000 years old. The class consists of several extinct orders of heavily armored fishes collectively called the **ostracoderms**, and the living lampreys and hagfishes belonging to the order **Cyclostomata**. As seen in the preceding chapter, members of this class are characterized by the absence of jaws and paired appendages.

Class Placodermi. All vertebrates above the Agnatha have jaws, and typically paired appendages. Having jaws, they are often referred to as **gnathostomes**. The earliest gnathostomes were members of the class **Placodermi**. Their jaws and paired appendages were rather primitive and also variable among the widely divergent fishes that compose the class. Unhappily, none have survived.

The other two classes of fishes, Chondrichthyes and Osteichthyes, arose independently from the placoderms. Both evolved "conventional" jaws and paired appendages.

Class Chondrichthyes. The class **Chondrichthyes** contains fishes with a cartilaginous skeleton. There are two groups—the subclasses **Elasmobranchii** and **Holocephali**. Contemporary elasmobranchs are the sharks and dogfishes of the order **Selachii**, and the skates and rays of the order **Batoidea**. Living

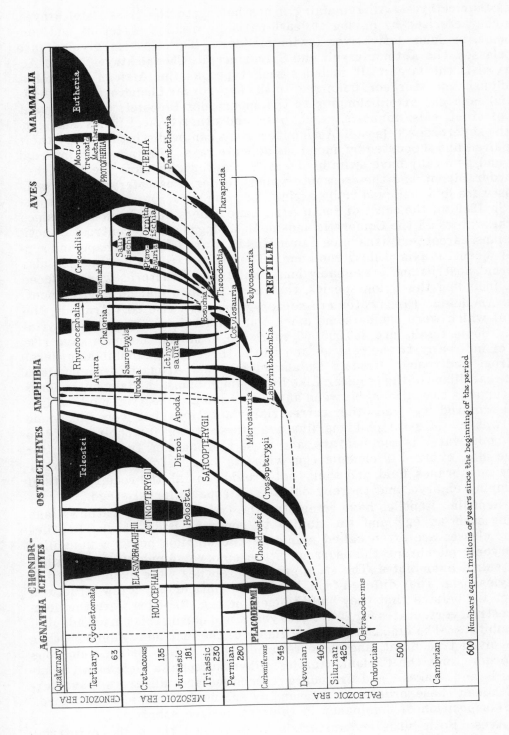

Figure 3-1. A phylogenetic tree of the vertebrates showing their distribution and relative abundance in time. (From Villee, Walker, and Smith, General Zoology; partly after Romer and Colbert.)

holocephalans are peculiar types known as chimaeras (order **Chimaerae**). With few exceptions, the cartilaginous fishes are marine.

Class Osteichthyes. All remaining fishes belong to the class **Osteichthyes**, and are characterized by having at least partially ossified skeletons and generally lungs, or lung derivatives (swim bladder). The class is divided into two subclasses, the **Actinopterygii** and **Sarcopterygii** (**Choanichthyes**). The Actinopterygii is the larger. It includes such types as the African **Polypterus**, the paddlefish, and sturgeon, belonging to the superorder **Chondrostei**; the bowfin (*Amia*), and gar pike, belonging to the superorder **Holostei**; and the great variety of trout, eels, minnows, perch, bass, and other typical fishes that constitute the superorder **Teleostei**. Actinopterygians can be distinguished by their fanlike paired fins supported by numerous delicate rays. They also lack internal nostrils and generally have swim bladders. Sarcopterygians are the lungfishes of the order **Dipnoi**, and the crossopterygians (order **Crossopterygii**). Crossopterygians were long believed to be extinct, but one surviving member was discovered in 1939 off the coast of South Africa and others have been found since then in the waters off the Comoro Islands near Madagascar. In contrast to actinopterygians, sarcopterygians have paired fins that are lobate in shape and supported by a central axis of flesh and bone. The extinct freshwater species also had internal nostrils and presumably lungs, but these structures were reduced in those, including the living species, that adapted to a marine environment.

Class Amphibia. Freshwater crossopterygians gave rise to tetrapods, the earliest of which were cumbersome amphibians collectively known as **labyrinthodonts**. Some fossils are 350,000,000 years old. Labyrinthodonts gave rise directly, or indirectly, to the reptiles and to the three living amphibian orders: **Anura** (frogs and toads), **Urodela** (salamanders), and **Apoda** or **Gymnophiona** (wormlike caecilians of the tropics). Like living amphibians, the labyrinthodonts were presumably transitional between an aquatic and a terrestrial mode of life. They had acquired legs and other terrestrial adaptations, but were not found far from water. Most amphibians live near fresh water for their ability to conserve body water is rudimentary, and most reproduce in this medium, for they have not evolved a cleidoic egg with its large store of yolk, extraembryonic membranes, fluid, and shell that provide for all the requirements of the developing embryo, and thereby obviate an aquatic larval stage.

Class Reptilia. Reptiles have completed the transition from water to land by evolving such an egg and the ability to conserve body water. Reptiles and higher classes are often called **amniotes**, the amnion being a fluid-filled extraembryonic membrane that surrounds the embryo; amphibians and fishes are often called **anamniotes**. The stem group of reptiles were members of the order **Cotylosauria**. They differed from labyrinthodonts in only a few structural details, but we believe that they had the cleidoic egg. Reptiles were the dominant terrestrial vertebrates for 200,000,000 years, and during this period adapted to many habitats. The dinosaurs became giants of the land and swamps. Other reptiles evolved true flight, and some returned to the sea. Of the numerous reptilian orders, only the **Chelonia** (turtles), **Rhynchocephalia** (*Sphendon*), **Squamata** (lizards and snakes), and **Crocodilia** (crocodiles and alligators) have survived. Although these orders are reasonably successful, mammals and birds now occupy a position of dominance in their respective spheres.

Class Aves. Both birds and mammals improved on the basic terrestrial adaptations of reptiles, becoming active and warm-blooded (homoiothermic). But birds, as a group, specialized for flight, evolving wings and feathers. Primitive birds had toothed jaws, clawed fingers, and long tails, and, aside from their feathers, were very similar to, and probably evolved from, certain small bipedal reptiles closely related to early dinosaurs. More recent birds have lost these reptilian features, but continue to lay reptilian type eggs. Contemporary species are usually arranged in 27 orders, the perching and song birds (order **Passeriformes**) being among the highest.

Class Mammalia. The line of evolution to mammals early diverged from the cotylosaurs and passed through two extinct reptilian orders (**Pelycosauria** and **Therapsida**) collectively referred to as the mammal-like, or **synapsid reptiles.** As the common name implies, these reptiles gradually came to resemble mammals, at least in their dentition and osteological features. The actual transition was made surprisingly early, even before the complete dominancy of the reptiles, but mammals remained inconspicuous in the fauna for millions of years. The most primitive of living mammals are the duckbilled platypus (*Ornithorhynchus*) and spiny anteater (*Tachyglossus*), belonging to the subclass **Prototheria**, order **Monotremata**. Like other mammals, they have hair, mammary glands, although no nipples, and a characteristic type of jaw joint and auditory ossicle mechanism. But they continue to lay a reptilian type of egg. Although practically unknown as fossils, monotremes must be an ancient group. It is now believed that they diverged from mammal-like reptiles independently of the other mammals.

All other mammals are placed in the subclass **Theria**, and reproduce viviparously. That is, they retain the embryo in the uterus, where its needs are provided for by some sort of placenta, until it is ready for birth. The earliest known mammal fossils probably belong in this subclass. Little is known of these archaic mammals, but they seem to have been rat-sized, semiarboreal forms that must have spent much of their time eluding the reptiles. The more important ones are placed in the infraclass and order **Pantotheria**.

With the extinction of the dominant reptilian orders about 65,000,000 years ago, mammals came into their own. Two lines of evolution diverged from the pantotheres. One led to the infraclass **Metatheria**, which includes but one order, the **Marsupialia**, Marsupials are pouched mammals such as the kangaroo and opossum. They generally have a poorly formed placenta, and so give birth to young at a relatively early stage. The young move into the pouch, or marsupium, where they attach to nipples of the mammary glands and complete their development. The period of intrauterine gestation is very short compared to that of pouch development. Opossums for example develop only 13 days in the uterus compared to 50 to 60 in the pouch.

The other line of evolution led to the infraclass **Eutheria**, or true placental mammals. Eutherians lack the marsupium, and have a more efficient placenta, for the embryos are retained in the uterus until a much more advanced stage. They are now the dominant mammals. The order **Insectivora**, which includes the shrews and moles, is the most primitive of the numerous orders, and the others diverged directly, or indirectly from it. Man belongs in the order **Primates**.

In studying the evolution of placental mammals, which is to be the main theme in this manual, one should ideally consider extinct groups through which the line of evolution passed—an ostracoderm, placoderm, crossopterygian, labyrinthodont, cotylosaur, pelycosaur, therapsid, pantothere, and primitive insectivore. This can be done, to some extent, in lectures on the skeleton, but in the laboratory one must rely on available, living species. These inevitably have many peculiar .features of their own, but if carefully selected, will at the same time show a number of primitive features characteristic of the early members of their group. The lamprey, representing a primitive, jawless fish, has already been examined. In the sections that follow, the evolution of vertebrate anatomy will be considered by following the changes in each organ system through a representative series of animals. For the most part, this can be done in three stages — jawed fish, primitive tetrapod, and mammal.

EXTERNAL ANATOMY

Aside from the skin or **integument** (**cutis**), most of the external features of vertebrates are simply manifestations of other organ systems. Nevertheless, it is worth while to study them along with the skin at this time, for this will bring out general changes in body shape and regions, appendages, openings, and special sence organs. The skin itself consists of two fundamental layers of tissue — an outer **epidermis** of stratified epithelial cells derived from the embryonic ectoderm, and an inner **dermis** (**corium**) of dense connective tissue derived from the mesoderm. The dermis is the thicker and also the more stable layer, for its basic structure remains much the same. But as vertebrates adapt to a terrestrial environment, the epidermis becomes thicker, and its outer cells become cornified (keratinized). That is, the outer cells, as they die, become filled with a horny, water insoluble protein called **keratin** that renders them less pervious to water. The cornified cells, which often form a distinct epidermal layer called the **stratum corneum** are continually being sloughed off, but they are replaced by mitosis in the basal cells of the epidermis, the **stratum basals**.

These changes cannot be seen grossly in the laboratory, but certain derivatives of the integument can be studied along with the general external features. Major derivatives are pigment, glands, and a variety of bony and horny structures such as scales, feathers, and hair. Although some pigment is contained within unspecialized epidermal cells in vertebrates, it is usually found within specific cells called chromatophores in fishes, amphibians, and reptiles. The chromatophores are stellate-shaped cells in which the pigment may move about. When the animal is dark, the pigment is dispersed throughout the cell; when the animal is light, the pigment is withdrawn to the center. The chromatophores are located in the dermis despite their ectodermal origin. All the glands are derivatives of the epidermis, but they have usually invaginated into the dermis. Feathers and hair are also epidermal derivatives, but scales (of which there are a variety of types) may come from either the epidermis or the dermis. The horny and bony derivatives of the skin (various types of scales and plates, feathers and hair) are sometimes referred to as an "exoskeleton," but they should not be confused with the invertebrate exoskeleton which represents a noncellular secretion superficial to the epidermis.

Fishes

Even though the Chondrichthyes are on an evolutionary side line (Fig. 3–1), they represent the primitive jawed-fish stage better than any available bony fish, for they lack lungs, or lung derivatives, and certain other peculiarities found in most living Osteichthyes. A favorable species for study is the spiny dogfish, *Squalus acanthias* of the North Atlantic and northern Pacific. The Pacific population is sometimes described as a separate species, *S. suckleyi*.

Dogfish are small sharks, and are placed with the sharks in the order Selachii. Adult males range in length from 2 to 3 feet; females are slightly larger. They prefer water temperatures ranging from 6° to 15°C., and hence migrate north in the spring and south in the fall. They are voracious and prey upon most species of fish smaller than themselves. They are considered a foodfish in Europe, but North American fishermen consider them only a nuisance. They drive more favorable fish away, and are rather destructive to gear and hooked or netted fish.

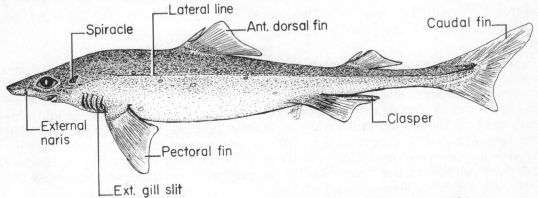

Figure 3–2. A lateral view of the dogfish, *Squalus acanthias*.

(A) General External Features

Examine a specimen, noting the streamlined, or **fusiform**, shape that enables the animal to move easily through the water. The body can be divided into a **head**, which includes the gill region, a **trunk** which continues to the cloaca, and a **tail** posterior to this. The **cloaca** is the chamber on the ventral side that receives the intestine and the urinary and genital ducts. The urinary ducts, and in the male the genital ducts as well, open at the tip of a **urinary papilla** which can be seen inside the cloaca. The cloaca opens at the surface by way of the **cloacal aperture**.[5] The body regions are not so well demarcated as in tetrapods, for all blend into one another. There is no neck, and the tail is a powerful organ of locomotion.

Note that the body is a dark color above, and light beneath. Such a distribution of pigment, referred to as **counter shading**, is common in vertebrates, especially aquatic forms. Optically, it tends to neutralize the effect of natural lighting, which highlights the back and casts a shadow on the belly and thus renders the organism less conspicuous.

There are two **dorsal fins** (anterior and posterior), with a large spine in front of each. The spines are defensive, and are associated with modified skin glands that secrete a slightly irritating substance. Large, paired **pectoral fins** will be seen just behind the head, and paired **pelvic fins** at the posterior end of the trunk. Males have stout, grooved copulatory

---

[5] See footnote 4, p. 17.

organs, called **claspers**, on the medial side of their pelvic fins. The tail ends in a large **caudal fin** of the **heterocercal** type. That is, the fin is asymmetrical, for the body axis turns up into its dorsal lobe and most of the fin rays are ventral to the axis. All the fins are supported by fibrous fin rays (**ceratotrichia**) which lie in the skin on each surface of the fin. Deep cartilages, which will be seen later, lie along the base of each fin and provide further support. In well preserved specimens, a lateral keel will be seen on the trunk on each side of the base of the caudal fin.

Most primitive groups of fish have a heterocercal tail. This type of tail is associated with heavy bodied fishes that have large pectoral fins and are somewhat flattened along the anteroventral part of the body. When a fish of this type moves through the water, its anterior end tends to veer up. The movement of the caudal fin with its flexible sculling blade ventral to a stiff axis tends to give a compensatory lift to the posterior end, thereby keeping the fish on an even keel. More progressive fishes with lungs or swim bladder and more mobile paired fins, have better control over their stability, and the caudal fin tends to become symmetrical.

The **mouth**, which is of course supported by jaws, is located on the underside of the head, and is bounded laterally by deep **labial pockets**. There is a **labial fold**, containing a cartilage, between the mouth and pocket. A pair of large **eyes** will be seen set in deep sockets on each side of the head. The rim of each socket forms immovable **eyelids**. Paired nostrils (**external nares**) are on the underside of the pointed snout. The opening of each is partially subdivided by a little flap of skin which separates the stream of water that flows through the nostril into, and out of, each **olfactory sac**. Probe to determine whether or not the olfactory sacs communicate with the mouth cavity.

A row of five **external gill slits** is located in front of the pectoral fin. Posterior to the eye you will see a large opening called the **spiracle**. Little parallel ridges, representing a reduced gill called the **pseudobranch**, can be seen on a fold of tissue that is separated from the anterior wall of the spiracular passage by a deep recess. Probe to determine the extent of the recess. This fold of tissue is a **spiracular valve**, which can be closed to prevent water from leaving this way.

The spiracle is really the reduced, and modified, first gill slit of gnathostomes. Most fishes respire by taking water into the pharynx through the mouth and discharging it through the gill slits. When a spiracle is present, water also enters through it, and in the bottom-dwelling skates and rays most of the water enters this way.

If you look at the top of the head between the spiracles with a hand lens, you will see a pair of tiny **endolymphatic pores**, one pore on each side of the midline. They communicate with the inner ear, which in most fishes is an organ of both equilibrium and hearing. Other vibrations and movements in the water are detected by the lateral line system. The position of one canal of this system (the **lateral line** in a restricted sense) is indicated by a fine, light colored, horizontal stripe extending along the side of the body. It is nearer the dorsal than the ventral surface. You will also see patches of pores on the head through which a jelly-like substance extrudes if the area is squeezed. They are the openings of the **ampullae of Lorenzini**. These ampullae are closely related to the lateral line system.

(B) Integumentary Derivatives

Major derivatives of the fish integument are chromatophores; glands, usually in the form of simple, scattered mucous cells; and hardened, dermal structures. Most of the last are bony scales. In the dogfish, these are minute **placoid scales**, often called **dermal denticles**, which cover the animal and can be felt by moving your hand anteriorly over the surface. To see them, you will have to use a hand lens, or, better still, observe a special microscopic preparation. (**Chromatophores** may be seen at the same time.) Each denticle consists of a basal plate imbedded in the dermis from which a spine perforates the epidermis and projects caudad (Fig. 3-3). At one time, elasmobranch skin, sold as shagreen, was used as an abrasive for polishing wood.

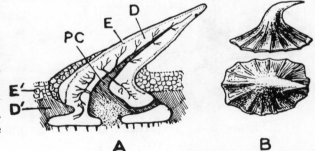

Figure 3-3. Placoid scales of a shark, A, vertical section through a scale; B, side and top view of a scale. Abbreviations: D, dentine; D',dermis; E,enamel-like vitrodentine; E', epidermis; PC, pulp cavity. (From Romer, The Vertebrate Body. After Dean.)

Histologically, a dermal denticle resembles a tooth in many ways. It contains a **pulp cavity**, a thick layer of **dentine**, and a surface covering of enamel-like material. The enamel-like material, however, is actually a very hard dentine (**vitrodentine**), for the denticle is solely of dermal origin. True enamel is derived from the epidermis.

Primitive fishes had an extensive armor of bony scales and plates, which consisted of basal layers of dermal bone overlaid first by dentine-like (**cosmine**) and then by enamel-like (**ganoine**) material. Depending on the relative amounts of cosmine and ganoine, and details of blood supply to the scale, they were called either **cosmoid** or **ganoid scales**. The denticles of the Chondrichthyes represent essentially the surface parts of cosmoid scales. Living bony fishes have very thin bony scales called **cycloid** if they are smooth, or **ctenoid** if they have tiny processes on their surface. These represent part of the bony, basal layers of the ganoid or cosmoid scale. Primitive actinopterygians had a ganoid scale; primitive sarcopterygians a cosmoid scale.

Other hard derivatives of the integument are the ceratotrichia and spines already observed. The ceratotrichia are composed of a fibrous material elaborated in the dermis. The spines are essentially enlarged denticles. In the Osteichthyes, the fins are supported by bony rays, called **lepidotrichia**, which are also modified scales.

Primitive Tetrapods

Labyrinthodonts and cotylosaurs were, of course, the true primitive tetrapods. It is difficult to find a living amphibian or reptile that is a good representative of this stage of evolution, but any one of the species of mudpuppy (*Necturus*), which belongs to the class Amphibia, order Urodela, is reasonably satisfactory in most ways. However, it is unusual in one respect, for it is a permanent larva. That is, it reaches sexual maturity while in the larval stage and never completes its metamorphosis, a condition referred to as **neoteny**. This fact must be discounted in studying its anatomy. *Necturus* is distributed throughout most of the eastern half of the United States. The most widespread species is *N. maculosus*. It is most abundant in clear waters of lakes and larger streams, but it is also found in weed-choked, turbid, and smaller bodies of water. It tends to be most active at night when it forages for small fish, crayfish, aquatic insect larvae, and mollusks.

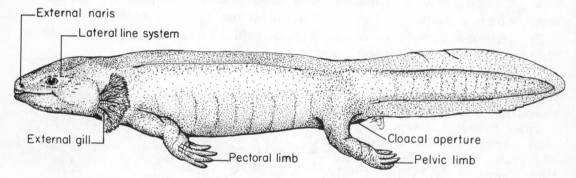

Figure 3–4. A lateral view of the mudpuppy, *Necturus maculosus*.

(A) General External Features

Examine a specimen of *Necturus*, and compare it with *Squalus*. The body is elongate with a modest-sized, flattened **head**; an incipient **neck** region; a long **trunk**; and a powerful, laterally compressed **tail**. There are no median fins, except for traces on the tail, and these lack fin rays. The paired fins of fish have become transformed into pectoral and pelvic limbs. Each limb consists of three segments. In the pectoral appendage these are upper arm (**brachium**), forearm (**antibrachium**), and hand (**manus**). The **elbow joint** is between the brachium and antibrachium; the wrist (**carpus**) is in the proximal part of the manus. Corresponding parts of the pelvic appendage are thigh, shank (**crus**), and foot (**pes**); corresponding joints are the **knee** and ankle (**tarsus**). Only four toes are present, the most medial probably being homologous to the second toe of amniotes. If the entire leg is pulled out to the side at right angles to the body with the palm of the hand, or sole of the foot, facing ventrally, the anterior border is said to be **preaxial**; the posterior, **postaxial**. This is best seen in the hind leg.

Locomotion is still partly fishlike with a lateral undulation, or squirming, of the trunk and tail playing an important role. The legs are used, more so in terrestrial salamanders than in *Necturus*, but are rather weak. When used, they tend to be held in the primitive tetrapod position. The proximal segment of the leg projects horizontally at right angles to the body, the distal segment extends vertically downward, the manus points forward, the pes more or less laterally. Note that in such a position, the preaxial border of the entire hind leg is still anterior, and that there is a simple hinge joint at the knee. In the front leg, the preaxial border of the brachium is anterior, but there has been a torsion, or rotation, at the elbow, which brings the foot forward, so the preaxial border of most of the antibrachium is medial. The position of limbs in a primitive tetrapod is such that the body is not raised far off of the ground, and the humerus and femur move back and forth in the horizontal plane. In short, the limbs are rather inefficient mechanically for support and locomotion.

The **mouth** is terminal, and is bounded by fleshy **lips**. Just above the upper lip, you will find a pair of widely spaced external nostrils (**external nares**). As will be seen later, they communicate with the front of the buccal cavity by way of internal nares, thus permitting air to be taken into the mouth. Small **eyes** are present, but are devoid of lids. The absence of eyelids is a larval feature not found in metamorphosed Amphibia. Most tetrapods have movable lids that help to protect, cleanse, and moisten the eye. Unlike the situation in most other amphibians and reptiles, salamanders also lack an external eardrum. There is an internal ear, however, and vibrations reach it by way of skull bones, or, oddly, the front leg (p. 59)! Another sense organ is the **lateral line system** which will appear, when the specimen has dried a bit, as rows of depressed dashes above and below the eyes, on the cheek, and on the ventral surface of the head. A less obvious row of dashes also extends caudad along the side of the trunk. The lateral line system, too, is a larval feature that would be lost at metamorphosis.

There are three pairs of prominent **external gills** at the posterior end of the head. Although some gas exchange takes place through the highly vascular skin, and the animal occasionally comes to the surface to gulp air, these gills are the major respiratory organ. It should be emphasized that external gills are larval structures, and are different from the internal gills, located inside the gill pouches, of adult fish. *Necturus* also has **gill slits** which can be seen at the base of the external gills. How many are there? The fold of skin extending across the ventral surface of the head between the gills is called the **gular fold**.

Finally, observe the **cloacal aperture** at the posterior end of the trunk. It is bounded by lips bearing tiny papillae in the male, and by lips bearing small folds in the female.

(B) *Integumentary Derivatives*

As will be seen, certain of the deeper parts of the bony scales of fish become associated with the skeleton proper. Aside from this, and traces on the belly of labyrinthodonts, bony scales have been lost in most amphibians, leaving the skin naked. Only the peculiar caecilians have dermal scales hidden in folds of the skin. The skin is not well protected against drying. Numerous alveolar mucous glands help keep the skin moist, but most amphibians cannot wander far from water without danger of desiccation. Chromatophores are abundant.

Since the reptilian stage of the evolution of the integumentary derivatives is essential to understand those of birds and animals, it will be considered at this point. Chromatophores continue to be present, but most of the skin glands are lost, only a few scent glands being retained. The reduction of glands is correlated with an extensive cornification of the epidermis, so that the skin is covered with **horny scales**. Sometimes little nodules of bone lie beneath the horny scales. A lizard is a good example. On the head and outside of the lips, the horny scales are enlarged to form plates. The horny scales are connected to each other by less heavily keratinized areas of the stratum corneum. In turtles, the scales are modified, and form large horny plates that overlie plates of bone. Most of the bony plates represent a redevelopment of bone within the dermis, but some of the ventral plates (plastral plates) include remnants of the original bony armor of primitive fish. The tips of the toes of reptiles bear **claws** which are also keratin structures.

Mammals

The evolutionary line of chief interest to us culminates in placental mammals. Although the house cat (*Felis domestica,* order **Carnivora**) will be used as the primary example, directions for the rabbit (*Lepus* sp. order **Lagomorpha**) are included for those who care to make comparisons between divergent members of the same evolutionary level. This can be done by providing certain students with cats and others with rabbits. Such a comparison illustrates the essential uniformity of mammalian anatomy, and, at the same time, reveals important differences that have been superimposed upon a common structural plan through the long independent evolution of these species, and their adaptation to carnivorous and herbivorous modes of life.

(A) *General External Features*

Examine either a cat or a rabbit, and compare it with *Necturus*. The diagnostic **hair** of mammals is at once evident, and it will be seen that the evolutionary trends which began in primitive tetrapods have continued. The **head** (**corpus**) is large and separated from the **trunk** by a distinct and movable **neck** (**cervix**). The trunk itself can be divided into an anterior thoracic (**thorax**) and a posterior lumbar-abdominal region. The thorax is enclosed by the ribs and sternum. The **lumbar** region is dorsal to the **abdomen**, or belly. A **tail** (**cauda**) is typically present in mammals, but, except in the whales and their allies, is greatly reduced in size in comparison with that of a primitive tetrapod. In some terrestrial mammals it is of use, often as a balancing organ in semiarboreal carnivores, in some other mammals it is essentially vestigial (rabbit), and in a few (man) it has been lost as an external structure.

The paired appendages consist of the usual parts — **brachium**, **antibrachium**, and **manus**, in the pectoral; **thigh**, **crus**, and **pes**, in the pelvic. In both the cat and rabbit, the most medial, or first toe, of the manus is vestigial, and the corresponding toe of the pes completely lost. Observe either on a mounted specimen or on a skeleton, that a cat walks on its toes with the wrist and heel raised off the ground. This method of locomotion is referred to as **digitigrade**, in contrast to **plantigrade** (man) in which the entire sole of the foot is flat on the ground, or **unguligrade** (ungulates such as the horse) in which the animal walks on the tips of its toes. The rabbit is also digitigrade in respect to the front feet, but the hind legs are modified for jumping. Just before a leap, the pes is in the plantigrade position. In both animals, the terminal segment of each toe has a **claw**. In the cat, this segment is hinged in such a way that the claw can be retracted or extended.

Note that the limbs no longer occupy the primitive sprawled position, but have rotated in such a way that they are directly beneath the body and the legs move back and forth in the vertical plane. In the hind leg, the change results from a 90° forward rotation. The knee and pes point anteriorly, and the original preaxial border is medial instead of anterior. In the front leg, there has been a 90° backward rotation, so that the elbow points posteriorly. Due to a continued torsion at the elbow, seen beginning in primitive tetrapods, the manus still points anteriorly. The original preaxial border of the brachium is now lateral, instead of anterior but that of the antibrachium continues to be primarily medial. In order, to understand these changes, try to visualize your own appendages in the primitive tetrapod position, and then slowly rotate them into the quadruped mammal position. The new limb posture is more efficient for support. It also increases speed, for the legs are now capable of a powerful and rapid fore and aft drive. Walking on the toes, with the wrist and heel raised, increases the length of the stride. Although the trunk and tail are flexible, lateral undulations play no significant part in locomotion.

Turning to the head, notice that the **mouth** (**os**) is bounded by fleshy **lips** (**labia**), the upper one being deeply cleft in the rabbit (harelip). The paired **external nares** are close together on the **nose**. The **eyes** (**oculi**) are large and protected by movable upper and lower **eyelids** (**palpebrae**). Spread these lids apart, and observe a third lid, called the **nictitating membrane**, in the medial corner of the eye. The nictitating membrane can be drawn across most of the eye, thus helping to moisten and cleanse this organ. Mammals have a prominent external ear consisting of a conspicuous external flap, called the **auricle** or **pinna**, and an external ear canal (**external acoustic meatus**) that extends into the head from the base of the auricle. The eardrum (**tympanum**) is located at the bottom of the meatus. It will not be seen at this time. That part of the head that includes the jaws, mouth, nose, and eyes is referred to as the **facial region**; the rest, containing the brain and ears, the **cranial region**.

The cloaca of primitive vertebrates has become divided in therian mammals, so that the intestine and urogenital ducts open independently at the surface. The opening of the intestine, called the **anus**, will be found just ventral to the base of the tail. If the animal is a female, the combined opening of the urinary and reproductive ducts will appear as a second hole, bounded by small folds, ventral to the anus. This immediate region is called the **vulva**. If the animal is a male, the urogenital ducts open at the tip of a **penis**. Associated with this, you will see the sac-shaped **scrotum** containing the testes. The testes may be retracted in the rabbit. The entire area of anus and external genitals is called the **perineum** in both sexes. Further discussion of the details of the external genital organs will be deferred.

Carefully feel along the ventral surface of the thorax and abdomen on each side of the midline, and you will find two rows of **teats** (**papillae mammae**) hidden in the fur. These bear the minute openings of the mammary glands. The teats are more prominent in females, but rudiments can sometimes be found in the males. There are about five pairs in the cat and six in the rabbit, but the number is subject to variation. How many are there on your specimen?

(B) *Integumentary Derivatives*

The integument of mammals is rich in derivatives, many of which can be seen, or demonstrated, in the laboratory. Chromatophores are largely absent, but pigment granules are present in the epithelial cells. There are many glands, but these fall into three categories — **mammary glands**, tubular **sweat glands** (**sudoriferous galnds**), and alveolar **sebaceous glands** usually associated with the hair follicles. The mammary glands resemble sweat glands in having contractile myo-epithelial cells peripheral to the secretory cells, and for this reason are usually considered to be modified sweat glands. Wax and scent glands seem to be modified sebaceous glands. The openings of sweat glands can be seen with a hand lens on one's finger tips. The openings of the sebaceous glands and hair follicles are the more familiar "pores" of the skin.

The most conspicuous integumentary derivative is the insulating covering of **hair** (**capillus**). Hair replaces the horny scales of reptiles in most mammals, but scales may still be found on the tails of certain rodents, and they have redeveloped over the bony plates of the armadillo shell. Although hair is composed of keratinized cells, it is a new development of the epidermis. It is not considered to be homologous with either horny scales or feathers, for details of its embryonic development are different. Moreover, the distribution of hair, as seen, for example, on the back of one's hand, leads to the conclusion that hairs evolved in small clusters between the scales, and then the scales were lost. The simultaneous presence of scales and hairs can be seen on the tails of some rodents. In most mammals, the hair forms a dense fur over the body, being modified in certain places such as the **eyelashes** (**cilia**) and tactile whiskers (**vibrissae**) on the head of the cat and rabbit. But there are many departures from this pattern. Hair is reduced in man, and lost in the adults of such highly aquatic mammals as the whale. In some other mammals, the hair has become adapted for very specialized purposes. The quills of a porcupine are a case in point, and the "horn" of a rhinoceros resembles a compact mass of hair.

Reptilian **claws** are retained in most mammals, but have been transformed into **nails** (**ungues**) in certain primates, and into **hoofs** in the ungulates. Other common integumentary derivatives are the **foot pads** on the feet of most mammals. These are simply thickenings of the stratum corneum. The **friction ridges** (finger prints) of primates are their homologue.

Aside from the widely distributed structures mentioned above, some mammals have still other derivatives. The redevelopment of **bone** in the dermis of the armadillo, the keratinized **whalebone plates** about the mouth of the toothless whales, and the horny covering of the **horns** of sheep, antelope and cattle belong in this category. The horns of these animals consist of a core of bone which arises from the skull and is covered with very heavily keratinized epidermis, i. e., horn. Horns of this type should not be confused with **antlers** found in the deer group. Antlers are bony outgrowths of the skull that are covered with skin (the velvet) only during their growth. In contrast with horns, they generally are restricted to the male, branch, and are shed annually.

4 / The Axial and Visceral Skeleton

THE NEXT ORGAN systems to be studied will be a group concerned with the general functions of support and locomotion, namely, the skeleton, muscles, sense organs, and nervous system. The skeleton may appropriately be considered first, as it is a fundamental system about which the body is built.

Divisions of the Skeleton

The vertebrate skeleton consists of two basic parts — the **dermal skeleton** and the **endoskeleton**. Although these two become united in various degrees, they are distinct in their ontogenetic and phylogenetic origins. The dermal skeleton consists of bone that develops embryologically directly from the mesenchyme in, or just beneath, the dermis of the skin. This type of bone is called either **dermal** or **membrane bone**, It follows that the dermal skeleton is superficial. Bony scales and plates, and their derivatives, are dermal in nature. Among their derivatives are the dermal plates in the head region, teeth, and the dermal portions of the pectoral girdle. In addition, dermal bone has evolved independently of bony scales in the dermis of certain animals. Portions of the shell of the turtle and the armadillo are familiar examples.

The endoskeleton, on the other hand, arises in deeper body layers, and consists of cartilage, or bone that develops in association with cartilaginous rudiments. Although such bone has the same histological structure as dermal bone, it is convenient to differentiate it as **cartilage replacement bone**. The endoskeleton may be subdivided into visceral and somatic portions. The **visceral skeleton**, as the name implies, is associated with the "inner tube" (gut) of the body. It consists of skeletal arches (visceral arches) that form in the wall of the pharynx, and primitively supported the gills. The **somatic** portion of the endoskeleton is associated with the "outer tube" of the body (body wall and appendages). It may be further broken down into axial and appendicular subdivisions. The **axial skeleton** includes those parts of the somatic skeleton located in the longitudinal axis of the body — vertebrae, skeleton of the median fins, ribs, sternum, and those portions of the brain case composed of cartilage, or cartilage replacement bone. The **appendicular skeleton** consists of the more laterally placed portions of the somatic skeleton — the skeleton of the paired appendages and those portions of their girdles composed of cartilage, or cartilage replacement bone.

Evolutionary Tendencies in the Dermal Skeleton and Endoskeleton

At one time it was thought that the endoskeleton was more primitive than the dermal skeleton, but paleontological studies have shown that this is not the case. Such primitive fishes as the ostracoderms and placoderms had an extensive dermal skeleton consisting of thick, bony scales of the cosmoid type over the trunk and tail, and larger bony plates over the head. The endoskeleton, although present, was neither completely ossified nor so conspicuous. From these early ancestors, the evolutionary tendency has been one of reduction of the dermal skeleton, and increase of the endoskeleton. As seen in the preceding chapter, the primitive, heavy, bony scales have become much thinner in living fishes, and are lost as such in tetrapods. However, the deeper parts of the original cephalic plates persist in Osteichthyes and tetrapods, and become associated with the endoskeleton as integral parts of the head skeleton and pectoral girdle.

One must be continually aware of the difference between the endoskeleton and the dermal skeleton. But in studying the evolution of the entire skeleton, this dichotomy cannot always be followed. Bony scales were studied with the integument, and the rest of the dermal skeleton will be considered along with those portions of the endoskeleton with which it becomes associated. In approaching the endoskeleton, it is convenient to study the axial and visceral portions together to some extent because they are combined in the formation of the head skeleton. After their evolution has been traced, that of the appendicular skeleton will be followed.

FISHES

In using the skeleton of *Squalus* as representative of the jawed-fish stage of evolution, one must remember that its skeleton is atypical in two major respects. For one thing, the endoskeleton is entirely cartilaginous. This represents the retention of an embryonic, not a primitive adult, condition. For another, the dermal skeleton is lost except for the small denticles. In view of the latter shortcoming, the bowfin, *Amia* (class Osteichthyes, superorder Holostei), will be used to illustrate the dermal cephalic bones that were present in primitive fishes ancestral to tetrapods.

Postcranial Axial Skeleton

(A) *Relationships of the Vertebral Column*

Make a fresh cross section of the tail of your specimen, and observe that the vertebral column lies at the intersection of several connective tissue septa which separate the surrounding muscles. A **dorsal skeletogenous septum** extends from the top of the vertebrae to the middorsal line of the body; a **ventral skeletogenous septum**, from the bottom of the vertebrae to the midventral line; and a **horizontal skeletogenous septum** from each side of the vertebrae to the lateral surface of the body. The last is the hardest to see, and it may be confused with portions of the more or less transverse **myosepta** which separate the muscle segments. The horizontal septum forms an arc that reaches the skin at the position of the lateral line, which appears as a small hole in the skin.

Although it will not be seen at this time, the relationships in the trunk are much the same. The only difference is that the ventral septum has split, so to speak, and passes on each side of the body cavity (coelom).

(B) *Vertebral Regions and Caudal Vertebrae*

The vertebral column of fishes consists of only **trunk** and **caudal vertebrae**, but these become somewhat modified at the attachment of the chondrocranium and in the vicinity of the median fins. Study the structure of the caudal vertebrae of *Squalus* on a special preparation, or by dissecting the tail of your own specimen. If you dissect the tail, cut out a piece about two inches long, and carefully expose the vertebrae by cleaning away all of the surrounding muscle and connective tissue. Do not select a piece adjacent to either the posterior dorsal or caudal fin.

Make a cross section through the joint between two vertebrae near one end of your piece, and examine it along with the lateral aspect of other vertebrae. Each vertebra consists of a cylindrical, central portion, called the **centrum**, which bears a dorsal and ventral arch of cartilage. The dorsal arch, which protects the spinal cord, is the **neural arch**; its cavity, the **vertebral canal** (Fig. 4–1). The ventral arch, which surrounds and protects the caudal artery and vein, is the **hemal arch**; the passage through it, the **hemal canal**. You may see parts of the vessels. The artery lies dorsal to the vein. The top of each arch may be seen extending a short distance into the dorsal and ventral septa as **neural** and **hemal spines**.

Make a sagittal section through several vertebrae, and study it along with the lateral aspect of others. It will be seen that each side of the neural arch is composed of two roughly V-shaped blocks of cartilage. That block located directly on top of the centrum, and having its apex pointing dorsally, is called the **neural plate**. The other block located above the joint between the centra, and having its apex directed ventrally, is the **dorsal intercalary plate**. Each plate of the neural arch is perforated by a small hole (**foramen**) for either the dorsal or ventral root of a spinal nerve. A continuous neural arch is peculiar to cartilaginous fishes; such a condition is possible because cartilage is much more flexible than bone. In other vertebrates there is some space between successive neural arches. The neural arch of other vetebrates may be considered homologous to the neural plate, for it develops primarily from a comparable group of cells.

The hemal arch is composed, on each side, of but one block (**hemal plate**). A small foramen for intersegmental branches of the caudal artery and vein can be found between the bases of the hemal arches.

Examine the centrum in the sagittal section. It is shaped like a biconcave spool, a shape termed **amphicelous**. The concavity at each end contains a gelatinous substance which is **notochordal tissue** that has persisted from the embryonic stage. A small strand of notochord also perforates each centrum. The cartilage immediately adjacent to the noto-chord has become calcified, and thus appears white. Calcification of cartilage involves the deposition of calcium salts in the matrix. This strengthens the cartilage, but it is not the same as ossification. The area of calcified cartilage is hourglass shaped.

(C) *Trunk Vertebrae*

Compare the caudal vertebrae with a special preparation of the trunk vertebrae (Fig. 4-1). The structure is basically the same, but the trunk vertebrae lack hemal arches. Instead, they have short, ventrolateral proc-esses (**basapophyses**) that project from the sides of the centrum. Basa-pophyses are serially homologous to at least the proximal part of the hemal arches. Small **ventral intercalary plates**, lie between successive basapophyses, but they are often lost in preparing the vertebrae. They are the ventral counterparts of the dorsal intercalary plates, and, like them, are secondary developments.

Each vertebra develops embryonically from sclerotomal mesenchyme cells that migrate from the segmented, embryonic somites and accumulate around the nerve cord and notochord in an intersegmental position. A vertebra thus comes to lie between two segmental myomeres, which in turn have developed directly from the somites. Much of the vertebral mesenchyme forms car-tilaginous vertebral arches, **arcualia**. Certain of these form the neural arch around the nerve cord; others, the hemal arch around the caudal vessels. The vertebral centrum develops between these arches and largely replaces the notochord. Its mode of development is extremely variable in vertebrates, but in fishes it appears to derive in part from arcualia bases, and in part from the direct deposition of cartilage, or in some cases of bone, in concentric rings around the notochord and even within the notochordal sheath.

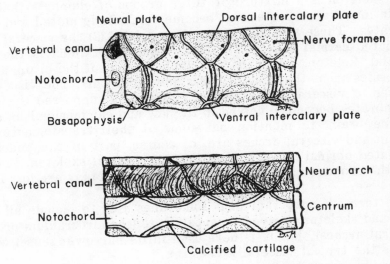

Figure 4-1. Trunk vertebrae of *Squalus*. Upper, lateral view; lower, sagittal section.

(D) Ribs

On a mounted skeleton it can be seen that short ribs articulate with the basapophyses, and extend into the horizontal skeletogenous septum at the point where it intersects with the myosepta. Some fishes have **dorsal ribs** which lie in the horizontal septum and **ventral ribs**, which lie on each side of the coelom and are serially homologous to the hemal arches.

There is some doubt as to the nature of the ribs of *Squalus,* but the consensus is that they are ventral ribs, for they attach onto the basapophyses. Their extension into the horizontal septum is believed to have resulted from a secondary displacement.

(E) Median Fin Supports

Look at the dorsal fins on a mounted skeleton, and notice that each is supported proximally by several cartilages and distally by fibrous rods, the **ceratotrichia**, described earlier (p. 32). The cartilages, which collectively may be called **pterygiophores**, are in two series. Those that rest on the vertebral column are **basals**; the more distal ones, **radials**. There is some question whether the pterygiophores develop phylogenetically from neural spines of the vertebrae, or independently within the dorsal septum. The caudal fin, which is of the heterocercal type (p. 32), is supported proximally by enlarged neural and hemal spines; distally by ceratotrichia.

Head Skeleton

The head skeleton is a mixture of three groups of elements that are distinct in certain lower vertebrates, but become confusingly united and mixed in other vertebrates. These are (1) the **chondrocranium**, (2) the **visceral skeleton**, and (3) associated **dermal bones**. The chondrocranium[6] which is the anterior end of the axial skeleton, surrounds a variable amount of the brain, and forms protective capsules about the olfactory sacs and inner ears. The visceral skeleton is composed of visceral arches, which primitively supported the gills, but in other vertebrates become involved in the jaw and ear mechanism, and even help encase the brain, to mention but some of their transformations. The chondrocranium and visceral arches are, of course, part of the endoskeleton, but the associated dermal bones are part of the dermal skeleton. Primitively they covered the chondrocranium and visceral arches. Many are lost in higher

---

[6] The term chondrocranium is sometimes used to include all parts of the head skeleton derived from cartilage (chondrocranium proper and visceral arches), but it is here used in its narrower sense, which excludes the typical visceral arches.

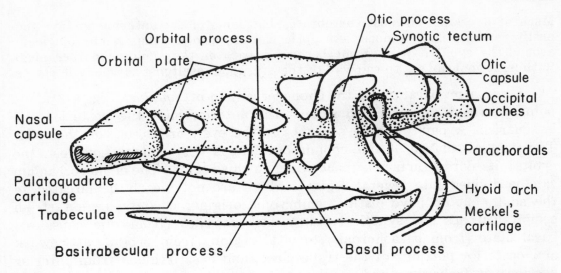

Figure 4–2. Lateral diagram of the embryonic chondrocranium and mandibular arch of a tetrapod to show the components of the chondrocranium and the autostylic articulation of the palatoquadrate. Basipterygoid process is a common synonym for basitrabecular process. (Modified after Goodrich, Studies on the Structure and Development of Vertebrates, Macmillan Company.)

vertebrates, but some persist to help form the braincase, the jaws, and the facial portion of the skull. [7]

(A) Composition and Structure of the Chondrocranium

The chondrocranium is a complex box of cartilage, or cartilage replacement bone, that can best be understood by reciting the major features of its embryonic development (Fig. 4–2). The basic elements in its formation in any vertebrate are two pairs of longitudinal cartilages that lie beneath the brain. The posterior pair, called **parachordals**, are located on each side of the anterior end of the notochord. (The notochord, itself, only extends as far forward as the pituitary gland.) The anterior pair, called **trabeculae**, are situated in front of the notochord. The parachordals develop from mesenchyme derived from the sclerotome and hence are closely related to the vertebrae. But the trabeculae, like the visceral arches, develop from mesenchyme derived from the neural crest, and are regarded as representing the remnant of a visceral arch that, in ancestral vertebrates, was located anterior to the first arch of gnathostomes. To these are added various other cartilages. An **otic capsule** develops around each inner ear; **a synotic tectum** connects the two otic capsules dorsally; **a nasal capsule** develops about each olfactory sac; a complex **orbital plate**, or latticework, forms medial to each eye; and a variable number of anterior vertebral elements (**occipital arches**) are added to the back of the otic capsules. The expansion and fusion of all these components forms the chondrocranium.

---

[7] The term skull is one of those words in common usage that is difficult to define precisely for all vertebrates. It may be used in a broad sense for the entire skeletal complex of the head, but is more often restricted to those parts of the head skeleton immovably united with each other, i.e., the braincase (cranium), facial region, and upper jaw.

Much of it ossifies in most vertebrates, but some of the anterior portion generally remains cartilaginous. An optic capsule also develops embryologically around the eye, but it does not become a part of the adult chondrocranium. Rather, it contributes to the wall of the eye, often ossifying as sclerotic plates.

Study a preparation of the chondrocranium of *Squalus* (Fig. 4–3). The pointed, trough-shaped **rostrum** is at the anterior end; the posterior end is squarish. A pair of large sockets for the eyes, **orbits**, lies on the sides. The ventral surface of the chondrocranium is very narrow between the orbits; its dorsal surface is much wider in this region. The chondrocranium can be divided into several regions, which are not clearly demarcated in the adult, but do have distinct embryonic origins: occipital region (from occipital arches and synotic tectum), otic capsules (from otic capsules), basal plate (from parachordals), orbital region (from orbital latticework and posterior portions of the trabeculae), and nasal region (from anterior portions of trabeculae and nasal capsules).

The **occipital region** is the very posterior portion of the chondrocranium lying in the midline. It surrounds a large hole, the **foramen magnum**, through which the spinal cord enters the **cranial cavity** within the chondrocranium. A pair of bumps, the **occipital condyles**, will be seen ventral to the foramen magnum on each side of a centrum-like area. They develop from a pair of basapophyses of a vertebra that has been incorporated in the occipital region, and they help to articulate the chondrocranium with the vertebral column. A double occipital condyle of this type is unusual in fish; most have a single rounded condyle located directly ventral to the foramen magnum. There is little movement at the occipito-vertebral joint in the dogfish.

The paired **otic capsules** are the large, squarish, posterior corners of the chondrocranium that extend from the occipital region to the orbits. Between them, on the dorsal side, is a large depression called the **endolymphatic fossa**. Within the fossa, you will see two pairs of openings which communicate with the inner ear. The smaller, anterior pair is for the **endolymphatic ducts**; the larger, posterior pair is for the **perilymphatic ducts**. On each capsule you may find ridges, two dorsally and one laterally, beneath which lie the **semicircular canals** of the ear. Finally, there are two large foramina on the posterior edge of the chondrocranium lateral to the occipital condyles. The more medial is the **vagus foramen** for the vagus nerve; the more lateral the **glossopharyngeal foramen** for a nerve of the same name.

Ventrally, the otic capsules are connected by the flat, broad **basal plate**. Some of the notochord persists in the chondrocranium of the adult dogfish, and its position is indicated by a white strand of calcified cartilage which can be seen through the clear cartilage along the midventral line of the basal plate. The small hole in the midline, anterior to the strand of calcified cartilage, is the **carotid foramen** for the passage of internal carotid arteries.

The **optic region** is that area which includes and lies between the orbits. The anterior, dorsal, and posterior portions of each orbit are formed by walls of cartilage called the **antorbital process, supraorbital crest**, and **postorbital process**, respectively. Ventrally, the orbit is open. Most of the floor of the chondrocranium is narrow between the two orbits, but near the posterior part of the orbit the floor is wider and bears a pair of prominent lateral bumps called the **basitrabecular (basi-pterygoid) processes**. As will be seen later, orbital processes of the upper jaws have a movable articulation with the floor of the chondrocranium just anterior to these processes. Note that the roof of the chondrocranium between the orbits is complete in *Squalus*. Primitively it was incomplete. The small, median hole near the anterior end of the roof is the **epiphyseal foramen**. In life, it contained a stalk of the same name which represents a vestige of the pineal eye. The series of foramina (one large, many small) that perforate the supraorbital crest are the **superficial ophthalmic foramina** for the passage of a similarly named nerve trunk and its branches. A number of other foramina can be seen in the medial wall of the orbit. The large anterior one is the **optic foramen** for the optic nerve; the large posterior one the **trigemino-facial foramen** for the trigeminal and facial nerve. The remaining are for smaller cranial nerves and vessels.

All the chondrocranium anterior to the antorbital processes may be called the **nasal region**. It consists of a pair of round **nasal capsules** attached to the front of each antorbital process, and a long, median **rostrum** which helps to support the snout. The wall of each nasal capsule is very thin, and is generally broken. If complete, you will be able to see its external opening (**external naris**). The opening within the capsule is for the passage of the olfactory tract. The rostrum is trough-shaped dorsally, and keeled ventrally. Its dorsal concavity, called the **precerebral cavity**, communicates with the **cranial cavity** by way of a large opening, the **precerebral fenestra**. In life, the precerebral cavity is filled with a gelatinous material. Two other large openings (**rostral fenestrae**) into the cranial cavity will be seen ventrally lying between the nasal capsules on each side of the rostral keel.

(B) Sagittal Section of the Chondrocranium

Examine the inside of the cranial cavity in a sagittal section of the chondrocranium, if such a preparation is available. Certain structures seen earlier can be noted again in this view, but there are additional features of interest. Notice the depression in the floor dorsal to the basitrabecular processes. This is the **sella turcica**, a recess for the hypophysis. The large opening in the medial wall of the otic capsule, just posterior to the trigemino-facial foramen, is the **internal acoustic meatus**. Part of the glossopharyngeal nerve enters here to pass beneath the ear and emerge through the glossopharyngeal foramen, but most of the passage is occupied by the statoacoustic nerve from the inner ear.

(C) Visceral Skeleton

In primitive agnathous vertebrates, the visceral skeleton is composed of **visceral arches** that pass between the gill slits and help to support the gills. These are generally fused to each other, and to the chondrocranium, forming a compact unit. Recall the branchial basket of the lamprey. The arches are also numerous. *Petromyzon* has traces of nine, two being anterior to the first of the seven gill slits, and ostracoderms had even more. In gnathostomes, there has been a reduction of the number of gill slits and arches, and a mobilization of those arches that remain. No more than the last seven of primitive series persist in higher fishes, and the first of these (the mandibular arch) forms, or helps to form, the jaws. Teeth, which are derivatives of the dermal skeleton, become attached to the jaws, and also frequently to the roof of the mouth and to other visceral arches.

In placoderms, there may have been a complete gill slit posterior to the jaws, but in fishes above this level, the second visceral arch (the hyoid arch) has moved close to the jaws, sometimes aiding in their support. Thus the original first gill slit of gnathostomes is lost, or reduced to a spiracle, in fishes above the placoderms. *Squalus* is a good example of this stage in the evolution of the visceral skeleton.

Examine a preparation of the visceral skeleton of *Squalus* (Fig. 4-3). The seven **visceral arches** of which it is composed show clearly. The first is modified to form the upper and lower jaws, and it therefore is called the **mandibular arch**. The second, called the **hyoid arch**, extends from the otic capsule to the angle of the jaws and ventrally into the floor of the mouth. The spiracle would be located just anterior to the dorsal portion of the hyoid arch. The last five visceral arches, called **branchial arches**,

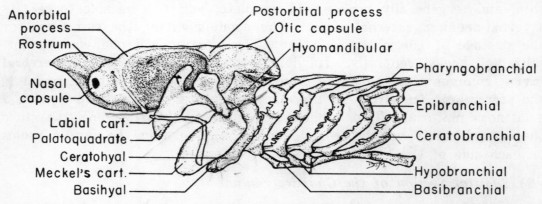

Figure 4–3. Lateral view of the chondrocranium and visceral skeleton of *Squalus*.

[8] The tissue lying between successive gill slits supports the gill filaments and contains a number of structures associated with the gills — the skeletal arches, muscles, nerves, and blood vessels. There is some confusion as to what the entire complex should be called. It has often been called gill, branchial, or visceral arch, using the term arch in a broad sense. To avoid confusion, I prefer to limit the term arch to the skeletal elements, and use the term interbranchial septum for the entire complex.

support the interbranchial septa,[8] and hence pass between the remaining gill slits. Note that the third visceral arch and first branchial arch are synonymous. In addition to the arches, small **labial cartilages** may be seen on the lateral surface of the jaws. They are located in the labial folds previously noted (p. 32). Some consider that they represent vestiges of a premandibular visceral arch, but this is doubtful, for there is good evidence that the chondrocranial trabeculae represent vestiges of such an arch.

Study the mandibular and hyoid arches in more detail. Each side of the upper jaw is formed by a **palatoquadrate (pterygoquadrate) cartilage**; each side of the lower jaw by **Meckel's cartilage**. Both cartilages bear several rows of sharp teeth which are loosely attached to the surface of the jaws and are similar to each other (Fig. 9-7, p. 236). A dentition in which the teeth are essentially the same is referred to as **homodont**. The teeth of selachians differ from those of other fishes primarily in their triangular shape. The teeth are conical in primitive fishes. Two prominent processes extend dorsally from the palatoquadrate. The one above the angle of the jaw is for the attachment of mandibular muscles. The one that extends up into the orbit (**orbital process**) passes lateral to the braincase and just anterior to a basitrabecular process. It helps to stabilize the upper jaw. The single midventral piece of the hyoid arch is called the **basihyal**. A **ceratohyal** extends from this element to the angle of the jaw, and a **hyomandibular** continues to the otic capsule. Ligaments unite the angle of the jaw and hyoid arch, so this arch provides the primary support for the jaws.

The type of jaw suspension seen in the dogfish, in which the palatoquadrate is loosely articulated to the chondrocranium by an orbital process and is buttressed posteriorly by the hyoid arch, is termed **hyostylic**. In primitive fishes ancestral to tetrapods, the palatoquadrate articulated to the underside of the braincase by two or three processes — an anterior orbital process, a middle basal process connecting with a basitrabecular process, and a posterior otic process articulating on the otic capsule (Fig. 4-2). This self-supporting condition is called **autostylic**. The hyoid arch, although situated close behind the jaws, was not involved, and, with the loss of gills in tetrapods, was available for other purposes.

Each of the branchial arches ideally consists of a midventral **basibranchial (copula)**, and a chain of four additional elements extending dorsally on each side — a short, ventral **hypobranchial**; a longer **ceratobranchial** extending to the height of the angle of the jaws; an **epibranchial** continuing beyond this; and finally a **pharyngobranchial** which overlies the pharynx and points posteriorly, but does not unite with the vertebrae. All these elements can be seen in the first branchial arch, but there has been some fusion and loss of certain of the elements in the posterior arches. Also notice that each branchial arch, and the hyoid arch as well, bears a number of laterally projecting **gill rays** which stiffen the interbranchial septa.

Certain of the relationships of the visceral skeleton to other organs of the head can be seen clearly in a sagittal section of the head (Fig. 10-7, p. 279).

(D) Dermal Bones

The third component of the head skeleton is the dermal bones associated with the anterior portions of the endoskeleton as far posteriorly as the pectoral girdle. They have been lost in the evolution of the Chondrichthyes, but were present in primitive fishes, and in those ancestral to tetrapods. Some appreciation of the extent of the dermal bones in primitive forms can be gained by studying them in the bowfin, *Amia*.

Examine the skeleton of the head of *Amia*, noting the groups, or series, of dermal bones that sheathe the chondrocranium and visceral arches (Fig. 4-4). In some types of preparations, the dermal bones have been removed on one side, exposing the endoskeletal structures. A **dermal roof** covers the top of the head, the cheek region, and the lateral surface of the partly ossified palatoquadrate. It is pierced by tiny external nares (two on each side), and by the large orbits. There is also a large gap, which is not found in more primitive fishes, between the cheek and upper jaw. The dermal bones on the margin of the jaw bear a single row of small, conical **teeth**, which are essentially similar to each other (**homodont**).

The posterior part of the palatoquadrate cartilage ossifies as the **quadrate**, but the part of it anterior to this is lost in the adult. A **palatal series** of dermal bones, on some of which there are several raws of teeth, forms much of the roof of the mouth in the general region of the missing portion of the palatoquadrate. A third "series" (actually

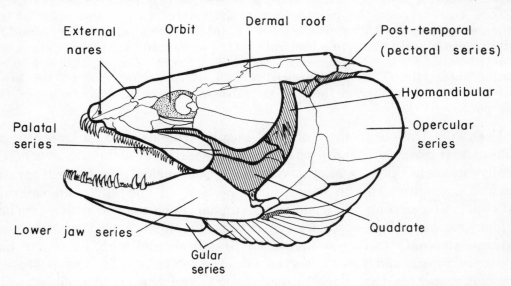

Figure 4–4. Lateral view of the skull of *Amia calva*. Series of dermal bones is shown in heavy outline. The quadrate and hyomandibular are parts of the visceral skeleton. (Modified after Goodrich, Vert. Craniata.)

composed of but a single element, the **parasphenoid**) lies in the midline of the roof of the mouth directly beneath the chondrocranium. It bears a number of very small teeth.

A **lower jaw series** covers the lateroventral and medial surfaces of the partly ossified Meckel's cartilage. As with the upper jaw and palate, there is a single row of marginal teeth, and several rows of more medial teeth. The medial teeth in both the palate and lower jaw occupy the same general position as the teeth on the mandibular arch of *Squalus*, and are considered to be their homologues.

The remaining visceral arches are covered laterally by an **opercular series**, and ventrally by a **gular series**, of dermal bones. Finally, the endoskeletal portions of the pectoral girdle are covered laterally by a **pectoral series** of dermal bones.

PRIMITIVE TETRAPODS

Since *Necturus* is neotenic, its skeleton is not as thoroughly ossified as is the skeleton in the ancestral tetrapods. This is particularly true for the head region and appendicular skeleton. It is therefore desirable to supplement the study of *Necturus* by examining these portions of the skeleton in another reasonably primitive tetrapod. The snapping turtle, *Chelydra serpentina*, is a good example.

Postcranial Axial Skeleton

The vertebrae of tetrapods vary considerably, but the most primitive type seems to have been the **rhachitomous vertebrae** of primitive labyrinthodonts and cotylosaurs (Fig. 4–5). Such a vertebra consisted of a neural arch beneath which was a notochord partly constricted by three blocks of bone that made up the centrum. A single **intercentrum** (sometimes called hypocentrum), lay beneath the notochord at the front of each vertebra. A pair of small **pleurocentra** was located dorsal and posterior to the intercentrum. Hemal arches (chevron bones) were also present in the caudal region where they were fused with the intercentra.

Numerous attempts have been made to compare the primitive tetrapod vertebra with that of fishes, and the subject has been recently reviewed by Williams (1959). The consensus is that the neural arch is homologous to the neural plate of the dogfish, and the tetrapod hemal arch represents the hemal arch of fishes. The central elements of the primitive tetrapod vertebra are believed to have developed independently of the arches from perichordal sclerotomal mesenchyme.

In the evolution from primitive amphibians through reptiles (Fig. 4–5) to birds and mammals, the intercentrum decreases in size, and the pleurocentrum enlarges to become the definitive centrum. The reduced intercentrum may contribute to the **intervertebral disc** when one is present. The notochord disappears in the adult, except for traces that lie within the disc.

The ribs of tetrapods are generally considered to be dorsal ribs. Each usually bears two processes that articulate with the vertebrae: a dorsal **tuberculum**, which articulates with the transverse process (**diapophysis**) on the neural arch,

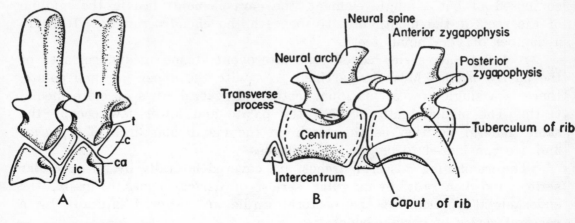

Figure 4–5. Diagrams to show; *A*, the adult rhachitomous vertebrae of a primitive labyrinthodont; *B*, the vertebrae and rib attachment of a primitive reptile. All are lateral views with anterior toward left. Abbreviations: *c*, pleurocentrum; *ca*, attachment of caput of rib; *ic*, intercentrum; *n*, neural arch; *t*, transverse process (diapophysis). (From Romer, The Vertebrate Body.)

and a ventral **caput**, which primitively articulates with the intercentrum (Fig. 4–1, *B*). Sometimes there is a small lateral process in the latter region, called a **parapophysis**. With the reduction and loss of the intercentrum, the caput has an intervertebral articulation.

Primitively, ribs were found on all the vertebrae from the atlas to the base of the tail, those in the anterior trunk region being the longest. But in later tetrapods, many of these free ribs disappear as such by fusing onto the sides of the vertebrae to form a more complex transverse process known as a **pleurapophysis**. The ribs in the anterior half of the trunk, however, always remain free. Typically, most of the free ribs in amniotes connect ventrally with a median **sternum** formed of cartilage, or cartilage replacement bone. Anteriorly, the sternum also becomes associated with the coracoid region of the pectoral girdle. An unossified sternum was probably present in labyrinthodonts, but whether or not it connected with the ribs is unknown. It does not connect with the short ribs of living amphibians.

(A) *Vertebral Regions*

Examine the vertebral column and ribs on a skeleton of *Necturus* and note that there is only slightly more differentiation into regions than in the skeleton of *Squalus*. The most anterior vertebra, called the **atlas**, lacks free ribs and is otherwise specialized for articulation with the skull. It is, of course, located in the incipient neck region and is the only one that can be called a **cervical vertebra**. **Trunk vertebrae** extend from here to the level of the pelvic girdle. All have free ribs. A single **sacral vertebra** (usually the nineteenth), and a pair of free ribs, is modified for the support of the pelvic girdle. The remaining are **caudal vertebrae**; they lack free ribs, and most bear **hemal arches** (**chevron bones**).

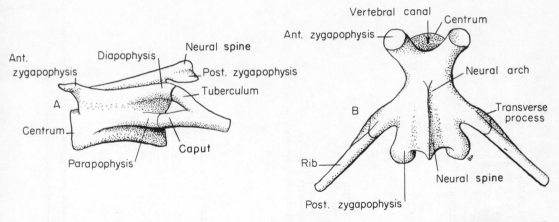

Figure 4–6. *A*, lateral, and *B*, dorsal view of a trunk vertebra and rib of *Necturus*.

(B) *Trunk Vertebrae*

Study a trunk vertebra on the mounted skeleton, and also examine an isolated specimen, if possible (Fig. 4-6). Dorsally it consists of a **neural arch** that overlies the spinal cord located in the **vertebral canal**. A small **neural spine** projects caudad from the top of the arch. Two pairs of small processes for the articulation of successive vertebrae project laterally from the front and back of the neural arch. Two anterior pair of processes, whose smooth articular facets face dorsally, are called the **anterior zygapophyses**; the posterior pair, whose facets face ventrally, the **posterior zygapophyses**. Note how the zygapophyses of adjacent vertebrae overlap. Zygapophyses strengthen the vertebral column, and are a characteristic feature of tetrapod vertebrae. In fishes, the body is partly supported by water, and there are no zygapophyses.

Ventrally, the vertebra consists of a biconcave (amphicelous) **centrum**. Notochordal tissue persists in the concavities. A prominent **transverse process** projects laterally and posteriorly from each side of the neural arch and centrum. Observe that it is rather high from its dorsal to ventral edge. The dorsal part represents a **diapophysis**; the ventral part a **parapophysis**.

(C) *Ribs*

The ribs of *Necturus* are shorter than the ribs of labyrinthodonts or cotylosaurs. but have the two heads (bicipital condition) characteristic of terrestrial vertebrates. The ventral head (**caput**) attaches to the parapophysis; the dorsal head (**tuberculum**) to the diapophysis. Which is the more anterior? The distal portion of the rib is its **shaft**.

(D) *Sternum*

Necturus does not have the small, cartilaginous sternum found in

TABLE 1

Components of the Tetrapod Skull and Lower Jaw

The components of the skull and lower jaw of a labyrinthodont, together with the part of the skeleton to which they belong, are shown in the left hand column. The homologies between these elements and those of certain other tetrapods are shown in the right hand columns. An X indicates that an element is present; O, that it is absent; Unos., that the region is unossified. All of the elements are paired unless indicated to the contrary by a number in parentheses: (1) indicates a median element; (2) or (3) that two or three of these are present on each side.

LABYRINTHODONT	NECTURUS	TURTLE	ALLIGATOR	MAMMAL
Skull				
Chondrocranium (Cartilage Replacement Bone)				
Basioccipital (1)	Unos.	X (1)	X (1)	X (1) ⎫ Occipital
Exoccipital	X	X	X Fused with opisthotic	X ⎬
Supraoccipital (1)	Unos.	X (1)	X (1)	X (1)
Opisthotic	X ⎱ Operculum	X	X	X ⎫ Periotic part
Prootic	X ⎰	X	X	X ⎬ of Temporal
Basisphenoid (1)	Unos.	X (1)	X (1)	X (1) Basisphenoid ⎫ part of
				X Presphenoid (1) ⎬ Sphenoid
Sphenethmoid (1)	Unos.	Unos.	X Laterosphenoid	X Orbitosphenoid ⎰ Sphenoid
Unossified Ethmoid Region	Unos.	Unos.	Unos.	X Mesethmoid
Unossified Nasal Capsule	Unos.	Unos.	Unos.	X Turbinals
Visceral Arches (Cartilage Replacement Bone)				
Palatoquadrate				
Quadrate	X	X	X	X Incus
Epipterygoid	O	X	O	X Alisphenoid part of Sphenoid
Hyomandibular				
Stapes	X	X	X	X
Dermal Bones				
Roof				
Tooth-bearing Marginal Bones				
Premaxilla	X	X	X	X
Maxilla	O	X	X	X
Median Series				
Nasal	O	O	X	X
Frontal	X	X	X	X
Parietal	X	X	X	X
Postparietal	O	O	O	X Interparietal, often a part of Occipital

TABLE 1 (Continued)

Components of the Tetrapod Skull and Lower Jaw

LABYRINTHODONT	NECTURUS	TURTLE	ALLIGATOR	MAMMAL
Circumorbital Series				
Lacrimal	o	o	X	X
Prefrontal	o	X	X	o
Postfrontal	o	o	o	o
Postorbital	o	X.	X	o
Jugal		X	X	X
Temporal Series				
Intertemporal	o	o	o	o
Supratemporal	o	o	o	o
Tabular	o	o	o	X ? Part of Occipital
Cheek Bones				
Squamosal	X	X	X	X Squamous part of Temporal
Quadratojugal	o	X	X	o
Palate and Underside of Chondrocranium				
Parasphenoid	X	X Fused with Basisphenoid	X Reduced	X ? Part of Sphenoid
Vomer	X	X (1)	X	X (1)
Palatine	o	X	X	X
Ectopterygoid	o	o	X	X ? Part of Sphenoid
Pterygoid	X	X	X	X Pterygoid process of Sphenoid
Lower Jaw				
Visceral Arches (Cartilage Replacement Bone)				
Meckel's Cartilage				
Articular	Unos.	X	X	X Malleus
Dermal Bones				
Lateral Series				
Dentary	X	X	X	X
Splenials (2)	X	o	X	o
Surangular	o	X	X	o
Angular	X	X	X	X Ectotympanic part of Temporal (Entotympanic a new cartilage replacement bone)
Medial Series				
Coronoids (3)	o	X	X	o
Prearticular	o	X	X Fused with articular	X Anterior process of Malleus

most urodeles just posterior to the pectoral girdle. However, traces of cartilage, which may represent an incompletely developed sternum, are located in the ventral portions of the myosepta dorsal and posterior to the pectoral girdle, and in the midventral connective tissue septum. They may be noticed during the dissection of the muscles (p. 124).

Head Skeleton of Primitive Tetrapods and Necturus

The three groups of elements present in the head region of fishes (chondrocranium, visceral skeleton, and dermal bones) are represented in amphibians and reptiles, but the opercular and gular series of dermal bones are lost, and the visceral arches are greatly reduced. These changes are correlated with the loss of the gills and the beginning of neck formation. The elements that remain may be grouped, for purposes of description, into four units: (1) the skull in its restricted sense, (2) the lower jaw, (3) teeth, (4) the hyoid, or hyobranchial apparatus.

The skull (Fig. 4-7) is composed of the chondrocranium, the palatoquadrate of the visceral skeleton, and the dermal bones that more or less encase these two. The chondrocranium covers the back, the underside, and most of the lateral surfaces of the brain. It does not cover the top of the brain, and originally there was a gap in each of its lateral walls. It is usually ossified to a large extent, as shown in Figure 4-7 (see p. 57) and Table 1, but the anterior ethmoid (nasal) region and nasal capsules were primitively unossified. As in most fishes, but not *Squalus*, the occipital condyle was originally single. The chondrocranium is perforated by foramina for nerves and vessels, and by an **oval window** (or fenestra ovalis), located on the lateral surfaces of the otic capsule, for the stapes (see below).

The posterior portion of the palatoquadrate ossifies as the **quadrate**, and continues to serve for the articulation with the lower jaw. The portion of it adjacent to the basitrabecular process of the chondrocranium generally ossifies as the **epipterygoid**. This bone articulates with the braincase, and helps to fill in the gap in its side (Fig. 4-16, *A*). The rest of the palatoquadrate is lost in the adult.

The dermal bones constitute the largest component of the skull. A great many were present in labyrinthodonts and cotylosaurs, but the evolutionary tendency since then has been one of reduction. There were also a number of general features of interest in the dermal roof and palate of very primitive forms (Fig. 4-7). The **external nares** were widely separated, and the **internal nares** entered the mouth at the very front of the palate. A foramen was present between the parietal bones for the pineal or parietal eye. An **otic notch**, which may have held a tympanic membrane, was located on the posterior part of the skull roof. Aside from this notch, the temporal portion of the roof was solid (**anapsid** condition). But posteriorly, a pair of **posttemporal fenestrae** passed between the dermal roof and otic capsule. Mandibular muscles arose from the underside of the roof on each side and passed through a pair of lateral palatal openings (**subtemporal fenestrae**) to insert on the lower jaw. The palate was movably articulated with the basitrabecular processes by way of the epipterygoids, and was separated from the anterior portion of the parasphenoid by **interpterygoid vacuities**. A cranioquadrate passage, which transmitted important vessels, nerves and possibly an auditory tube, was situated just posterior to the basal articulation between the braincase and pterygoid.

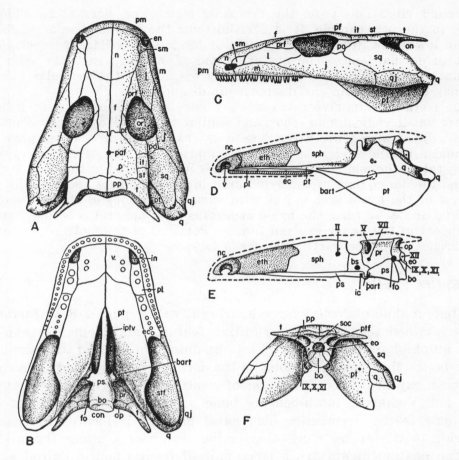

Figure 4–7. The skull of an ancestral land vertebrate, based primarily on the Carboniferous labyrinthodont, *Palaeogyrinus*. *A*, dorsal view of the dermal skull roof; *B*, palate; *C*, lateral view; *D*, lateral view with dermal skull roof removed (outline in broken line); palatal bones—dermal and endoskeletal—of left side are shown; deep to them, the braincase. The hatched area is the sutural surface of palatal bones against the maxilla. *E*, lateral view, and *F*, posterior view, of braincase. Roman numerals, foramina for cranial nerves. Abbreviations: *ab*, auditory bulla; *as*, alisphenoid; *bart*, basal articulation between palate and basitrabecular process of braincase; *bo*, basioccipital; *bs*, basisphenoid; *con.* condyle; *do*, postparietal; *e*, epipterygoid; *ec*, ectopterygoid; *en*, external naris; *eo*, exoccipital; *eth*, ethmoid region; *f*, frontal; *fm*, foramen magnum; *fo*, oval window; *ic*, foramen for internal carotid; *in*, internal naris; *iptv*, interpterygoid vacuity; *it*, intertemporal; *j*, jugal; *l*, lacrimal; *m*, maxilla; *n*, nasal; *nc*, nasal capsule; *on*, otic notch; *op*, opisthotic; *or*, orbit; *p*, parietal; *paf*, pineal or parietal foramen; *pf*, postfrontal; *pl*, palatine; *pm*, premaxilla; *po*, postorbital; *pp*, postparietal; *pr*, prootic; *prf*, prefrontal; *ps*, parasphenoid; *pt*, pterygoid; *ptf*, posttemporal fenestra; *q*, quadrate; *qj*, quadratojugal; *sm*, septomaxilla; *soc*, supraoccipital; *sph*, sphenethmoid; *sq*, squamosal; *st*, supratemporal; *stf*, subtemporal fenestra; *t*, tabular; *v*, vomer. (From Romer, The Vertebrate Body.)

To these major components of the skull may be added the **stapes**, or **columella**. This is a small bone, found in most amphibians and reptiles, that transmits sound vibrations from the tympanic membrane, across the middle ear cavity, to the otic capsule. It is derived from the hyomandibular of fishes.

The lower jaw (Fig. 4–10) consists of Meckel's cartilage surrounded by a sheath of dermal bones. The posterior end of Meckel's cartilage (the region articulating with the quadrate) generally ossifies as the **articular** bone. The rest may either remain cartilaginous, or disappear.

The **teeth** of primitive tetrapods are very similar to those of fishes, for they are small, conical, numerous, and similar to each other (homodont). They are usually arranged in two series—a lateral series on the margins of the jaws, and a medial series on the palate and medial side of the lower jaw.

As stated, the mandibular arch becomes incorporated in the skull and lower jaw, and the dorsal part of the hyoid arch (hyomandibular) becomes the stapes. The rest of the hyoid arch unites with variable portions of the more anterior branchial arches to form the **hyoid apparatus**. This apparatus is concerned with the support of the newly evolved tongue. Portions of the more posterior branchial arches form the cartilages of the larynx.

(A) *Entire Skull*

Study a skull of *Necturus,* comparing it with Figure 4–8. Unfortunately, it has retained fewer of the primitive features and elements than many other amphibians or reptiles. First examine the top of the skull. Most of the bones that you see belong to the dermal roof, although other groups have been exposed through the loss of some of the original elements of the roof. The V-shaped, tooth-bearing bone on each side of the front of the upper jaw is the **premaxilla**. The **nasal passage** is located in the notch posterior to its lateral wing. Continuing posteriorly along the middorsal line, the next elements are a large pair of **frontal** bones. Paired **parietals** extend from the frontals nearly to the **foramen magnum**. Each parietal also has a narrow process that extends anteriorly lateral to the frontals. The **orbits** are located ventral to these processes.

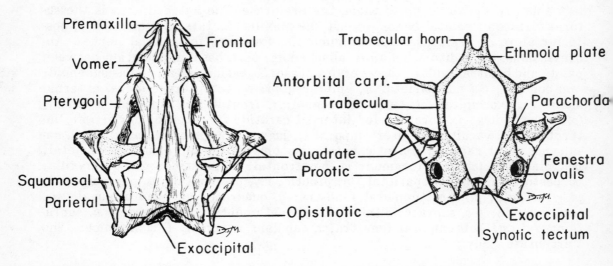

Figure 4–8. Skull of *Necturus*. Left, dorsal view of the entire skull; right, ventral view of chondrocranium. Cartilage is stippled.

Lateral to the posterior half of the parietal, you will see two small bones. One forms the very posterior angle of the skull, and continues forward to a tiny window of cartilage. The other is anterior to this window. These bones, the **opisthotic** and **prootic**, respectively, are a part of the chondrocranium. The thin sliver of bone on the margin of the skull, lateral to the otic bones, is the **squamosal** — the last element of the original dermal roof. Anterior and ventral to it, at the point where the lower jaw articulates, you will see the partly ossified **quadrate** — the only part of the palatoquadrate present in *Necturus*.

The posterior end of the skull, lateral and ventral to the foramen magnum, is formed by the paired **exoccipitals**. Each bears a condyle. There are therefore two **occipital condyles**, in *Necturus,* unlike the single one of more primitive tetrapods.

Look at the underside of the skull. The palate consists of two pairs of dermal bones. The anterior pair, **vomers**, are located posterior to the premaxillae, and, like the premaxillae, bear a row of teeth. The posterior pair, **pterygoids**, continue back from the vomers to the otic region. The front of the pterygoids also have a row of teeth. Portions of the vomers and pterygoids can be seen from the dorsal side. The large median bone on the ventral surface, lying between the vomers and pterygoids and continuing to the exoccipitals, is the dermal **parasphenoid**. A bit of the unossified **ethmoid plate** of the chondrocranium is exposed anterior to the parasphenoid. The two otic bones can also be seen in the ventral view lying posterior to the pterygoids, and lateral to the back of the parasphenoid. A cartilaginous area, containing the oval window, or **fenestra ovalis**, separates them. The oval window may be covered by a tiny disc-shaped bone bearing a little stem (stylus). This bone is generally called the stapes. It is connected by its stylus, and by ligaments, to the squamosal.

The sound-transmitting apparatus of urodeles is quite different from that of frogs. The tympanic cavity and tympanic membrane are absent, but a **stapes**, which connects either with the squamosal or with the quadrate, is present in larval stages. The aquatic larvae detect waterborne vibrations via this route. At about the time of metamorphosis, an **operculum** (no relation to the bony fish structure of the same name) develops from the wall of the otic capsule, and this eventually becomes connected with the pectoral girdle by way of an opercular muscle. With the development of the operculum, the stapes tends to disappear as a distinct and functional element, either by fusing onto the wall of the otic capsule, or onto the operculum. Terrestrial adults detect ground vibrations by way of the pectoral appendage, girdle and opercular muscle (p. 124), but in neotenic forms this development sequence is arrested at various intermediate stages. The auditory ossicle of *Necturus* is a fusion of the stapes and operculum, the opercular portion being the larger. But an opercular muscle is incompletely developed, so sound transmission is essentially larval by way of the stapedial portion of the bone and the squamosal.

(B) Chondrocranium

Although parts of the chondrocranium can be seen in the complete skull, it can be seen more clearly by examining a preparation in which the dermal bones have been removed. The chondrocranium of *Necturus* is not too good a representative of the primitive adult stage, as it is largely in the embryonic condition. (Compare what follows with the description of the development of the chondrocranium on page 45.) The pair of large, round, posterolateral swellings are the **otic capsules**. You will see two ossifications in each — an anterior **prootic**, and a posterior **opisthotic** (Fig. 4-8). The hole on the side of the capsule is the oval window. It may be covered by the stapes. The only other ossifications are a pair of ventral **exoccipitals** which form in the **occipital arch**. They are connected by a delicate cartilaginous bridge, the **basioccipital arch**. Dorsally, the otic capsules are connected by another bridge, the **synotic tectum**. Often the quadrate (a part of the mandibular arch) is left on preparations of the chondrocranium and will be seen anterior and lateral to the prootic.

The shelves of cartilage united with the medioventral edge of each otic capsule are the **parachordals**. The pair of cartilaginous rods that continue forward from the parachordals are **trabeculae**. They are united anteriorly to form an **ethmoid plate**, from which a pair of **trabecular horns** continue between very delicate nasal capsules. The nasal capsules are generally destroyed. A small **antorbital cartilage**, representing a part of the orbital latticework of other vertebrates, extends laterally from each trabecular cartilage.

(C) Lower Jaw

Compare the lower jaw of *Necturus* with Figure 4-9. Only three dermal bones cover Meckel's cartilage in this animal. The largest of these

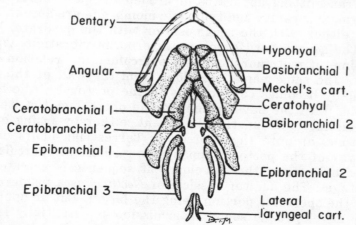

Figure 4-9. Ventral view of the lower jaw, hyoid apparatus and laryngeal cartilages of *Necturus*. Cartilage is stippled.

is the **dentary**, which forms practically all of the lateral surface of the jaw, and a small portion of the medial surface near the front of the jaw. It bears the long anterior row of teeth. The short posterior row of teeth is on the **splenial**, most of which is on the dorsomedial surface of the jaw, but a bit of the bone shows laterally. Most of the medial surface of the jaw is formed by the **angular**. This bone is widest posteriorly, and then tapers to a point which passes ventral to the splenial and continues anteriorly to the dentary. A small portion of the angular shows laterally at the posteroventral corner of the jaw. The surface of the lower jaw that articulates with the quadrate is formed by **Meckel's cartilage**, which is unossified in *Necturus*.

(D) *Teeth*

Notice that the teeth of *Necturus* are essentially fishlike. That is, they are small, conical, and all of the same type (**homodont**). They are also relatively numerous, and you can see parts of a **lateral** and **medial series**. The lateral series are on the premaxillae and dentaries; the medial on the palatal bones and splenials. The teeth attach loosely to the jaws.

(E) *Hyoid Apparatus*

Preserved skeletal material is better than dried for studying the hyoid apparatus. Notice that it is made up of parts of four visceral arches — the hyoid arch and the first three branchial arches (Fig. 4–9). The hyoid arch is the most anterior and the largest component. It consists on each side of a short **hypohyal**, which is just lateral to the midventral line, and a longer **ceratohyal**, which extends toward the angle of the jaw. A median **basibranchial 1** extends posteriorly from the hypohyals to the first branchial. This arch, too, is composed of two cartilages on each side, the more medial being **ceratobranchial 1**; the more lateral **epibranchial 1**. The next two arches (branchial arches numbers 2 and 3) are greatly reduced, and at first sight appear to consist only of **epibranchials 2 and 3**. On closer examination, however, you will see a small cartilage (**ceratobranchial 2**) that connects them with the first branchial arch. The branchial arches support the three external gills. The two gill slits pass on either side of the second branchial arch. A small, median **basibranchial 2**, which is generally partly ossified, extends caudad from the base of the first branchial arch.

The last two branchial arches do not contribute to the hyoid apparatus, and will not be seen. There is some doubt as to their fate, but traces of them may persist. A raphe in one of the branchial muscles, which contains a few cartilage cells, may represent a part of the fourth branchial (sixth visceral) arch. It is also probable that the rest of this arch, plus the last arch, have contributed to the lateral cartilages of the laryngotracheal chamber.

Head Skeleton of the Reptile

In many ways the skulls of certain living reptiles are better examples of a primitive tetrapod skull than is that of *Necturus*. The skull of the snapping turtle, *Chelydra,* is described and illustrated below (Figs. 4–11 to 4–13) and an illustration of an alligator's skull (Fig. 4–10) is included for comparison.

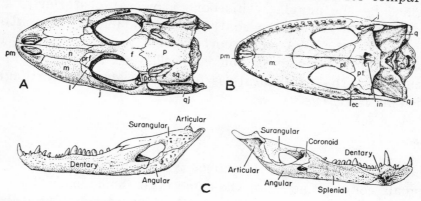

Figure 4–10. The alligator skull. *A,* dorsal view; *B,* ventral view; *C,* lateral and medial views of the lower jaw. (From Romer, The Vertebrate Body.)

(A) General Features of the Skull

Examine a skull of *Chelydra.* Teeth are absent, and are replaced functionally in life by a horny beak ensheathing the jaw margins. The **dermal roof** and **palate** can be seen surrounding the small, partly ossified **chondrocranium**. The originally paired **external nares** have united to form one opening at the surface of the dermal roof, but the nasal passages remain distinct and open into the mouth by paired **internal nares** located at the front of the palate. The **orbits** are situated far forward, for the snout is short. The dermal roof in the temporal area has been "eaten away," or emarginated, from behind in most turtles, but is complete in the sea turtles. Although this emargination is related to the bulging of the powerful jaw muscles, it is not comparable in its relationship to the temporal fenestrae of other reptiles. Turtles are usually considered to have the complete (**anapsid**) temporal region of primitive amphibians and reptiles. In any case, one can easily visualize the nature of such a roof by looking at the turtle. A pair of large **posttemporal fenestrae** can be seen in a posterior view of the sea turtle skull between the dermal roof and otic capsule. They are present in other turtles too, but the loss of the overlying dermal roof in this region makes them less apparent.

The palatal bones have united solidly with the underside of the braincase, so interpterygoid vacuities are absent. The pair of large **subtemporal fenestrae** can be seen between the palatal bones and the lateral margins of the dermal roof. Two pairs of **infraorbital fenestrae** can be found in the palate beneath the orbits.

Look at the back of the skull, and note the position of the middle ear. It is quite close to the jaw joint. It has also been partly encased and constricted by bone, so that it appears as an hourglass-shaped cavity. The expanded lateral portion of the cavity lying dorsal to the jaw joint is the **middle ear**, or **tympanic**, **cavity** proper; the expanded medial portion, known as the **tympanic recess**, contains an extention of the perilymphatic system of the inner ear (Fig. 4-11). The reduced cranioquadrate passage lies in the floor of the tympanic recess.

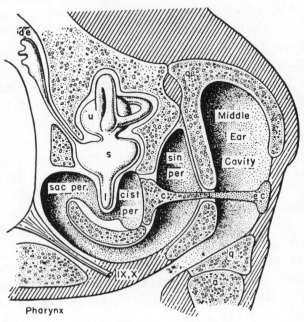

Figure 4–11. Diagrammatic vertical section through the ear and adjacent parts of a turtle skull. Abbreviations: *a*, articular; *c*, columella or stapes; *cist per*, perilymphatic cistern; *de*, endolymphatic duct and sac; *ec*, extracolumella; *l*, lagena; *q*, quadrate; *s*, sacculus; *sac per*, perilymphatic sac; *sin per*, pericapsular sinus; *u*, utriculus; *IX, X*, cranial nerves. (From Romer, Osteology of Reptiles, University of Chicago Press.)

(B) *Composition of the Skull*

The dermal elements along either lateroventral margin of the roof are a very small **premaxilla** ventral to the external naris, a large **maxilla** continuing beneath the orbit, a **jugal** posterior to this, and a **quadratojugal** just anterior to the middle ear cavity. The jugal enters only the posteroventral corner of the orbit. The middorsal elements are a pair of large **prefrontals** posterior to the external naris, and dorsal to much of the orbit; a pair of small **frontals** that do not enter the orbit in *Chelydra*; and a pair of large **parietals** that extend to the prominent, middorsal, occipital crest. Each parietal also sends a wide flange ventrally that covers a part of the brain not covered by chondrocranial elements. Two other dermal elements complete that portion of the roof situated between the dorsal and marginal bones. A large **postorbital** (sometimes considered to be a **postfrontal**) lies between the orbit and temporal emargination, and a **squamosal** forms a cap on the extreme posterodorsal corner of the skull.

Dermal bones also make up the palate. The very front of this region is formed by palatal processes of the premaxillae and maxillae. A median **vomer** (a fusion of originally paired elements), is located posterior to the premaxillae, and **palatines** lateral to the vomer. The internal nares enter on each side of the front of the vomer in most turtles. But in the sea turtles, and a few others, the vomer and palatines have sent out flanges that unite with each other to form a small **secondary palate** ventral to the **primary palate**. This pushes the openings of the internal nares caudad. The rest of the palate is formed by a pair of large **pterygoids**.

The chondrocranium surrounds the base, some of the sides, and the posterior portion of the brain, but it fails to cover the rest of the brain, which is covered instead by dermal bones of the roof, and by the epipterygoid (see below). Four bones surround the **foramen magnum**—dorsally the **supraoccipital**, which forms the occipital crest; laterally the paired **exoccipitals**; and ventrally the **basioccipital**. The **occipital condyle** has begun to shift from its primitive position on the basioccipital to the exoccipitals. Since distinct portions of it are borne on all three of these bones, the condyle is tripartite. In a dorsal view, an **opisthotic** can be seen extending laterally from the supraoccipital and exoccipital to the squamosal. A **prootic** is situated anterior to this. The only other ossification of the chondrocranium is the **basispheniod**, which can be seen ventrally lying between the pterygoids anterior to the basioccipital. The dermal **parasphenoid** has united with it.

The posterior part of the palatoquadrate has ossified as the **quadrate**. This bone occupies the posteroventral corner of the skull, extending from the jaw joint dorsally to the squamosal. Most of it passes anterior to the middle ear, but a part of it goes posterior to the ear. A very slender, rodlike **stapes** may be seen extending from the tympanic cavity proper, to the otic capsule. It passes through a small hole in the quadrate and crosses the tympanic recess. Its median end is enlarged to form a foot plate that fits into the **round window**. The central portion of the palatoquadrate has ossified as the **epipterygoid**. This bone helps to fill in a gap in the side of the chondrocranium. It is difficult to see, but occupies a small triangular area between the pterygoid and ventral flange of the parietal anterior to a large **prootic foramen** for certain cranial nerves (the trigeminal and abducens).

(C) *Lower Jaw*

Compare the lower jaw of *Chelydra* with Figures 4-12 and 4-13. The posterior end of Meckel's cartilage has ossified as the **articular**, and can be recognized by its smooth articular surface. The rest of the cartilage

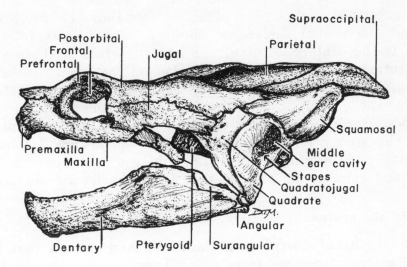

Figure 4-12. Lateral view of the skull and lower jaw of the snapping turtle, *Chelydra*.

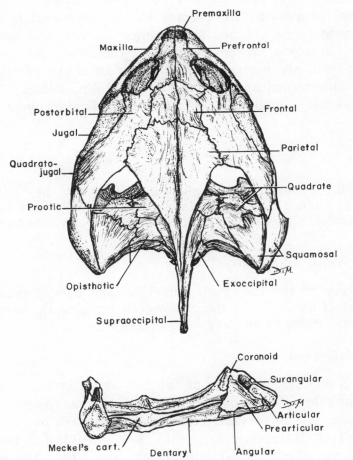

Figure 4-13. Snapping turtle skull and lower jaw. Upper, dorsal view of the skull; lower, posteromedian view of the lower jaw.

remains unossified, but it can often be found in the adult lying in a groove on the medial surface of the jaw. The largest of the dermal elements is the **dentary**. Most of the lateral surface of the jaw is formed by the dentary, but a small **surangular** lies on the lateral surface between the dentary and articular. The front half of the medial surface is also formed by the dentary, but three additional elements form the medial surface posteriorly — a **coronoid** dorsally; a **prearticular** ventral to it, and bridging the groove for Meckel's cartilage; and an **angular** beneath the prearticular, and forming the ventral border of the jaw. A bit of the coronoid and angular can also be seen from the lateral surface.

(D) *Hyoid Apparatus*

Study the hyoid apparatus of the turtle, and observe that it consists of a median body (**corpus**) from which horns (**cornua**) project laterally. The corpus develops from the fusion of the basihyal and first two basibranchials. Most of it ossifies, but the front remains cartilaginous. A slender **lingual process** projects from its anterior end into the tongue. The two pairs of large, ossified horns represent the first and second ceratobranchials. In some cases, a trace of the first epibranchial may be found. Ceratohyals are represented by a pair of short, cartilaginous horns arising from the corpus anterior to the ceratobranchials. More posterior visceral arches have doubtless contributed to the laryngeal cartilages.

MAMMALS

The following directions are based on the cat, but they could be applied to many other mammals. If material is available, it would be desirable to compare the skeleton of the cat with that of man and other species.

Postcranial Axial Skeleton

The structure of the individual vertebrae and ribs of mammals is very similar to that of some of the more advanced reptiles discussed earlier. Review the general remarks on page 51.

(A) *Vertebral Groups*

Examine a mounted skeleton of the cat, and a string of disarticulated vertebrae. There are five groups of vertebrae in mammals. The most anterior is, of course, the **cervical vertebrae**. There are more of them than in very primitive tetrapods, for, with few exceptions, mammals have seven cervical vertebrae, all of which lack free ribs. **Thoracic vertebrae** bearing free ribs follow the cervical, and **lumbar vertebrae** follow the thoracic. The lumbar vertebrae lack free ribs, but have very large transverse processes. The number of vertebrae in these two regions varies

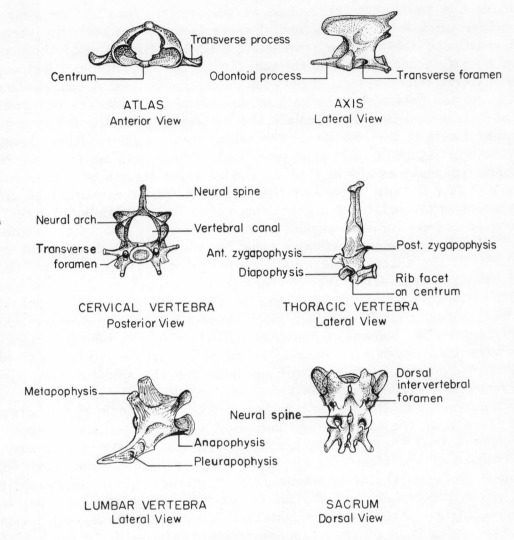

Figure 4-14. Drawings of selected vertebrae of the cat.

among mammals. The cat normally has 13 thoracic and seven lumbar vertebrae. **Sacral vertebrae** follow the lumbar and are firmly fused together and form a solid point of attachment, the **sacrum**, for the pelvic girdle. The number of vertebrae contributing to the sacrum varies between species. Most mammals, including the cat, have three; some, more than this. The remaining are **caudal (coccygeal) vertebrae**. Their number varies with the length of the tail, but all mammals have some. Even in man there are three to five vestigial caudal vertebrae that are fused together to form a **coccyx**.

(B) Thoracic Vertebrae

The thoracic vertebrae should be studied first as they are more similar to the vertebrae of *Necturus* than are the vertebrae in other regions.

Examine one from near the middle of the series. Identify the **neural (vertebral) arch** with its long **neural spine** (**spinous process**), the **vertebral canal**, and the **centrum** (**corpus**), Fig. 4-14. Each end of the centrum is flat, a shape termed **acelous**. In life, small, fibrocartilaginous **intervertebral discs** are located between successive centra. Articulate two vertebrae, and notice how the back of one neural arch overlaps the front of the one behind it. Disarticulate the vertebrae, and look for smooth articular facets in the area where the neural arches came together. These are the zygapophyses (articular processes). The facets of the pair of **anterior zygapophyses** are located on the anterior surface of the neural arch, and face dorsally; those of the pair of **posterior zygapophyses** are on the posterior surface of a neural arch, and face ventrally. A lateral **transverse process** (or **diapophysis**) projects from each side of the neural arch. Notice the **smooth articular facet** for the tuberculum of a rib on its end. In the anterior and middle portion of the thoracic series, the head of a rib articulates between vertebrae, so part of the facet for a head is located on the front of one centrum, and part on the back of the next anterior centrum. Why does the head have an intervertebral articulation? The somewhat constricted portion of the neural arch between the diapophysis and centrum is called its **pedicle**. When several vertebrae are articulated, you will see holes for the spinal nerves, the **intervertebral foramina**, between successive pedicles.

Compare one of the thoracic vertebrae from the middle of the series with those near the anterior and posterior ends of the thoracic region, and observe that the neural spines, zygapophyses, transverse processes, and position of the facets for rib heads differ in details. For example, the most posterior thoracic vertebrae lack transverse processes, for their ribs lack a tuberculum, and the facets for their rib heads are entirely on one centrum. Articulate the last two thoracic vertebrae, and notice how the articulation of the zygapophyses is reinforced by a process of the pedicle that extends caudad lateral to the zygapophyses. This is an **anapophysis** (**accessory process**). A skilled observer can find differences among the vertebrae sufficient to identify each one precisely as the third, sixth, or eighth thoracic, etc. However, it will be sufficient for students to be able to distinguish the major group to which a vertebra belongs. All thoracic vertebrae have at least one articular facet for a rib. No other vertebrae have such facets.

(C) Lumbar Vertebrae

Lumbar vertebrae are characterized by their large size, and by having prominent, bladelike transverse processes. The transverse process is a **pleurapophysis**, for a rib has united with the diapophysis. Lumbar vertebrae also have a small bump for the attachment of ligaments and tendons dorsal to the articular surface of each anterior zygapophysis.

This is called a **metapophysis** (**mamillary process**). Traces of metapophyses can also be seen on the more caudal thoracic vertebrae. Most of the lumbar vertebrae also have anapophyses.

(D) Sacral Vertebrae

The sacral vertebrae can easily be distinguished by their fusion into a single piece (the sacrum), and by the broad surface they present for the articulation with the pelvic girdle. Examine the sacrum closely and you will be able to detect the neural spines, zygapophyses, pleurapophyses, etc., of the individual vertebrae of which it is composed. Notice how the distal ends of the pleurapophyses have fanned out and united with each other lateral to the intervertebral foramina. This has produced separate dorsal and ventral foramina for the respective rami of spinal nerves.

(E) Caudal Vertebrae

Caudal vertebrae are characterized by their small size and progressive incompleteness. The more anterior ones have the typical vertebral parts, but soon there is little left but an elongated centrum. V-shaped **chevron bones**, which protect the caudal artery and vein, are found in some mammals. The cat has traces of such bones on the anterior caudal vertebrae, but they are usually lost in a mounted skeleton. However, the points of articulation of a chevron bone will appear as a pair of tubercles, **hemal processes**, at the anterior end of the ventral surface of a centrum.

(F) Cervical Vertebrae

Most of the cervical vertebrae can be recognized by their characteristic transverse processes. The transverse process is a pleurapophysis, but, unlike those in the lumbar region, it is perforated in all the cervical vertebrae, except the most posterior one, by a **transverse foramen** through which the vertebral blood vessels pass. The last cervical vertebra normally lacks this foramen, and, aside from the absence of rib facets, is much like the first thoracic vertebra. The transverse processes of many of the cervical vertebrae (the fifth or sixth is a good example) also have two parts — a dorsal, pointed transverse portion comparable to a diapophysis, and a ventral, platelike costal portion comparable to a rib. Most of the cervical vertebrae also have low neural spines, and wide neural arches.

The first two cervical vertebrae, the **atlas** and **axis**, respectively, are very distinctive. The atlas is ring-shaped with winglike transverse processes perforated by the transverse foramina. Its neural arch lacks a spine, and is perforated on each side by an **atlantal foramen** through which the vertebral artery enters the skull, and the first spinal nerve leaves the spinal cord. Anteriorly, the neural arch has facets which articulate with the two occipital condyles of the skull; posteriorly, it has facets that articulate with the centrum of the axis. The centrum of the atlas is reduced to a thin transverse rod.

The axis is characterized by an elongated neural spine that extends over the neural arch of the atlas; very small transverse processes; rounded articular surfaces at the anterior end of its centrum; and a median, finger-like **odontoid process** (**dens**) that projects from the front of the centrum of the axis into the atlas.

Although an atlas is present in amphibians, an axis does not appear until reptiles. Together these two vertebrae form a kind of universal joint which permits the freer movement of the head characteristic of amniotes. The composition of the atlas and axis is also somewhat different from that of other vertebrae. The atlas lacks the true centrum of amniotes (pleurocentrum), having instead a small intercentrum. The centrum of the axis is a typical pleurocentrum, but its odontoid process is made up of the intercentrum of the axis and the missing pleurocentrum of the atlas.

(G) *Ribs*

Study the **ribs** (**costae**) of a cat from disarticulated specimens and on a mounted skeleton. Most of them have both heads characteristic of tetrapods — a proximal **head** (**caput**) articulating with the centrum and a more distal **tuberculum** articulating with the transverse process. But the last three ribs have only the head. The portion of the rib between its two heads is its **neck**; the long distal part, its **shaft**. A **costal cartilage** extends from the end of the shaft. Those ribs whose costal cartilages attach directly on the sternum are called **vertebrosternal ribs**; those whose costal cartilages unite with other costal cartilages before reaching the sternum are **vertebrocostal ribs**; those whose costal cartilages have no distal attachment are **vertebral ribs**. Sometimes vertebrosternal ribs are referred to as **true ribs**, and all the rest are referred to as **false ribs**. A floating rib is an alternate name for a vertebral rib. The cat normally has nine vertebrosternal ribs, three vertebrocostal ones, and one vertebral rib.

(H) *Sternum*

The **sternum** is composed of a number of ossified segments called the **sternebrae**. The first of these constitutes the **manubrium**; the last, the **xiphisternum**; and those between, the **body** of the sternum. A cartilaginous **xiphoid process** extends caudad from the xiphisternum. Where do the costal cartilages attach in relation to the sternebrae? In many mammals, but not in the cat, the clavicles of the pectoral girdle articulate with the anterior end of the manubrium.

The vertebral column of mammals is the primary girder of the skeleton.

Not only does it carry the body weight and transfer this to the appendages, but, through its extension and flexion, it participates in the locomotor movements of the body. In order to facilitate interpreting its structure, some authors have compared the vertebral column to a cantilever girder, but Slijper (1946) and others have taken a broader view and have compared the entire trunk skeleton and associated muscles to a bowstring bridge, or to an archer's bow (Fig. 4–15). The vertebral centra represent the main supporting arch, but this is a dynamic arch that has a tendency to straighten out because of the elasticity of certain dorsal ligaments and the tonus of the dorsal muscles. This is prevented by the "bow string"—the sternum and ventral abdominal muscles such as the rectus abdominis. The "bow" and "string" are connected posteriorly by the pelvic girdle, and anteriorly by the stout anterior ribs, which are held in place by the action of such muscles as the scalenus. This hypothesis has the merit of viewing the vertebral column as a dynamic girder capable both of providing support and of participating in the movements of the body.

The neural spines are viewed as muscle lever arms acting on the vertebral centra. The direction of their inclination tends to be perpendicular to the major muscle forces acting upon them. Since several muscles may attach onto a single spine, the situation becomes quite complex, but in carnivores and many other mammal quadrupeds the tips of the spines of the lumbar vertebrae point toward the head, partly in response to a very powerful longissimus muscle (Fig. 6–18, p. 154), whereas the spines of the thoracic vertebrae slope in the opposite direction, partly in response to a powerful splenius. The vertebra near the middle of the trunk, where the angle of inclination of the spines reverses, is called the **anticlinal vertebra**.

If the neural spines are lever arms, an increase in their height increases the length of the lever arms and the mechanical advantage of muscles acting

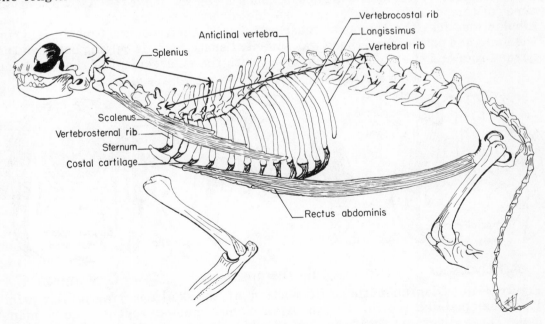

Figure 4–15. Diagram of a lateral view of a cat's skeleton to illustrate Slijper's views on the biomechanics of the trunk. The scapula has been omitted. Heavy arrows indicate the line of action of certain muscles upon the neural spines. (Modified after Slijper.)

upon them. The great length of the anterior thoracic neural spines is considered to be related to the support of a relatively heavy head by the splenius and other muscles.

Head Skeleton

Many changes occur during the evolution of the mammalian skull from the labyrinthodont stage. Although certain of the details are set forth in Table 1 and will be seen during the examination of the cat skull, it is desirable to summarize the major transformations at this point. Correlated with a great increase in brain size, we find that the chondrocranium only suffices to cover the back and underside of the brain. The rest of the braincase is formed by other bones, as will be seen. All the original elements of the chondrocranium are represented, but there has been considerable fusion in the adult. Moreover, a **mesethmoid** ossifies from the originally unossified anterior ethmoid region, and delicate **turbinal bones**, or cartilages, develop from the nasal capsules. The median occipital condyle of primitive vertebrates becomes divided, so that there is a condyle on each side of the foramen magnum.

The major changes in the dermal roof include a loss of the pineal foramen, otic notch, and posttemporal fenestrae; a medial migration of originally widely separated external nares; a reduction in the number of elements through loss and fusion; a ventral extension of certain elements to help complete the braincase (Fig. 4–16); and the appearance of a pair of **temporal fenestrae**. To elaborate on the last, the temporal fenestrae make their appearance in the mammal-like reptiles (Figs. 4–16 and 4–17), and are of the **synapsid** type. That is, there is one on each side of the skull and it was originally laterally placed with the postorbital and squamosal bones meeting above it. As with other types of temporal fenestrae, these provide room for the bulging of certain mandibular muscles (the temporal muscles) during their contraction. In labyrinthodonts and cotylosaurs, there was sufficient space for this bulging between the dermal roof and the chondrocranium, but with the enlargement of the brain and

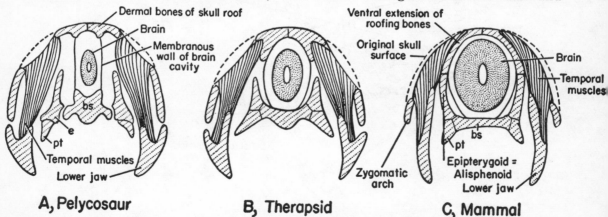

Figure 4–16. Diagrammatic cross section of the skull and jaws of *A*, a primitive mammal-like reptile; *B*, an advanced mammal-like reptile; and *C*, mammal. These diagrams show the way in which the temporal fenestrae "erode" the dermal roof, and how part of the mammalian braincase is formed by the ventral extension of roofing bones, and by the epipterygoid. Abbreviations: *bs*, basisphenoid; *e*, epipterygoid; *pt*, pterygoid. (From Romer, The Vertebrate Body.)

the development of a more powerful jaw apparatus, the problem became critical. At first the temporal fenestrae were small, but they gradually enlarged. In primitive mammals, most of the original lateral surface of the dermal roof in the temporal area has disappeared, and the temporal fenestra and orbit merge. What appears to be the lateral dermal roof in a mammal skull is really the inward extension of flanges of dermal bones that cover the brain. All that is left of the original dermal roof in this region is a strip of bone in the middorsal line, another strip bordering the occipital region of the skull, and the handle-like **zygomatic arch**.

In the palatal region, there is also a reduction in the number of elements. Those that remain unite firmly with the underside of the chondrocranium, so that there is no movement between the palate and braincase. In addition, a **secondary palate** evolves from processes of the premaxillary, maxillary, and palatine bones that extend toward the midline and unite with each other ventral to the original palate. This displaces the openings of the internal nares posteriorly, thus permitting simultaneous respiration and manipulation of food within the mouth. The anterior portion of the primitive palate, now situated dorsal to the secondary palate, regresses to some extent, and this area is occupied by enlarged nasal cavities.

Of the elements of the mandibular arch, the epipterygoid becomes the **alisphenoid** portion of the sphenoid bone and helps to form the lateral wall of the braincase, and the articular and quadrate have become additional auditory ossicles—the **malleus** and **incus**, respectively.

The latter change is closely correlated with changes in the jaw apparatus which gradually appeared during the evolution of the later mammal-like reptiles. Teeth in the lower jaw became restricted to the dentary bone, and the entire jaw apparatus became more powerful. The zygomatic arch bulged laterally, permitting an expansion of the temporal musculature and its extension onto the lateral surface of the dentary. The dentary also enlarged and extended dorsally and posteriorly into the temporal fossa, thereby permitting muscle fibers that attached upon it to have more favorable lines of action. Dentary and squamosal were soon in close juxtaposition. The parts of the lower jaw posterior to the dentary had little importance in the jaw apparatus, other than bearing the joint, and they became quite small. Certain of these bones (quadrate and articular), however, were closely associated with the auditory apparatus. There is evidence that a tympanic cavity and membrane were situated directly behind the jaw joint (cf. Figure 7–6, p. 175), and some investigators have suggested that these bones were part of a bone conduction system carrying sound waves from the lower jaw to the ear. The final change was the development of a new joint between dentary and squamosal, and this made it possible for certain of the posterior jaw elements to become specialized as parts of the auditory apparatus. The change in jaw joint and auditory ossicles is the osteological feature diagnostic of mammals, but this change evolved independently several times during the evolution of different groups of extinct mammals.

The delicate auditory ossicles of mammals (malleus, incus, stapes) are protected by the formation of a plate of bone beneath the tympanic, or middle ear, cavity. Two elements contribute to this bony encasement in most instances—a cartilage replacement **entotympanic** having no homologue in lower vertebrates, and a dermal **tympanic**, or **ectotympanic**), homologous to the angular of the lower jaw of amphibians and reptiles.

(A) *General Features of the Skull*

Examine the skull of a cat (Figs. 4–18 and 4–19). As with other vertebrates, it can be divided into an anterior **facial region** containing the jaw, nose, and eyes, and a posterior **cranial region** housing the brain and ear. Notice the **external nares** at the anterior end of the skull, and the large, circular **orbits**. Although the two external nares may appear contiguous, they are separated in life by a fleshy and cartilaginous septum.

The **foramen magnum** can be seen in the occipital region at the posterior end of the skull. There is an **occipital condyle** on each side of it. The large, round swelling on the ventral side of the skull, anterior to each occipital condyle, is the **tympanic bulla**, which encases the underside of the tympanic cavity. In life, the tympanum is lodged in the opening (**external acoustic meatus**) on the lateral surface of the bulla. The tongue-shaped bump of bone on the lateral surface of the bulla just posterior to the external auditory meatus is the **mastoid process**. A similarly shaped **paraoccipital** (**jugular**) **process** is located on the posterior surface of the bulla.

The handle-like bridge of bone on the side of the skull, extending from the front of the orbit to the external auditory meatus, is the **zygomatic arch**. All of it is included within the facial portion of the skull. A **mandibular fossa** for the articulation of the lower jaw appears as a smooth groove on the ventral surface of the posterior portion of the arch.

A large **temporal fenestra** (**temporal fossa**) is situated dorsal to the back half of the zygomatic arch and posterior to the orbit. Although temporal fenestra and orbit merge to some extent, they tend to be separated by two **postorbital processes**, one of which projects down from

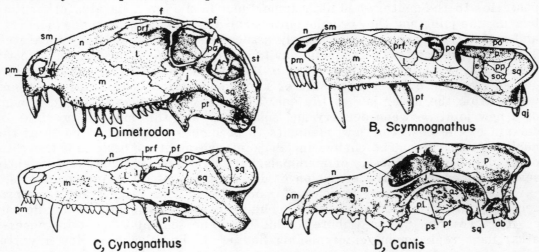

Figure 4–17. Lateral views of the skull of *A,* a primitive mammal-like reptile; *B* and *C,* advanced mammal-like reptiles; *D,* a dog. The enlargement of the temporal fenestra, and its union with the orbit, show particularly well in this series. Abbreviations as in Figure 4–7 except in *D; ps,* presphenoid. (From Romer, The Vertebrate Body, *B,* after Watson; *C,* after Broili and Schroeder.)

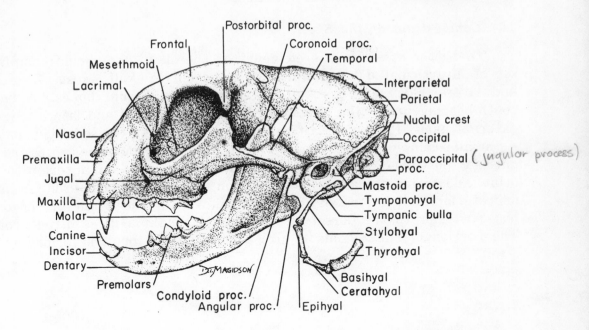

Postorbital proc.
Frontal
Coronoid proc.
Mesethmoid
Temporal
Lacrimal
Interparietal
Parietal
Nuchal crest
Nasal
Occipital
Premaxilla
Paraoccipital (jugular process) proc.
Jugal
Mastoid proc.
Maxilla
Tympanohyal
Molar
Tympanic bulla
Canine
Stylohyal
Incisor
Thyrohyal
Dentary
Basihyal
Premolars
Ceratohyal
Condyloid proc.
Angular proc.
Epihyal

Figure 4–18. Lateral view of the skull, lower jaw, and hyoid apparatus of the cat.

the top of the skull, the other up from the zygomatic arch. The temporal fenestra extends from the orbit and zygomatic arch dorsally and posteriorly over much of the side of the skull. Its posterior border is the **nuchal crest**, a ridge of bone that extends from the mastoid process on one side, over the back and top of the skull, to the mastoid process on the other side. Cervical muscles attach on the posterior surface of the nuchal crest, and mandibular muscles on the anterior surface. The dorsal boundary of the fenestra is the **sagittal crest**, a middorsal ridge of bone extending anteriorly from the middle of the nuchal crest, and a faint ridge (**temporal line**) that curves from the anterior end of the sagittal crest toward the dorsal postorbital process. What is the significance of the temporal fenestra? Are the bones that one sees through the fenestra on the side of the cranium part of the original dermal roof?

The shelf of bone that extends across the palatal region, from the teeth on one side to those on the other, is the **secondary palate** (**bony palate**). In life, it is continued posteriorly by a fleshy **soft palate**. The remains of the primitive, primary palate lies dorsal to the secondary. **Internal nares,** or **choanae,** will be seen dorsal to the posterior margin of the secondary palate. A nearly vertical plate of bone, the **pterygoid process,** continues posteriorly from each side of the secondary palate. The point of bone at its posterior end is called the **hamulus.** A small **pterygoid fossa,** for the attachment of certain mandibular muscles, appears as an elongated groove just lateral and posterior to the hamulus.

(B) Composition of the Skull

To better visualize the elements of the skull, one should examine a set of disarticulated bones along with the entire skull. Most of the top and lateral sides of the skull are formed by dermal bones. A small, tooth-bearing **premaxilla (incisive)** is located ventral and lateral to each external naris. It also contributes a small process to the secondary palate. A **maxilla** completes the upper jaw. It, too, contributes a process to the anterior portion of the secondary palate, sends one up anterior to the orbit, and forms the anteromost portion of the zygomatic arch. A small, delicate **lacrimal** is located in the medial wall of the front of the orbit just posterior to the dorsal extension of the maxilla. It is often broken. The portion of the zygomatic arch ventral to the orbit is formed by the

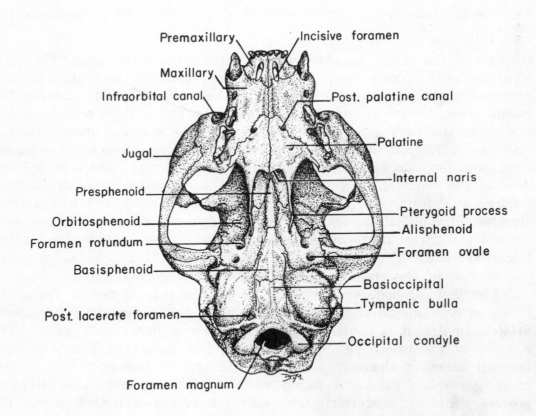

Figure 4-19. Ventral view of the skull of a cat.

jugal (**zygomatic**). A large **temporal** forms the posterior portion of the arch, the adjacent cranial wall, and encases the internal and middle ear.

The temporal bone is compounded of a number of elements fused together (Table 1). Its zygomatic process, and that portion of it that helps to encase the brain (**squamous portion**), represent the squamosal. Of course, the part encasing the brain is an inward extension of the original squamosal of the dermal roof. The mastoid process, and its inward extension which will be seen in the sagittal section of the skull, represent the opisthotic and prootic of the chondrocranium. This part of the temporal may be called the **periotic** (**petrous**). The inner ear is contained within it. The thick part of the tympanic bulla adjacent to the external auditory meatus is the **tympanic**, and is homologous with the dermal angular of the lower jaw. Finally a new cartilage replacement **entotympanic** forms the rest of the bulla, and completes the encasement of the tympanic cavity.

You may be able to get glimpses of the auditory ossicles by looking in the external acoustic meatus, but you should examine a special preparation to see them clearly. The **malleus** is roughly mallet shaped. It has a long, narrow handle that attaches to the tympanic membrane, and a rounded head that articulates with the incus. The **incus** is shaped like an anvil. It has a concave surface for the reception of the malleus, and two processes that extend from the main surface of the bone. One of these articulates with the head of the stirrup-shaped **stapes**. A pair of narrow columns of bone extend from the head of the stapes to a flat, oval-shaped foot plate which fits into the oval window of the otic capsule. A stapedial artery passes between the columns of the stapes embryologically, and also in the adult of some species.

Dermal bones along the top of the skull are a pair of small **nasals** dorsal to the external nares, a pair of large **frontals** dorsal and medial to the orbits, and a pair of large **parietals** posterior to the frontals. The parietals have a long suture with the squamous portion of the temporals. As was the case with squamosal, much of each parietal and frontal represents flanges of the original roofing bones that grew down median to the temporal muscles and helped to cover the brain.

A large, median **occipital** surrounds the foramen magnum and forms the posterior surface of the skull. The paraoccipital processes and most of the nuchal crest are on this element. Ventrally, the occipital forms the floor of the braincase between the tympanic bullae. It is a compound bone containing the four separate occipital elements of the chondrocranium of primitive tetrapods and in most mammals certain dermal elements (Table 1). In the skulls of young cats, a triangularly shaped **interparietal** will be seen in the middorsal line in front of the occipital and between the posterior part of the two parietals. It represents a fusion of the paired postparietals of the dermal roof. In older individuals, it often fuses with the occipital and parietals.

The occipital bone forms the posterior portion of the chondrocranium, and the otic elements are incorporated in the temporal. The rest of the primitive chondrocranium is represented by a part of the sphenoid, the mesethmoid, and the turbinal bones. Although a tiny portion of the **mesethmoid** (**ethmoid**) can sometimes be seen in the medial wall of the orbit posterior to the lacrimal, most of this bone, and the turbinals too, can be seen best in a sagittal section described below.

The **sphenoid** can be observed in the entire skull. Roughly it forms the floor and part of the sides of the braincase anterior to the tympanic bullae and posterior to the secondary palate. In the cat, it is divided into separate anterior and posterior halves. These should be seen in a disarticulated skull to appreciate their extent. The posterior half includes the plate of bone on the underside of the skull just anterior to the ventral (basioccipital) portion of the occipital; the posterior portion of the pterygoid process; the hamulus; the three posterior foramina of a row of four at the back of the orbit; and a winglike process extending dorsally between the squamous portion of the temporal and the frontal. The midventral plate of bone is homologous to the primitive **basisphenoid**, That portion of it which includes the three foramina and dorsal extension is called the **alisphenoid**, and it is homologous to the epipterygoid of more primitive vertebrates. There is some doubt as to the homologies of the pterygoid process and its hamulus, but the consensus is that this region includes the pterygoid of the primitive palate, plus either the ectopterygoid or parasphenoid.

The anterior half of the sphenoid includes a narrow midventral strip of bone lying between the bases of the pterygoid processes, and lateral extensions that pass dorsal to the pterygoid processes to enter the medial wall of the orbits. The lateral portion contains the anterior foramen in the row of four referred to above, and has a common suture with the ventral extension of the frontal bone. The anterior half of the sphenoid is homologous to the sphenethmoid of primitive tetrapods, but in many mammals it ossifies from three centers — a median **presphenoid**, and a pair of lateral **orbitosphenoids**. In some mammals, including man, the anterior and posterior portions of the sphenoid are united and form a single bone.

Certain of the original dermal palatal bones are incorporated in the sphenoid; two others can be seen in more anterior parts of the skull. The originally paired vomers have united to form a single element (the **vomer**) which appears as a midventral strip of bone anterior to the presphenoid, and dorsal to the secondary palate. You will have to look into the internal nares to see it, for it helps to separate the nasal cavities. Paired **palatine** bones form the posterior portion of the secondary palate, the anterior portion of the pterygoid processes, and a bit of the medial wall of the orbits posterior and ventral to the lacrimals.

(C) Sagittal Section of the Skull

Examine a pair of sagittal sections of the cat skull cut in such a way that the nasal septum shows on one half, and the turbinals on the other. Note the large **cranial cavity** for the brain. It can be divided into three parts—a **posterior cranial fossa** in the occipital-otic region for the cerebellum; a large **middle cranial fossa** anterior to this for the cerebrum; and a small, narrow **anterior cranial fossa**, just posterior to the nasal region, for the olfactory bulbs of the brain. In the cat, a partial transverse septum of bone (the **tentorium**) separates the middle and posterior cranial fossae. The internal part of the periotic portion of the temporal, containing the inner ear, can be seen in the lateral wall of the posterior cranial fossa dorsal to the tympanic bulla. It is perforated by a large foramen, the **internal acoustic meatus**. The fossa dorsal to the internal acoustic meatus lodges a lobule of the cerebellum. The saddle-shaped notch (**sella turcica**) in the floor of the cerebral fossa lodges the hypophysis (Fig. 9–14, p. 249). What bone is it in? A large **sphenoidal air sinus** can be seen in the presphenoid region beneath the anterior portion of the cerebral fossa, and a **frontal air sinus** in the frontal bone dorsal to the cerebral and olfactory fossae.

The anterior wall of the anterior cranial fossa is formed by a sievelike plate of bone whose foramina communicate with the nasal cavities. This plate of bone, the **cribriform plate** of the **mesethmoid**, can also be seen by looking through the foramen magnum of a complete skull. A **perpendicular plate** of the mesethmoid, which will be seen on the larger section, extends from the cribriform plate anteriorly between the two nasal cavities. It connects with the nasal bones dorsally and with the vomer ventrally, and forms much of the nasal septum. The very front of the septum is cartilaginous. Most of the mesethmoid is shaped like the letter T, the top of the T being the cribriform plate, and its stem the perpendicular plate. In addition, the mesethmoid has little lateral processes that often can be seen in the medial wall of the orbits posterior to the lacrimal.

Each nasal cavity is largely filled with thin, complex scrolls of bone called the **turbinals (conchae)** (Fig. 9–14). They will show best in the section that lacks the perpendicular plate of the mesethmoid. Although the turbinals ossify from the nasal capsules, they become attached to one of the bones surrounding the nasal cavities. It is difficult to see the details, but there are three major scrolls in each cavity—a small **nasoturbinal** located for the most part dorsal to the anterior end of the cribriform plate, a large **ethmoturbinal** extending anteriorly from the cribriform plate and sphenoidal sinus, and a small **maxilloturbinal** located ventral and lateral to the anterior portion of the ethmoturbinal. Much of the air entering the nasal cavities continues directly back through an uninterrupted passage (**inferior nasal meatus**) between the turbinals and secondary palate, but some of it filters through the turbinals, which are covered with the nasal epithelium and thus serve to increase the surface available for olfaction and air conditioning (cleansing, etc.).

(D) Foramina of the Skull

Numerous foramina for nerves and blood vessels perforate the skull. These may be considered at this time, or their consideration may be postponed until the nerves have been studied. In any case, the foramina should be studied on a sagittal section of the skull so that the student can see both their external and internal aspects. Certain of the passages should be probed to determine their course.

First study the foramina for the 12 cranial nerves. The first cranial nerve, the olfactory, consists of many fine subdivisions that enter the cranial cavity through the **olfactory foramina** in the cribriform plate of the mesethmoid. The **optic canal**, for the second cranial nerve, is the most anterior of the row of four foramina in the posteromedial wall of the orbit. Next posterior, and largest in this row, is the **anterior lacerate foramen (orbital fissure)** for the third, fourth, and sixth nerves going to the muscles of the eyeball, and for the ophthalmic division of the fifth nerve. The third and smallest foramen in this row (**foramen rotundum**, Fig. 4-19) transmits the maxillary division of the fifth nerve, and the last in this row (**foramen ovale**) the mandibular division of this nerve. The **internal acoustic meatus**, seen in the periotic when you studied the sagittal section, transmits both the seventh and eighth nerves. The eighth (vestibulocochlear) nerve comes from the inner ear, but the seventh (facial nerve) continues through a facial canal within the periotic and emerges on the undersurface of the skull through the **stylomastoid foramen** located beneath the tip of the mastoid process. The ninth, tenth, and eleventh nerves, in company with the internal jugular vein, pass through the **posterior lacerate foramen (jugular foramen)** which can be seen internally in the floor of the posterior cranial fossa posterior to the internal acoustic meatus. The posterior lacerate foramen opens on the ventral surface of the skull beside the posteromedial edge of the tympanic bulla. The **hypoglossal canal** for the twelfth cranial nerve can be seen in the floor of the posterior cranial fossa posterior and medial to the posterior lacerate foramen. Probe to see that the hypoglossal canal extends anteriorly and emerges on the ventral surface in common with the posterior lacerate foramen.

As the ophthalmic and maxillary divisions of the fifth (trigeminal) nerve are distributed to various parts of the head, certain of their branches pass through other foramina, often in company with blood vessels. A branch of the ophthalmic division traverses a small **ethmoid foramen**, or series of foramina, in the medial wall of the orbit to enter the nasal cavity. The ethmoid foramen lies in the frontal bone very near its suture with the orbitosphenoid. Turning to the maxillary division, one of its branches emerges through an **infraorbital canal** located in the anterior part of the zygomatic arch. But before reaching the infraorbital canal, this branch has subsidiary branches that pass through several small foramina near the front of the orbit to supply the teeth of the upper jaw. A second branch of the maxillary division leaves the anterior portion of the orbit through a **sphenopalatine foramen** to enter the ventral part of the nasal cavity. The sphenopalatine foramen is the larger, and more medial, of two foramina that lie close together in the orbital process of the palatine bone. After entering the nasal cavity, part of this nerve continues forward through the nasal passages and finally drops down to the roof of the mouth through an **incisive foramen**, one of which is located on each side of the midline at the anterior end of the secondary palate. In certain mammals, the incisive foramina also carry nasopalatine ducts that lead from the mouth to the vomeronasal organs (Jacobson's organs, p. 173) located in the nasal cavities. The small foramen lateral and anterior to the sphenopalatine is the posterior end of the **posterior palatine canal**. A third branch of the maxillary division runs through this canal to the roof of the mouth. The anterior end of the canal can be seen on the ventral surface of the secondary palate in, or near, the suture between the palatine and maxillary bones. A fourth branch of the maxillary backtracks to enter the orbital fissure. It then passes into a small **pterygoid canal** whose anterior end may be seen in the floor of the anterior lacerate foramen. The posterior end of the canal appears as a tiny hole on the ventral side of the skull between the basisphenoid and the base of the pterygoid process of the sphenoid. After emerging from the pterygoid canal, this branch enters the tympanic cavity through the osseous portion of the **auditory tube**—a large opening at the anterior edge of the tympanic bulla.

The major foramina that remain do not carry nerves. A **nasolacrimal canal** for the lacrimal or tear, duct will be seen in the lacrimal bone extending from the orbit into the nasal cavity. You may also be able to find parts of the small **carotid canal** for the vestigial (in the cat) internal carotid artery. The posterior end of the canal appears as a tiny hole in the anteromedial wall of the posterior lacerate foramen. From here the canal extends forward, dorsal to the tympanic bulla, and enters the cranial cavity through the **middle lacerate foramen**. This foramen can be seen in the posterior cranial fossa anterior to the periotic and ventral to the tentorium. A **condyloid canal**, for a small vein, can be found in the posterior cranial fossa dorsal to the hypoglossal canal.

If you have access to a specimen in which the tympanic bulla has been removed, you will be able to see two openings on the underside of the periotic. The more dorsal is the **fenestra vestibuli** or **ovalis** for the stapes, the more ventral the **fenestra cochleae** or **rotunda** for the release of vibrations from the inner ear.

(E) Lower Jaw

With the transfer of certain of the original lower jaw bones to the ear region, and the loss of others, the pair of enlarged dentaries are left as the sole elements in the mammalian lower jaw.

Examine the lower jaw, or **mandible**, of the cat (Fig. 4–18) and notice that the **dentary bones** are firmly united anteriorly by a **mandibular symphysis**. The horizontal part of the dentary that bears the teeth is its **body**; the part posterior to this, its **ramus**. A large, triangular-shaped depression, **coronoid fossa**, occupies most of the lateral surface of the ramus. Part of the masseter, one of the mandibular muscles, inserts here. Posteriorly, the ramus has three processes — a dorsal **coronoid process**, to which the temporal muscle attaches; a middle, rounded **condyloid process** for the articulation with the skull proper; and a ventral **angular process** to which other mandibular muscles attach.

A large **mandibular foramen** will be seen on the medial side of the ramus, and two small **mental foramina** on the lateral surface of the body near its anterior end. A branch of the mandibular division of the trigeminal nerve, supplying the teeth and the skin covering the lower jaw, enters the mandibular foramen and emerges through the mental foramina. Blood vessels accompany the nerve.

(F) Teeth

The teeth of mammals are quite different from those of lower vertebrates for they are limited to the jaw margins, are set in deep sockets (**thecodont**), and are differentiated into various types (**heterodont**). Most adult mammals have in each side of each jaw a series of nipping incisors at the front, a large canine behind these, a series of cutting premolars behind the canine, and finally a series of chewing, or grinding, molars. The number of each kind of teeth present in a particular group of mammals may be expressed as a dental formula. For primitive placental mammals this was $\frac{3.1.4.3}{3.1.4.3} \times 2 = 44$. Such an animal had three incisors, one canine, four premolars, and three molars in each side of each jaw. The number of teeth, and their structure, are adapted to the animal's mode of life. The molars especially are subject to much divergence among the groups of mammals.

Examine the teeth of the cat (Figs. 4–18 and 4–19). In each side of the upper jaw there are normally three **incisors** at the anterior end followed by one **canine**, three **premolars**, and one very small **molar**. In the lower jaw there are three incisors, one canine, two premolars, and one large molar. What would the dental formula be? In which bones are the teeth of the upper jaw located?

During its evolution, the cat has lost the first premolar in the upper jaw, the first two in the lower, and all molars posterior to the first. The gap left between each canine and the premolars is called a **diastema**. The first of the remaining premolars, and the molar of the upper jaw, are more or less vestigial. But the last premolar of the upper jaw (phylogenetically premolar number four), and the lower molar have become large and complex in structure. Articulate the jaws and note how these two teeth, which are known as the **carnassials**, intersect to form a specialized shearing mechanism. Carnassials are restricted to carnivores, and in contemporary species have the formula Pm 4/M 1.

(G) *Hyoid Apparatus*

A hyoid apparatus is present in mammals, but is not as large as in more primitive tetrapods, for it is formed from only the ventral parts of the hyoid and first branchial arches. In so far as more posterior arches are present, they are incorporated in the laryngeal and possibly the tracheal cartilages (p. 251).

Study the hyoid apparatus of the cat (Fig. 4–18). It may be in place on the skeleton, or removed and mounted separately. It consists of a transverse bar of bone, the **body of the hyoid** or **basihyal**, from which two pairs of processes (horns) extend anteriorly and posteriorly. The **posterior horns** are the smaller, and each consists of but one bone, which, although it is called the **thyrohyal**, is derived from the first branchial arch. Each of the larger anterior horns, consisting of four little bones, curves anteriorly and dorsally to attach to the tympanic bulla just medial to the stylomastoid foramen. The components of an anterior horn are, from dorsal to ventral, the **tympanohyal**, **stylohyal**, **epihyal**, and **ceratohyal**. These are derivatives of the hyoid arch. The body of the hyoid develops partly from the hyoid arch and partly from the first branchial arch.

In some mammals, man for example, the hyoid apparatus consists of a single bone, the hyoid, having a body, a small anterior horn, and a large posterior horn. A stylohyoid ligament, which extends from a styloid process at the base of the skull to the anterior horn, replaces the chain of ossicles seen in the cat as the support for the apparatus. The styloid process itself represents a part of the hyoid arch fused onto the temporal bone.

5 / The Appendicular Skeleton

AS STATED in the preceding chapter, the appendicular skeleton consists of the bones of the paired appendages, and the girdles to which the appendages attach. Most vertebrates have both **pectoral** (shoulder) and **pelvic** (hip) **girdles** and **appendages**. The appendicular skeleton is basically a part of the endoskeleton, so it consists primarily of cartilage or cartilage replacement bone. However, dermal bones have become intimately associated with the cartilaginous elements of the pectoral girdle in the majority of vertebrates. That portion of the girdle formed of cartilage, or cartilage replacement bone, is called the **endo-skeletal girdle**; that portion formed of dermal bone, the **dermal girdle**.

FISHES

An appendicular skeleton is usually absent in primitive agnathous vertebrates—recall *Petromyzon*—but is typically present in gnathostomes. Although some ostracoderms had pectoral flaps, paired appendages were not commonly found until the placoderms. They were exceedingly variable in structure and number among the members of this class, but in cartilaginous and bony fishes they are in the form of pectoral and pelvic fins that help to keep the fish in the horizontal plane, and aid in steering.

The Chondrichthyes, and the actinopterygian group of Osteichthyes, have fins that are somewhat fan shaped. Primitively, such fins had a wide base and were supported by numerous parallel bars (Fig. 5–1, *A*), but in living members of these groups, the base of each fin has become constricted, so that the basal elements are crowded together and reduced in number (Fig .5–1, *B*). In the sarcopterygian group of bony fish, the fin is elongate and lobate in shape. It is supported by a single basal element and a central axis from which side branches arise (Fig. 5–1, *C*). This type of fin, called an **archipterygium**, is probably derived from the wider-based fin, and, in turn, is the type of fin from which the tetrapod limb evolved (Fig. 5–1, *D*).

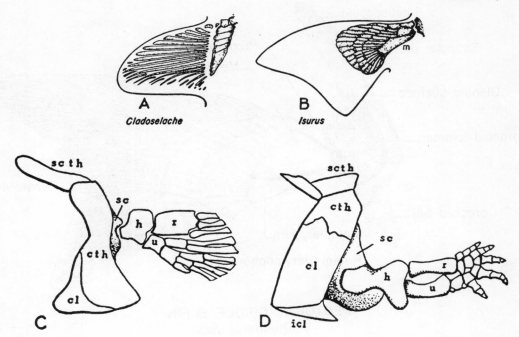

Figure 5-1. Pectoral fin of *A*, a primitive fossil shark, and *B*, a modern shark; pectoral fin and shoulder girdle of *C*, a crossopterygian, and *D*, a labyrinthodont. All figures are of the left side. Abbreviations: *cl*, clavicle; *cth*, cleithrum; *h*, humerus; *icl*, interclavicle; *m*, metapterygium: *r*, radius; *sc*, scapulocoracoid (endoskeletal girdle); *scth*, supracleithrum; *u*, ulua. The metapterygium of the tribasic selachian fin is believed to be homologous to the single basal of the archipterygium fin of crossopterygians. (From Romer, The Vertebrate Body; *A*, after Dean; *B*, after Mivart; *C*, after Gregory.)

Pectoral Girdle and Fin

(A) *Squalus*

The appendicular skeleton of *Squalus,* which will be used as the chief example of the fish condition, is reasonably representative of the type found in most living fishes. But, like other parts of the skeleton, it is atypical in being entirely cartilaginous. Also, the dermal bones that early became associated with the pectoral girdle are absent, but can be seen in one of the primitive bony fishes such as *Amia*.

Examine the pectoral girdle and fin on a skeleton of the dogfish (Fig. 5-2). The girdle is located just posterior to the visceral skeleton, for there is no neck region in fishes. It consists of a U-shaped bar of cartilage, which represents the endoskeletal pectoral girdle. A dermal girdle, as stated, is absent. At the top of each limb of the U there is a separate **suprascapular cartilage** about one half an inch long. The rest of the girdle is formed of a single piece, the **scapulocoracoid**. In some sharks, but not in *Squalus*, it is obvious that the scapulocoracoids are paired elements that have fused in the midventral line. That part

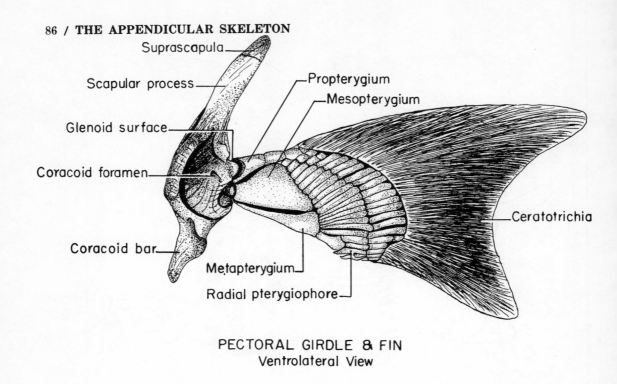

PECTORAL GIRDLE & FIN
Ventrolateral View

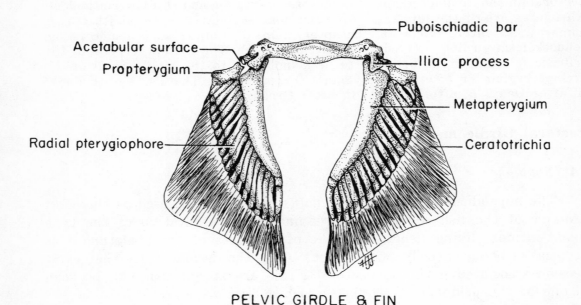

PELVIC GIRDLE & FIN
Dorsal View

Figure 5–2. Drawings of the girdles and fins of *Squalus*.

of the scapulocoracoid which is located ventral to the point where the fin attaches (**glenoid surface**) may be called the **coracoid bar**, for it has the same position as the coracoid of tetrapods. The rest of it may be called the **scapular process**. A small **coracoid foramen** for vessels and nerves can be found just anterior to the glenoid surface.

The pectoral fin is narrow at the base, but widens distally. It is supported proximally by a series of cartilages, collectively called the **pterygiophores**, and distally by fibrous fin rays (**ceratotrichia**) that develop in the dermis on each surface of the fin. The three large pterygiophores that articulate with the girdle are the **basals**; the rest are **radials**. The three basals are, from anterior to posterior, the **propterygium, mesopterygium,** and **metapterygium**. Note that the metapterygium is the longest.

In many cartilaginous fishes and actinopterygians, the metapterygium forms the axis of the fin. It is probable that the metapterygium is the element that persists as the single "basal" in the archipterygium fin of sarcopterygians.

(B) Amia

Examine a skeleton of the bowfin, *Amia*. Notice that the **endoskeletal girdle** consists, on each side, of a small area of unossified cartilage which lies between the pectoral fin and a conspicuous arch of bone posterior to the **operculum** (dermal gill covering). This arch of bone, which is of dermal origin, consists of four elements. A large, ventral **cleithrum** begins ventral to the posterior visceral arches and continues dorsally a bit beyond the fin and endoskeletal girdle. A **supracleithrum** continues from the cleithrum nearly to the roof of the skull, and a **posttemporal**, which is a part of the girdle even though it has surface sculpturings similar to the skull bones, joins the supracleithrum and skull. Finally, a small **postcleithrum** is located posterior to the junction of cleithrum and supracleithrum. A clavicle proper is absent in *Amia* although such a bone was found in crossopterygian fishes ancestral to the tetrapods.

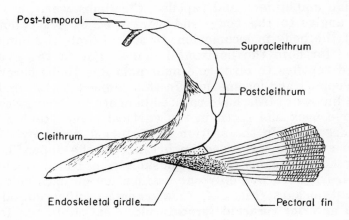

Figure 5–3. A lateral view of the pectoral girdle and fin of *Amia*; cartilage is stippled.

Pelvic Girdle and Fin

The pelvic girdle of *Squalus* consists of a simple transverse rod of cartilage, the **puboischiadic bar,**[9] located in the ventral abdominal wall just anterior to the cloaca. Each end of it extends dorsally as a very short **iliac process,** but these processes do not reach the vertebral column. There are often two small nerve foramina near the base of each iliac process.

Like the pectoral fin, the pelvic consists of a series of proximal cartilaginous pterygiophores and distal ceratotrichia. There are only two **basal cartilages** in *Squalus* — a long **metapterygium** extending posteriorly from the girdle, and clearly forming the main support of the fin, and a short anterior **propterygium** projecting laterally from the girdle. A number of **radials** extend into the fin from the two basals. In males, a **clasper,** the skeleton of which is formed by enlarged and modified radials, extends caudad from the posterior end of the metapterygium. The fin attaches to the **acetabular surface** of the girdle.

PRIMITIVE TETRAPODS

In the evolution from crossopterygian fishes to tetrapods, the archipterygium is transformed into a pentadactyl limb (Fig. 5–1, *C* and *D*), consisting of three segments. In the pectoral appendage these segments are the **brachium** (containing the **humerus**), **antibrachium** (containing the **radius** and **ulna**), and **manus** (**carpals, metacarpals,** and **phalanges**). Corresponding segments and bones in the pelvic appendage are **thigh** (**femur**), **shank** (**tibia** and **fibula**), and **pes** (**tarsals, metatarsals,** and **phalanges**). The number of phalanges present in the toes of different groups of tetrapods varies. In early reptiles (cotylosaurs) there were two phalanges in the first toe, three in the second, four in the third, five in the fourth, and either three or four in the fifth toe. This can be expressed as a **phalangeal formula,** i.e., 2-3-4-5-3 (or 4).

In primitive amphibians and reptiles, the limbs were in a sprawled position at right angles to the body, and were used as a supplement to lateral undulations of the body in locomotion. In such a limb, the humerus and femur move back and forth in the horizontal plane. But in the evolution through more advanced reptiles to birds and mammals, the limbs became increasingly important in support and locomotion. In this connection the limbs of mammals, and the pelvic limbs of birds, have rotated beneath the body so that the humerus and femur move back and forth in the vertical plane. Review the changes in limb posture described in the section on External Anatomy (p. 37).

Correlated with the increased importance of the appendages, the girdles became larger and stronger. In the endoskeletal pectoral girdle of labyrintho-donts (Fig. 5–1, *D*) there was only one ossification on each side (a **scapulocoracoid**), but in cotylosaurs (Fig. 5–4, *A*) a **scapula** ossified dorsal to the glenoid cavity, and an **anterior coracoid** (**procoracoid**) ventral to the fossa. Although these elements were large and platelike, the endoskeletal girdles of opposite sides neither united with each other ventrally nor connected with the vertebral column. Correlated with the evolution of a distinct neck, the dermal pectoral girdle lost its connection with the back of the skull. However, the cleithrum

---

[9] Ischium has four acceptable adjective forms — ischiac, ischiadic, ischial, ischiatic.

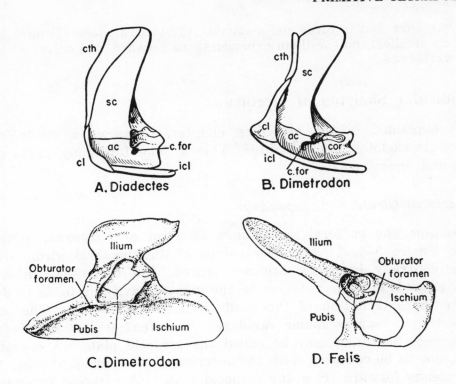

Figure 5-4. *A*, pectoral girdle of a cotylosaur; *B*, pectoral girdle of a mammal-like reptile; *C*, pelvic girdle of a mammal-like reptile; *D*, pelvic girdle of a mammal. All are lateral views of the left side. Abbreviations: *ac*, anterior coracoid; *c.for*, coracoid foramen; *cl*, clavicle; *cor*, posterior coracoid (or coracoid of mammals); *cth*, cleithrum; *icl*, interclavicle; *sc*, scapula. The acetabular bone, although not labeled, can be seen in the acetabulum of *Felis* above the pubis. The clavicle, interclavicle, and cleithrum are dermal elements. All others are cartilage replacement bone. (From Romer, The Vertebrate Body.)

and clavicle persisted (Fig. 5-4, *A*), and a new dermal element, the **interclavicle**, became associated with the girdle. The interclavicle is a median ventral element which connects the clavicles of opposite sides.

The endoskeletal pectoral girdle of living amphibians, reptiles, and birds is derived from the type just described and is usually close to it in essential pattern. But in the evolution toward mammals, a third ossification, the **posterior coracoid** (or simply coracoid) appeared in the endoskeletal girdle posterior to the anterior coracoid (Fig. 5-4, *B*). In mammals, the scapula area is greatly expanded, and the coracoid region is reduced. The anterior coracoid is completely lost in placental mammals, and the posterior coracoid is represented only by a small coracoid process of the scapula. In all living tetrapods the dermal portions of the girdle have become reduced, the cleithrum being completely lost, except in certain primitive frogs.

The pelvic girdle similarly enlarged during the transition from water to land, and in tetrapods typically consists of three cartilage replacement bones — a **pubis** and **ischium** ventral to the acetabulum, and an **ilium** that extends from the acetabulum to the sacral ribs and vertebrae (Fig. 5-4, *C* and *D*). The ventral elements of opposite sides unite with each other. Primitively the ventral elements formed a broad plate, and the smaller ilium connected with only a single sacral vertebra. But in birds and mammals, the ventral elements have become relatively smaller, and the ilium expands, turns forward, and unites with more sacral vertebrae.

Appendicular Skeleton of Necturus

As explained in the preceding chapter, *Necturus* is neotenic, and parts of its skeleton are unossified. This is true for many parts of the girdles and appendages.

(A) *Pectoral Girdle and Appendage*

Examine the pectoral girdle on a skeleton of *Necturus,* comparing it with Figure 5-4, *A*. The two halves of the pectoral girdle overlap slightly ventrally but they are not united. On each side an ossified **scapula** extends dorsally anterior to the **glenoid cavity** or **fossa** (a depression for the articulation of the girdle with the humerus). The scapula is capped by a **suprascapular cartilage**. The ventral part of the girdle remains unossified, and may be called the **coracoid plate**. A **procoracoid process**, not to be confused with the anterior coracoid bone of other tetrapods, extends forward from the coracoid plate. A **coracoid foramen**, for vessels and nerves, may be seen in the coracoid plate ventral to the scapula.

Study the pectoral skeleton, noting first the position of the different segments of the limb, and the preaxial and postaxial surfaces (see page 35). A single bone, the **humerus**, extends from the glenoid cavity to the elbow joint. Two bones of approximately equal size compose the forearm — a **radius** on the preaxial (medial) side, and an **ulna** on the postaxial (lateral) side. The manus consists of a group of six cartilaginous **carpals** in the wrist, four ossified **metacarpals** in the palm of the hand, and the ossified **phalanges** of the digits or toes. The individual carpals are usually not distinct in dried skeletons. Note that only four toes are present, a number characteristic of living amphibians. It is uncertain whether this is a retention of the condition found in many primitive amphibians, or whether it resulted from the loss of a toe present in certain five-toed, primitive amphibians. In any case, the consensus is that the toes present are homologous to the second through the fifth toes of amniotes. What is the phalangeal formula?

(B) *Pelvic Girdle and Appendage*

Each half of the pelvic girdle has a narrow ossified **ilium** that extends dorsally from the socket for the leg articulation (**acetabulum**) to attach on a single sacral rib and vertebra. Ventrally, there is a broad **puboischiadic** plate, which contains a pair of ossified **ischia** posteriorly and a **pubic cartilage** anteriorly. An **obturator foramen**, for a nerve of the same name, may be seen in the pubic cartilage. Note that the pelvic girdle, together with the sacral rib and vertebra, forms a ring of bone around the posterior end of the trunk. The passage through this ring is called the **pelvic canal**. Structures that lead to the cloaca necessarily pass through this canal.

Study the pelvic limb, noting its position and its preaxial and post-axial surfaces. A **femur** forms the upper segment of the limb; a **tibia** and **fibula** are contained within the shank (the former lying along the preaxial surface); and the pes consists of a group of six cartilaginous **tarsals** and ossified **metatarsals** and **phalanges**. The individual tarsals cannot be distinguished in dried skeletons. Many salamanders have five toes, but *Necturus* has only four, the most medial probably being homologous to the second toe of amniotes. What is the phalangeal formula?

Appendicular Skeleton of the Turtle

Although the appendicular skeleton of the turtle is specialized in some respects, certain primitive features can be seen better in it than in the appendicular skeleton of *Necturus*.

(A) *Pectoral Girdle and Appendage*

Examine a mounted skeleton and an isolated pectoral girdle of the turtle (Fig. 5-5). The endoskeletal pectoral girdle, which has an unusual

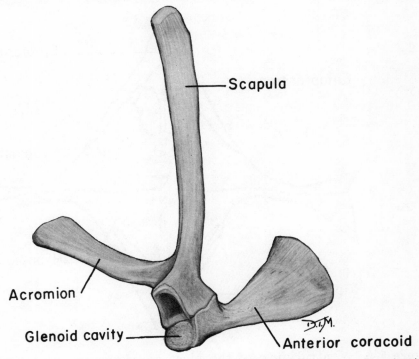

Figure 5-5. Lateral view of the left pectoral girdle of the snapping turtle, *Chelydra*. Anterior is toward the left.

triradiate shape, is located between the bottom (**plastron**) and top shell (**carapace**). Hence, it lies partly beneath the ribs. This is a feature peculiar to turtles. The dorsal prong of the endoskeletal girdle, which articulates with the carapace, represents the **scapula**; the anteroventral prong, a part of the scapula known as the **acromion**; and the expanded posteroventral prong, the **anterior coracoid**. In life, the acromion is connected by a ligament to the **entoplastral plate** of the plastron. An **acromiocoracoid ligament** may be seen extending between the tips of the acromion and coracoid. A **glenoid cavity** is present for the articulation of the arm.

Parts of the dermal girdle are present but they are incorporated in the anterior plates of the plastron. Examine a plastron in which the epidermal scutes (**lamina**) have been removed, and the **dermal plates** exposed (Fig. 5-6). The front of the plastron is formed by a pair of **epiplastra**, posterior to which are a median **entoplastron** and three additional paired plates — **hyoplastra**, **hypoplastra**, and **xiphiplastra**. All these plates represent in part an ossification in the dermis of the skin of the underside of the body. But during embryonic development the originally separate primordia of the clavicles and interclavicle become incorporated in the first three plates, which are therefore compound plates. The epiplastra include the paired **clavicles**; the entoplastron includes the **interclavicle**. The remaining plastral plates may include the **gastralia** of early amphibians and reptiles. In these primitive tetrapods, the gastralia were riblike rods of dermal bone found on the ventral abdominal wall. They

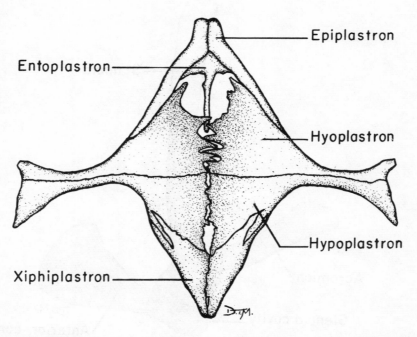

Entoplastron

Epiplastron

Hyoplastron

Hypoplastron

Xiphiplastron

Figure 5–6. Ventral view of the plastron of the snapping turtle. The epidermal scutes have been removed.

were remnants of the piscine dermal scales, and may have had a protective function. They are retained in a few living reptiles — *Sphenodon,* crocodiles, and, possibly, turtles.

Study the pectoral appendage (Fig. 5-7). How does the position of the limbs of a turtle compare with that of one of the early amphibians or reptiles (labyrinthodont or cotylosaur)? Identify the preaxial and postaxial surfaces. The long bones of the appendage are a humerus in the upper arm, and a **radius** and **ulna** in the forearm. The proximal end of the humerus has a round **head** that fits into the glenoid cavity and two prominent enlargements (processes) for the attachment of muscles. Of the forearm bones, the radius is the one on the preaxial (anterior or medial) surface. Both radius and ulna are about equal in size, but the ulna tends to extend over the distal end of the humerus, while the radius articulates on the underside of the end of the humerus. As in primitive tetrapods generally, both the radius and ulna articulate with the wrist bones, and there is no distal radioulnar joint.

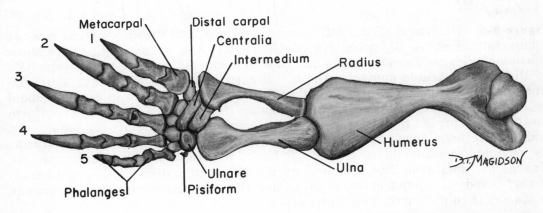

Figure 5-7. Dorsal view of the left pectoral appendage of the snapping turtle.

The manus consists of a group of **carpals** in the wrist, a row of five **metacarpals** in the palm, and the **phalanges** in the free part of the toes. There are five toes, the first being the most medial. What is the phalangeal formula?

The carpus of the turtle is very similar to that of more primitive tetrapods, and the individual components should be identified. The carpals can be grouped into a proximal and distal row. The proximal row consists of three bones — an **ulnare** adjacent to the ulna, an **intermedium** lying between the distal ends of the radius and ulna, and an elongated element distal to the radius and intermedium, which represents a fusion of two **centralia**. The **radiale** (Fig. 5-8, *B*), which one would expect to see distal to the radius, has been lost or reduced to a small nubbin of bone on the edge of the carpus medial to the fused centralia. The distal row consists of five distal carpals which are numbered according to the digit

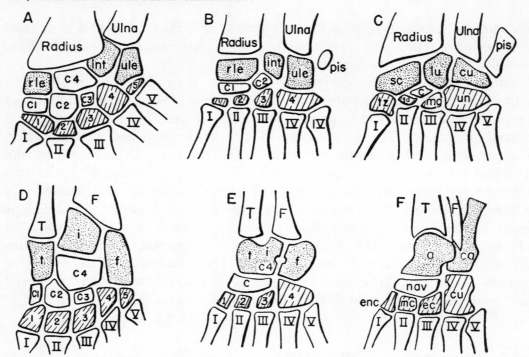

Figure 5–8. Diagrams of the left carpus (top row) and tarsus (bottom row) of a labyrinthodont (*A, D*), primitive reptile (*B, E*), and mammal (*C, F*), to show the homologies between the elements. Roman numerals indicate the metapodials of the digits; Arabic numerals the distal carpals and tarsals. Other abbreviations: *a*, astragalus; *c*, centralia; *ca*, calcaneus; *cu*, cuneiform in carpus, cuboid in tarsus; *ec*, external cuneiform (ectocuneiform); *enc*, internal cuneiform (entocuneiform; *f*, fibulare; *F*, fibula; *i, int,* intermedium; *lu*, lunar; *mc*, middle cuneiform (mesocuneiform); *mg*, magnum; nav, navicular; *pis,* pisiform; *rle*, radiale; *sc*, scaphoid; *t*, tibiale; *T*, tibia; *td,* trapezoid; *tz,* trapezium; *ule,* ulnare; *un*, unciform. The terminology for the mammalian carpus and tarsus is that used in comparative studies, not the Nomina Anatomica of human anatomy. (From Romer, The Vertebrate Body.)

to which they are related—distal carpal 1, distal carpal 2, etc. In addition there may be a small **sesamoid** bone on the lateral edge of the carpus. Sesamoid bones develop in the tendons of muscles, and they are rather variable. However, the one on the lateral edge of the wrist adjacent to the ulnare is consistent enough to be given a name—the **pisiform**.

(B) Pelvic Girdle and Appendage

Study the pelvic girdle of the turtle on a mounted skeleton and from an isolated specimen (Fig. 5-9). Each half of the girdle consists of a dorsal **ilium** that inclines posteriorly and articulates with two sacral ribs and vertebrae, an anteroventral **pubis**, and a posteroventral **ischium**. All three elements share in the formation of the **acetabulum**, the socket for the leg articulation. Pubis and ischium of opposite sides unite by a **symphysis**. An **epipubic cartilage**, which may be partly ossified, extends

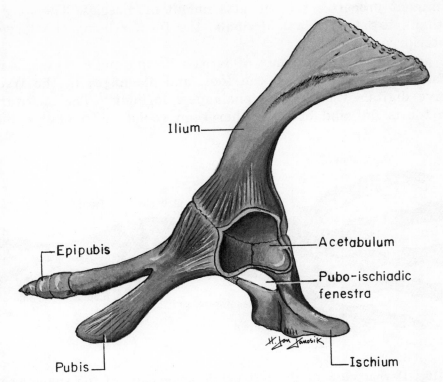

Figure 5-9. Lateral view of the left side of the pelvic girdle of the snapping turtle, *Chelydra*. Anterior is toward the left.

forward in the midventral line from the pubic bones. Both the pubis and ischium have a lateral process that is often directed ventrally and, in life, rests on the plastron. A large **puboischiadic fenestra**, which develops in association with the origin of a muscle, lies between the pubis and ischium on each side. A separate obturator foramen seen in many lower vertebrates is not present in turtles, for the obturator nerve also passes through the puboischiadic fenestra.[10]

Examine the pelvic limb, noting its position and its preaxial and postaxial surfaces (Fig. 5-10). The long bone of the thigh is the **femur**. Its proximal end bears a round **head** that fits into the acetabulum, and

[10] The terminology and homology of the various pelvic openings are unfortunately confused. In most amphibians and reptiles, an obturator foramen (pubic foramen) for an obturator nerve perforates the pubis anterior to the acetabulum (Fig. 5-4, *C*). The rest of the puboischiadic plate is solid. In some reptiles (lizards) an additional opening, known as the puboischiadic fenestra (thyroid fenestra), develops between the pubis and ischium in association with the origin of certain pelvic muscles. In mammals, the fenestration of the puboischiadic plate includes both the primitive obturator foramen and the puboischiadic fenestra. Such an opening is also termed an obturator foramen. The turtle would seem to parallel this condition, but most authorities call the opening a puboischiadic fenestra.

two prominent processes for the attachment of muscles. The long bones of the shank are the **tibia** and **fibula**, the former being the larger and more anterior or medial.

The pes consists of a group of **tarsals** in the ankle region, a row of five **metatarsals** in the sole of the foot, and **phalanges** in the free part of the five digits. What is the phalangeal formula? The metatarsal of the fifth toe is flat and broad, rather than round and elongate like the others.

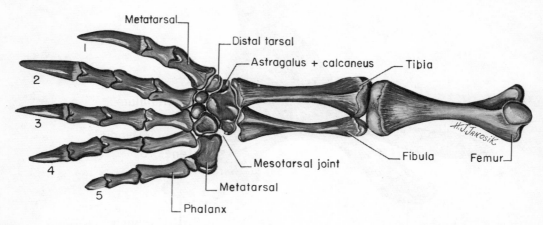

Figure 5–10. Dorsal view of the left pelvic appendage of the snapping turtle.

The individual tarsals should be studied and compared with Figure 5–8, *D*. There is a row of four **distal tarsals** next to the metatarsals. The fourth distal tarsal is larger than the others and is associated with the fourth and fifth toe. All the remaining elements of the tarsus tend to fuse into a single bone, but one can often see the lines of union between the three major elements: **astragalus** (tibiale, intermedium, and proximal centrale), **distal centrale**, and **calcaneus** (fibulare). Sometimes the calcaneus remains distinct. The main ankle joint of the turtle, as in many reptiles, is a **mesotarsal joint**, for it lies between the large proximal element(s) and the distal tarsals.

MAMMALS

The appendicular skeleton of mammals is much better suited than that of primitive tetrapods for terrestrial locomotion. Note the position of the limbs beneath the body on a mounted skeleton of the cat. How has the changed position of the limbs come about? Where are the original preaxial and postaxial surfaces now located? Primitive mammals walked with the soles of their feet on the ground (**plantigrade**). But in such mammals as the cat, additional leverage is provided by walking on the digits with the soles of the feet off the ground (**digitigrade**). Ungulates, such as the horse and cow, carry the tendency further and walk on the toe tips (**unguligrade**).

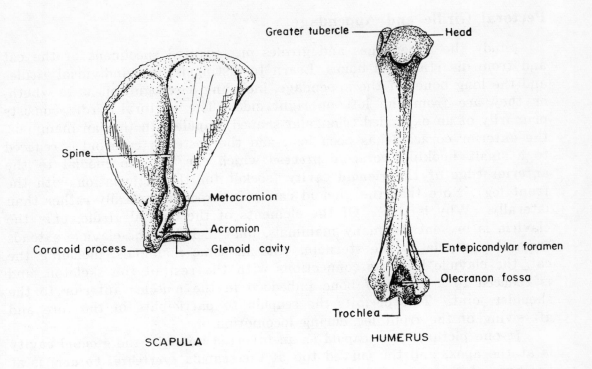

SCAPULA

HUMERUS

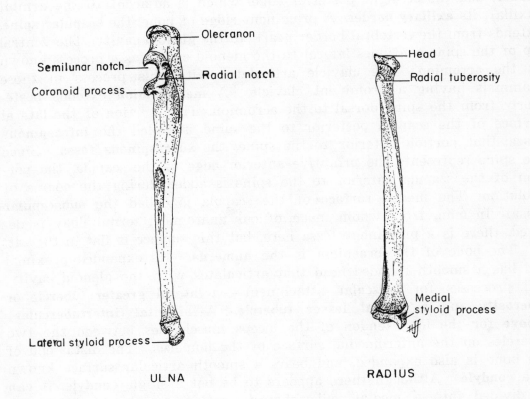

ULNA

RADIUS

Figure 5–11. Pectoral bones from the left side of the cat.

Pectoral Girdle and Appendage

Study the appendages and girdles on mounted specimens of the cat and from disarticulated bones. Learn to distinguish the individual girdles and the long bones of the appendage, including a recognition as to whether they are from the left or right side. The pectoral girdle consists primarily of an expanded triangular-shaped **scapula**. In therian mammals, the anterior coracoid has been lost, and the posterior coracoid is reduced to a small, hooklike **coracoid process** which can be seen medial to the anterior edge of the **glenoid cavity** (socket for the articulation with the front leg). Note that the glenoid cavity is directed ventrally rather than laterally. Why is this? Of the elements of the dermal girdle, only the **clavicle** is present. In many mammals, man included, the clavicle extends from the scapula to the sternum. But in some mammals, including the cat, the clavicle loses its connections with the rest of the skeleton, and is reduced to a sliver of bone imbedded in the muscles anterior to the shoulder joint. This permits the scapula to participate in the fore and aft swing of the front leg during locomotion.

If one pictures the scapula as an inverted triangle, the glenoid cavity is at the apex, and the curved top of the scapula (**vertebral border**) is at the base of the triangle. The anterior edge of the scapula is its **anterior border**, and the straight posterior edge, which is adjacent to the armpit (axilla), its **axillary border**. A prominent ridge of bone, the **scapular spine**, extends from the vertebral border nearly to the glenoid cavity. The ventral tip of the spine continues lateral to the glenoid cavity as a process known as the **acromion**. The clavicle articulates with this process in those mammals having a prominent clavicle. A **metacromion** extends posteriorly from the spine dorsal to the acromion. That portion of the lateral surface of the scapula posterior to the spine is called the **infraspinous fossa**; that portion anterior to the spine, the **supraspinous fossa**. Since the spine represents the primitive anterior edge of the scapula, the portion of the scapula anterior to the spine is added during the course of evolution. The medial surface of the scapula is called the **subscapular fossa**. In man, from whom much of our anatomical terminology is derived, there is a prominent fossa here, but this surface is flat in the cat.

The bone of the brachium is the **humerus**. Its expanded proximal end has a smooth rounded **head** that articulates with the glenoid cavity, and processes for muscular attachment — a lateral **greater tubercle** or **tuberosity** and a medial **lesser tubercle**. A **bicipital** (**intertubercular**) **groove** for the long tendon of the biceps muscle lies between the two tubercles on the anteromedial surface of the humerus. The distal end of the bone is also expanded, and bears a smooth articular surface known as a **condyle**. Although there appears to be but a single condyle, it can be divided into a medial pulley-shaped portion (the **trochlea**) for the ulna of the forearm (the bone which comes up behind the elbow), and a lateral rounded portion (the **capitulum**) for the radius. You may have to articulate the ulna and radius with the humerus to determine the extent of trochlea and capitulum. As was the case with the subscapular fossa, these features are more obvious on a human humerus than on the cat's. An **olecranon fossa**, for the olecranon process of the ulna, is situated proximal to the trochlea. The enlargements medial and lateral to the articular surfaces are the **entepicondyle** and **ectepicondyle**, respectively. An **entepicondylar foramen** for a nerve and vessel is located above the entepicondyle. This foramen is a primitive feature found in early reptiles, but lost in most mammals, including man. That portion of the humerus, or of any long bone, lying between its extremities is its **body** or **shaft**. The faint ridges and rugosities upon it mark the attachments of certain muscles.

The **ulna** is the longer of the two forearm bones, and a prominent **semilunar (trochlear) notch**, for the articulation with the humerus, will be seen near its proximal end. The end of the ulna lying proximal to the notch is the **olecranon** or "funny bone." A **coronoid process** forms the distal border of the notch, and a **radial notch**, for the head of the radius, merges with the semilunar notch lateral to the coronoid process. The bone terminates distally in a **lateral styloid process** which articulates with the lateral surface of the wrist. Note that the ulna and radius articulate distally, and that the ulna plays a relatively insignificant role in the formation of the wrist joint. In some mammals, the distal half of the ulna is lost.

The other bone of the forearm is the **radius**. The articular surfaces on its **head** are of such a nature that the bone can rotate on the humerus and ulna. Slightly distal to the head is a prominent **radial tuberosity** for the attachment of the biceps muscle. The distal end of the radius is expanded, has articular surfaces for the ulna and carpus, and a short **medial styloid process**.

In primitive tetrapods the ulna extended straight down the lateral edge of the forearm and the radius straight down the medial edge. The articulation of the radius with the humerus, besides being medial, was slightly anterior to that of the ulna. The manus pointed forward. If the elbow rotated posteriorly only 90 degrees during the evolution of mammals, the manus would extend laterally. To bring the manus forward, the radius rotated at the elbow. The end result is that the radius continues to be the more medial bone at the wrist, but is anterior and lateral to the ulna at the elbow. One can put the arm of a mounted human skeleton into the various positions to better visualize the changes.

Study the hand (**manus**) of the cat. Its first portion, the **carpus**, or wrist (Fig. 5-12), consists of two rows of small **carpal bones**. The proximal row contains a large medial **scapholunar**, which represents the radiale, intermedium, and a centrale fused; a smaller **cuneiform (triquetrum)**, which represents the ulnare and, on the lateral edge, a large,

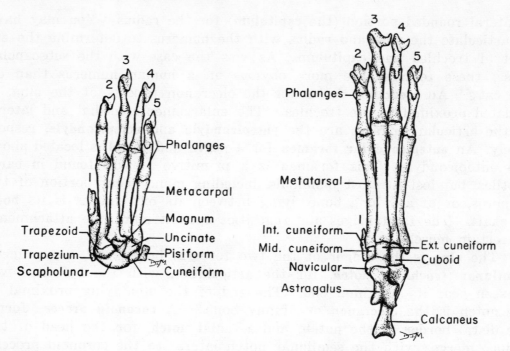

Figure 5–12. Dorsal views of the right manus (left drawing) and pes (right drawing) of the cat.

posteriorly projecting **pisiform** (a sesamoid bone). The four elements of the distal row are, from medial to lateral, the **trapezium**, representing distal carpal 1; **trapezoid**, representing distal carpal 2; **magnum (capitate)**, representing distal carpal 3; and **uncinate (hamate)**, representing distal carpal 4. Five **metacarpals** form the palm of the hand, and the free parts of the toes are composed of **phalanges**. The first toe is the most medial. Note that the number of phalanges has been reduced from that of primitive tetrapods. The terminal (**ungual**) phalanx of the catlike carnivores is articulated in such a way that it, and the claw which it bears, can be either extended or pulled back over the penultimate phalanx.

Pelvic Girdle and Appendage

The ilium, ischium, and pubis on each side of the pelvic girdle have fused together in adult mammals to form an **innominate bone** (**os coxae**), but they can be seen clearly in young specimens (Fig. 5-4, *D*). The **ilium** extends dorsally from the **acetabulum**, or socket for the hip joint, to the sacrum. Its dorsalmost border is its **crest**. Notice that the ilium inclines anteriorly, and unites with more sacral vertebrae than does the ilium in primitive tetrapods. The **ischium** surrounds all but the anterior portion, and some of the medial side, of the large opening (**obturator foramen**[11]) in the ventral portion of the girdle. The enlarged posterolateral portion

---

[11] See footnote 10, p. 95.

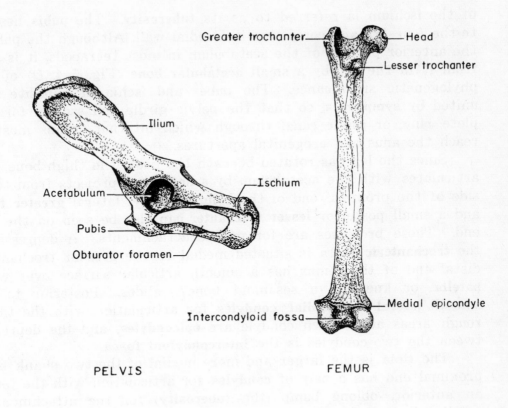

PELVIS

FEMUR

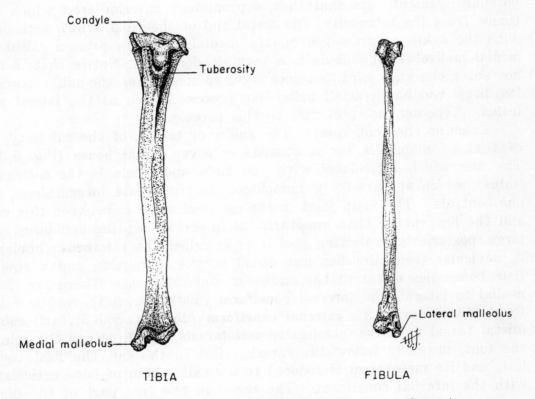

TIBIA

FIBULA

Figure 5-13. Pelvic bones from the left side of the cat.

of the ischium is referred to as its **tuberosity**. The **pubis** lies anterior to the fenestra and completes its medial wall. Although the pubis enters the anterior portion of the acetabulum in most tetrapods, it is separated from it in the cat by a small **acetabular bone** (Fig. 5-4, *D*) of unknown phylogenetic significance. The pubes and ischia of opposite sides are united by **symphyses**, so that the pelvic girdle and sacrum form a complete ring, or **pelvic canal** through which internal organs must pass to reach the anus and urogenital apertures.

Since the leg has rotated beneath the body, the thigh bone, or **femur**, articulates with the acetabulum by a **head** that projects from the medial side of the proximal end of the bone. A large, lateral **greater trochanter** and a small posterior **lesser trochanter** can also be seen on the proximal end. These processes are for muscle attachments. A depression called the **trochanteric fossa** is situated medial to the greater trochanter. The distal end of the femur has a smooth articular surface over which the **patella**, or kneecap (a sesamoid bone), glides. Posterior to this are smooth **lateral** and **medial condyles** for articulation with the tibia. The rough areas above each condyle are **epicondyles**, and the depression between the two condyles is the **intercondyloid fossa**.

The **tibia** is the larger and more medial of the two shank bones. Its proximal end has a pair of **condyles** for articulation with the femur, and an anterior, oblong bump (the **tuberosity**) for the attachment of the patellar ligament. Its shaft has a prominent anterior **crest** which continues from the tuberosity. The distal end of the tibia, which articulates with the ankle, is prolonged on the medial side as a process called the **medial malleolus**. The **fibula** is a very slender bone. Notice that it does not enter the knee joint, but does serve to strengthen the ankle laterally. Its distal end has a small pulley-like process known as the **lateral malleolus**. Tendons pass posterior to this process.

Examine the foot (**pes**). The ankle, or **tarsus**, of the cat is typical of that of mammals for it consists of seven **tarsal bones** (Fig. 5-12). The one which articulates with the tibia and fibula is the **astragalus** (**talus**), which appears to be homologous to the tibiale, intermedium, and one centrale. The main joint in the mammal ankle is between this bone and the leg, rather than mesotarsal as in certain reptiles and birds. The large, posteriorly projecting heel bone is called the **calcaneus** (fibulare). A **navicular** (centrale) lies just distal to the astragalus, and a row of four bones lies distal to the navicular and calcaneus. There are, from medial to lateral, the **internal cuneiform** (distal tarsal 1), middle cuneiform (distal tarsal 2), **external cuneiform** (distal tarsal 3), and **cuboid** (distal tarsal 4). Five enlongated **metatarsals**, which occupy the sole of the foot, normally follow the tarsals. But in the cat, the first toe is lost, and its metatarsal is reduced to a small nubbin of bone articulated with the internal cuneiform. The bones in the free part of the digits are the **phalanges**. What is the phalangeal formula?

6 / The Muscular System

Groups of Muscles

Continuing on the general theme of the organ systems concerned with support and locomotion, we will next consider the muscular system. In order to understand their evolution, the numerous muscles of vertebrates must be grouped in some way. Although the subdivisions of the muscular system according to histological structure (smooth vs. striated), or general type of innervation (involuntary vs. voluntary), are useful for certain types of work, the phylogenetically most natural method of subdivision appears to be according to their mode of embryonic development in lower vertebrates (Fig. 6–1). Using this basis, there are two major groups of muscles—the somatic and visceral. **Somatic**, or **parietal, muscles** develop from embryonic myotomes,[12] and are associated with the "outer tube" of the body. They are the more conspicuous muscles of the body and the appendages. All are striated and voluntary. **Visceral muscles** arise from the embryonic lateral plate, or hypomere, and are associated with the "inner tube" of the body. They are the muscles of the gut, including the wall of the pharynx and its visceral arches. Although we generally think of these muscles as being smooth and involuntary, those of the visceral arches are striated and to a large extent voluntary.

Both the somatic and visceral muscles can be further subdivided as shown in Table 2. The somatic muscles are broken down into axial and appendicular groups. The **axial muscles**, as the name implies, are the muscles in the axis of the body — the extrinsic muscles of the eyeball, the epibranchial and hypobranchial muscles of the neck region, and the muscles of the trunk and tail. In gnathostomes, the hypobranchial and trunk muscles are further subdivided. The hypobranchial musculature is divided at the level of the hyoid into prehyoid and posthyoid groups; the trunk musculature into epaxial muscles lying dorsal and lateral to the vertebral column, and into more ventral hypaxial muscles. The innervation of these, and of other groups, is shown in Table 2.

The **appendicular muscles** develop embryologically from mesenchyme that lies within the limb bud. In the lower vertebrates, at least, this mesenchyme is derived early in development from myotomic buds. Thus the appendicular muscles are closely related to axial muscles, but are sufficiently distinct to warrant separate treatment. Since the mesenchyme in the limb bud becomes divided into

---

[12] See footnote 2, p. 7.

Table 2. Major Muscles of Vertebrates

Showing the major muscles of the vertebrates studied in the laboratory, and the groups to which they belong. The main pattern of innervation and the probable homologies are also shown.

	Squalus	Necturus	Mammal

A. SOMATIC MUSCLES

I. Axial Muscles

1. Extrinsic Ocular Muscles

	Squalus	Necturus	Mammal
From First Somite (Oculomotor Nerve)	Superior rectus Inferior rectus Medial rectus Inferior oblique	Superior rectus Inferior rectus Medial rectus Inferior oblique	Levator palpebrae superioris Superior rectus Inferior rectus Medial rectus Inferior oblique
Second Somite (Trochlear Nerve)	Superior oblique	Superior oblique	Superior oblique
Third Somite (Abducens Nerve)	Lateral rectus	Lateral rectus Retractor bulbi	Lateral rectus Retractor bulbi

2. Epibranchial Muscles (Dorsal Rami of Occipital and Anterior Spinal Nerves)

	Squalus	Necturus	Mammal
	Epaxial portion of the myomeres in gill region	Anterior part of the Dorsalis trunci	Anterior part of the epaxial muscles

Table 2. Major Muscles of Vertebrates—Somatic (Axial) Muscles (Continued)

	Squalus	Necturus	Mammal

3. Hypobranchial Muscles (Ventral Rami of Spino-occipital or, in Amniotes, Hypoglossal Nerve and Cervical Plexus)

	Squalus	Necturus	Mammal
Prehyoid Muscles	Coracomandibular	Genioglossus Geniohyoid	Musculi linguae Genioglossus Hyoglossus Styloglossus Geniohyoid
Posthyoid Muscles	Rectus cervicis (or Coracoarcual + Coracohyoid) Coracobranchials	Rectus cervicis Omoarcuals Pectori-scapularis	Sternohyoid Sternothyroid Thyreohyoid

4. Axial Muscles of the Trunk (Spinal Nerves)

	Squalus	Necturus	Mammal	
Epaxial (Dorsal Rami)	Epaxial portion of myomeres	Interspinalis Dorsalis trunci	Interspinalis Intertransversarii Occipitals Multifidus Spinalis Semispinalis Longissimus Splenius Iliocostalis	Transverso-spinalis Longissimus Iliocostalis

Table 2. Major Muscles of Vertebrates—Somatic (Axial) Muscles (Continued)

	Squalus	Necturus	Mammal	
Hypaxial (Ventral Rami)	Hypaxial portion of myomeres	Subvertebralis	Longus colli Psoas minor Quadratus lumborum Levator scapulae	} Subvertebral
		Levator scapulae (Opercularis) Thoraci-scapularis		
		External oblique	Serratus ventralis (part) Serratus ventralis (part) Rhomboideus Rhomboideus capitis Serratus dorsalis Scalenus Transversus costarum External oblique External intercostals	} Lateral
		Internal oblique Transversus	Internal oblique Internal intercostals Transversus abdominis Transversus thoracis	
		Rectus abdominis	Diaphragm muscles Rectus abdominis	} Ventral

II. Appendicular Muscles (Ventral Rami of Spinal Nerves)

1. Pectoral Muscles

	Necturus	Mammal	
Dorsal Group — Abductor (Extensor)	Latissimus dorsi	Panniculus carnosus (part) Latissimus dorsi Teres major Subscapularis	}
	Subcoracoscapularis		
	Scapular deltoid	Spinodeltoid Acromiodeltoid Clavodeltoid	}
	Procoraco-humeralis	Teres minor	
	Triceps	Triceps brachii Epitrochlearis	}
	Forearm extensors	Forearm extensors	

Table 2. Major Muscles of Vertebrates—Somatic (Appendicular) Muscles (Continued)

	Squalus	Necturus	Mammal
1. Pectoral Muscles (Continued)			
Ventral Group	Adductor (Flexor)	Pectoralis	Panniculus carnosus (part) / Pectoralis complex
		Supracoracoideus	Supraspinatus / Infraspinatus
		Coracoradialis	Biceps brachii (part)
		Humeroanti- brachialis	Biceps brachii (part) / Brachialis
		Coracobrachialis	Coracobrachialis
		Forearm flexors	Forearm flexors
2. Pelvic Muscles			
Dorsal Group	Abductor (Extensor)	Iliotibialis	Sartorius / Iliacus / Psoas major / Pectineus / Vasti?
		Puboischio- femoralis internus	
		Ilioextensorius	Rectus femoris / Gluteus maximus
		Iliofibularis	Tensor fasciae latae? / Gluteus medius / Gluteus minimus
		Iliofemoralis	
		Shank extensors	Shank extensors

Table 2. Major Muscles of Vertebrates—Somatic (Appendicular) Muscles (Continued)

	Squalus	Necturus	Mammal
		2. Pelvic Muscles (Continued)	
Ventral Group	Adductor (Flexor)	Puboischiofemoralis externus (Adductor femoris. In Necturus this is not clearly separated from the preceding)	Obturator externus Quadratus femoris
			Adductor brevis et longus
		Pubotibialis	Adductor magnus
		Ischiofemoralis	Obturator internus Gemelli
		Caudofemoralis	Crurococcygeus (absent in some mammals) Piriformis
		Puboischiotibialis	Gracilis
			Semimembranosus
		Ischioflexorius	Semitendinosus Biceps femoris ? Tenuissimus ? (absent in some mammals)
		Shank flexors	Shank flexors

B. VISCERAL MUSCLES

I. Branchiomeric Muscles

	Squalus	Necturus	Mammal
Mandibular Muscles (Trigeminal Nerve)	Adductor mandibulae Levator palatoquadrati Spiracularis Preorbitalis	Levator mandibulae (3 parts)	Masseter Temporalis Pterygoids Tensor veli palati Tensor tympani
	Intermandibularis	Intermandibularis	Mylohyoid Anterior digastric

Table 2. Major Muscles of Vertebrates—Visceral (Branchiomeric) Muscles (Continued)

	Squalus	Necturus	Mammal
Hyoid Muscles (Facial Nerve)	Levator hyomandibulae Dorsal constrictor }	Depressor mandibulae Branchiohyoideus	Stapedius Platysma and Facial muscles }
	Ventral constrictor Interhyoideus }	Interhyoideus Sphincter colli	Posterior digastric Stylohyoid }
Branchiomeric Muscles of Remaining Arches (Glossopharyngeal, Vagus, and, in Amniotes, Spinal Accessory)	Cucullaris	Cucullaris	Trapezius complex Sternocleidomastoid complex }
	Interarcuals	Levatores arcuum	
	Branchial adductors	Dilatator laryngis	Intrinsic muscles of the larynx and certain muscles of the pharynx
	Superficial constrictors and Interbranchials	Subarcuals Transversi ventrales Depressores arcuum	

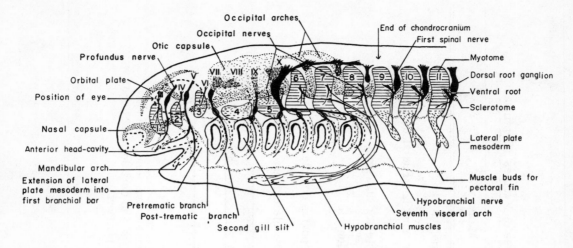

Figure 6–1. Diagram of the head of an embryo dogfish (*Scyllium*), to show the segmentation of the head, and the origin of the cranial muscles. The myotomes, which give rise to the somatic muscles, are hatched and numbered with Arabic numerals. The lateral plate mesoderm, which gives rise to the visceral musculature, is shown in outline ventral to the posterior myotomes and the gill slits. This part of the mesoderm contains the coelom, and sends processes up into each of the interbranchial septa. The visceral arches in these septa are shown by broken lines. The nerves are shown in solid black, and numbered with Roman numerals. The heavy stippling dorsally represents cartilage of the chondrocranium and vertebrae. (Slightly modified after Goodrich, On the development of the segments of the head of Scyllium, Quarterly Journal of Microscopical Science, Vol. 63.)

dorsal and ventral premuscular masses before giving rise to the muscles, the appendicular muscles may be further subdivided into dorsal and ventral groups. True appendicular muscles lie within the appendage, or on the girdle, or may grow from the appendicular skeleton back onto the trunk. They should not be confused with muscles that start their development in some other part of the body, and then grow over to the girdle.

The visceral muscles are subdivided into two categories: (1) the **branchiomeric**, or **branchial** muscles associated with the visceral skeleton, and (2) the remaining muscles of the gut tube, and associated structures. Only the branchiomeric group of visceral muscles will be considered. They, in turn, are grouped according to the arch with which they become associated — mandibular muscles, hyoid muscles, muscles of the third arch, etc. Branchiomeric muscles are innervated by the cranial nerves associated with the various visceral arches. The trigeminal nerve (V) is the nerve of the first arch and its muscles; the facial (VII) supplies the muscles of the second arch; the glossopharyngeal (IX), the third; and the vagus (X) and spinal accessory (XI), a derivative of the vagus present only in amniotes, supply the musculature of the remaining arches. How do the subdivisions of the muscular system compare with those of the endoskeleton?

Muscle Terminology

In describing the individual muscles, a number of terms will be used, and these may appropriately be defined at this time. The ends of muscles are attached to skeletal elements, or to connective tissue septa, and the intermediate part of the muscle (its belly) is free. One end of a muscle tends to be fairly stationary during contraction, and the other end moves the bone to which it is attached as the muscle shortens. The more stable end is the **origin** of the muscle; the opposite end is the **insertion**. In the case of limb muscles, the origin is usually proximal and the insertion distal. A muscle may have two or more points of origin, or two or more points of insertion. If the multiple points of attachment are segmentally arranged, they are called **slips**. The multiple origins of a muscle which are not segmented are sometimes called **heads**.

The attachment of a muscle to a bone is by its investing connective tissue, not by the actual muscle fibers. If the muscle fibers come very close to the bone, and the muscle appears to attach to it, we speak of a "fleshy" attachment; if a narrow band of connective tissue extends from the muscle to the bone, we speak of a **tendon**; if the muscle attaches by a broad, thin sheet of connective tissue, we speak of an **aponeurosis**.

The connective tissue of a muscle, besides implementing the attachment, forms a covering about the muscle, and spreads into the muscle where it surrounds bundles of muscle fibers, and even invests the individual muscle cells. The connective tissue covering of a muscle is a part of the **deep fascia** — a dense layer of connective tissue that forms a sheath for the individual muscles and also may hold groups of muscles together. Deep fascia is to be distinguished from **superficial fascia**, which is the layer of loose connective tissue beneath the skin. Superficial fascia generally contains fat; deep fascia does not.

Sometimes a transverse septum of connective tissue, called an **inscription** or **raphe**, is found in the middle of a muscle. Such an incription often represents either a persistent myoseptum, or the point at which two originally separate muscles have united.

Action of Muscles

Most muscles act in antagonistic groups. That is, the action of one muscle, or group of muscles, is offset by an antagonistic muscle or group. Various sets of terms describe these antagonistic actions. Unfortunately some of these have been used in different ways by different authors. The usage I shall follow is one proposed for quadruped mammals by Gray (1944) and slightly modified by Barclay (1946) to apply to all tetrapods except man. Barclay defines **protraction** and **retraction** of the humerus or femur as "movements causing the distal ends of these bones to move respectively forwards and backwards longitudinally." **Adduction** and **abduction** of the humerus and femur are defined as "movements causing the distal ends of these bones to be brought respectively nearer to and farther from the ventral median line." These terms can also be applied to movements of the scapula. The terms flexion and extension are here limited to movements of the distal parts of the appendage and certain movements of the trunk. **Flexion** is the movement of a distal segment of the limb toward the next proximal segment, as in the approximation of the antibrachium and brachium, or the hand and antibrachium. Flexion also describes the bending of the head or trunk toward the ventral surface. **Extension** is movement in the opposite directions. Lateral bending of the trunk is called lateral flexion. The terms flexion and extension have also been used to describe movements of the entire limb at the shoulder and hip, but their usage at the shoulder has been inconsistent. Some authors equate protraction and flexion, but others, especially human anatomists, use the term extension for a forward movement of the arm. It seems best to drop the terms flexion and extension here and use protraction and retraction. Rotation can be illustrated by the movement of the radius on the ulna, or the axis on the atlas. In rotation of the radius on the ulna, special terms are often used. Rotation of the forearm to a position in which the palm of the hand faces ventrally, or toward the ground, is called **pronation**; the opposite rotation, which brings the palm up, is **supination**.

The Study of Muscles

As methods of body support changed during the transition from water to land, and movements of the body and its parts became more complex during the evolution of vertebrates, the muscles became more numerous. It is not possible, in a course of this scope, to study all of them. One of two approaches may be taken. One could examine all the superficial muscles of the body, and omit the deeper ones; or one could examine certain muscle groups in more detail and omit other groups. I prefer the latter treatment for it gives a fairly complete evolutionary picture of at least certain groups. In this connection the details of the appendicular muscles of the distal portion of the limbs have been omitted, as have those of the tail and perineum. But other groups are described with reasonable completeness. In so far as possible, the muscles are described by natural groups. This will permit a further selection by the instructor of the muscles to be studied if time is short.

A few remarks concerning the dissection of muscles may be appropriate. In so far as possible confine your dissection of the muscles to one side of the body and cut open the body, when you do, on the opposite side. If *Necturus* is going to be studied, do the muscles on the left side, for specimens of *Necturus* have generally been cut open on the right in order to inject the blood vessels. Care must be exercised so as not to injure the underlying muscles when the skin is removed. It is best to try to peel the skin off by tearing underlying connective tissue with your fingers or with forceps. If you must cut with a scalpel, cut toward the underside of the skin, not toward the muscles. After the skin is off, the muscles must be carefully separated from each other. This involves cleaning off the overlying connective tissue with forceps until you can see the direction of the muscle fibers. Ordinarily the fibers of one muscle are held together by a sheath of connective tissue, and all run in the same general direction to a common tendon or attachment. The fibers of an adjacent muscle will be bound together by a different sheath, and will have a different direction and attachment. This will give you a clue as to where to separate one from the other. Separate muscles by picking away, or tearing, the connective tissue between them with forceps, watching the fiber direction as you do so. Do not try to cut muscles apart. If the muscles separate as units, you are doing it correctly; but if you are exposing small bundles of muscle fibers, you are probably separating the parts of a single muscle. It is best to expose and separate a few muscles of a given region before attempting to identify them. It will be necessary sometimes to cut through a superficial muscle to expose deeper ones. In such cases, you usually should cut across the belly of a muscle at right angles to its fibers, and turn back (reflect) its ends, rather than detach its origin or insertion. The dissection will be more meaningful if you have a skeleton before you on which to visualize the points of attachment of the muscles.

FISHES

Although advanced in some respects when compared with such agnathous forms as *Petromyzon,* the muscles of *Squalus* are a good example of the condition of the musculature in primitive vertebrates. The natural groups of muscles can easily be recognized and studied in the adult, for the groups have not lost their identity, as they have to some extent in higher vertebrates, through the migration of muscles.

Axial Muscles

(A) *Typical Myomeres of the Trunk and Tail*

The bulk of the musculature of fishes belongs to the axial group of somatic muscles, and the most conspicuous of these are the muscles of the trunk and tail. Remove a wide strip of skin from the posterior portion of the tail, and another strip from the front of the trunk between the pectoral and anterior dorsal fins (Fig. 6-2). These strips should extend from the middorsal to the midventral lines of the body. Try not to injure the underlying muscles when taking off the skin.

Notice that the trunk musculature consists of muscle segments, or **myomeres**, which develop from the embryonic myotomes, and that the segments are separated from each other by connective tissue septa, the **myosepta**, or **myocommata**. The myomeres are bent in a complex zigzag fashion and each is divided into dorsal and ventral portions by a longitudinal connective tissue septum (the **horizontal skeletogenous septum**), which lies deep to the lateral line. The dorsal portions of the myomeres constitute the **epaxial musculature**; the ventral, the **hypaxial musculature**. This division of the trunk musculature is found in all gnathostomes, but is absent in the Agnatha. In addition to their divisions into epaxial and hypaxial masses, the myomeres can be divided, at the apexes of the zigzags into **longitudinal bundles**. The muscle fibers within the two, somewhat darker bundles adjacent to the horizontal septum, extend longitudinally, but those in the others are somewhat oblique.

Each myomere is a complex entity, as can be seen in Figure 6-2. The internal part of each V-like fold of a myomere is cone shaped and extends further anteriorly or posteriorly than the fold does at the body surface. An analogous arrangement can be made by cutting a V-shaped notch out of the top half of several conical paper cups, and then fitting them together. The cone-within-a-cone nature of the folds can be seen if you make a cross section of the tail. In such a section, the overlapping V-shaped folds of several myomeres appear as a series of concentric rings (Fig. 6-5, p. 120). As described on page 41, the relationships of the myomeres to the vertebral column, and to the skeletogenous septa, also show in this view. In the abdominal region, the ventral band of connective tissue which separates the myomeres of opposite sides is called the **linea alba**.

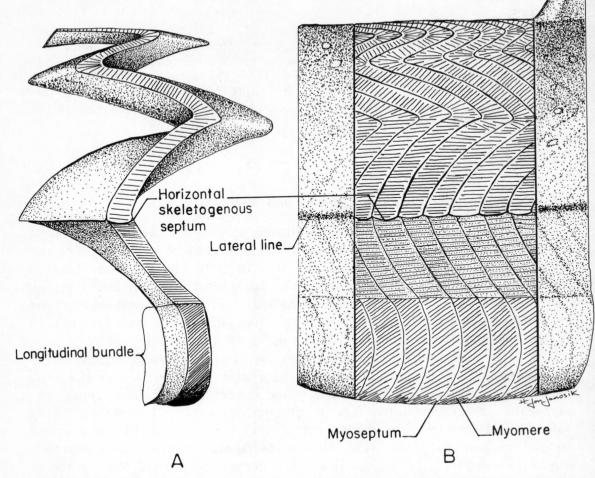

Ant. dorsal spine

Horizontal skeletogenous septum

Lateral line

Longitudinal bundle

Myoseptum

Myomere

A

B

Figure 6–2. Trunk muscles of the dogfish. *A*, diagram of a lateral view of an individual myomere; anterior is toward the left. *B*, lateral view of a group of myomeres. (*A* modified after Coles.)

The myomeres are the main locomotor organs of a fish, for they cause the lateral undulations of the trunk and tail whereby fish swim. Metachronal contractions are initiated anteriorly and pass toward the tail, affecting first one side of the body and then the other. These waves increase in amplitude as they move caudad. The large caudal fin, besides giving a lateral thrust, dampens the lateral displacement of the body and reduces the sinuosity of the fish's course through the water. As the body moves from side to side, its inclination to the vertical plane changes so that the thrusts it delivers against the water are directed downward and backward, not just laterally. The effect is very similar to the thrust given by a skulling oar.

How contraction of the myomeres brings about flexure of the body has been considered by Willemse (1959). As he points out, the fibers do not act directly on the vertebrae by way of the myosepta, as many investigators had thought. Rather, the contraction of the muscle fibers on one side of the body, being resisted by the incompressible vertebral column simply causes a bending of the body to this side. The situation is analogous to the bending, upon cooling, of a strip composed of two metals lying side by side. The shortening of one metal at a faster rate will be resisted by the other, and the strip will bend.

As Nursall has recently summarized (1962), the complex folding of the myomeres increases their fore and aft extension. This allows relatively short fibers to exert their effect over a considerable distance. Most of the fibers attach diagonally onto the myosepta, permitting the myomeres to shorten with a minimum of bulging, for, when the fibers shorten, they assume a more upright position in respect to the septum. The situation is analogous to an oblique parallelogram shortening into a rectangle of equal area. This is important in the fish, which must have a streamlined body surface. The darker (red) fibers that comprise certain of the longitudinal bundles are particularly rich in myohemoglobin. Although they contract less rapidly than the white fibers, they do not fatigue as rapidly.

(B) Epibranchial Muscles

The axial musculature is interrupted in the head by the gill region, but extends forward above and below the gills as the **epibranchial** and **hypobranchial** musculature, respectively (Fig. 6-3). Remove a strip of skin dorsal to the gill region, and the epibranchial muscles can be seen extending to the chondrocranium. Notice that the segmentation persists in this region. Certain of the deeper parts of this musculature, which are not easily seen, attach onto the tops of the branchial arches.

(C) Hypobranchial Muscles

To see the hypobranchial muscles, remove the skin on the underside of the head from the pectoral region forward to the lower jaw. It will be necessary in this case to skin both sides of the body. A broad sheet of transverse muscle fibers lies posterior to the jaws. This is part of the branchiomeric musculature, and will be studied presently. At this time, cut through it near the midventral line, and reflect it. Be careful not to injure a narrow, midventral, longitudinal muscle that lies just beneath the transverse layer.

The narrow midventral muscle (embryologically paired) thus exposed is called the **coracomandibular**. It arises anteriorly from the jaw and inserts posteriorly onto the surface of other muscles. It represents the **prehyoid** portion of the hypobranchial musculature, but in fishes this muscle commonly extends posterior to the hyoid. Cut through the posterior attachment of the coracomandibular and pull it forward. The dark, or pink, mass beneath its anterior end is the **thyroid gland**.

The paired muscles now exposed belong to the **posthyoid** portion of the hypobranchial musculature. The entire complex, which arises on the pectoral girdle and inserts on the hyoid arch, may simply be called the **rectus cervicis**. But it is sometimes considered to be two pairs of muscles, for the fibers of the posterior half converge toward the midline, while those of the anterior half extend longitudinally to the hyoid arch. The posterior pair is called the **coracoarcuals**, the anterior the **coracohyoids**. Make a cut from the ventral edge of the last gill slit to the pos-

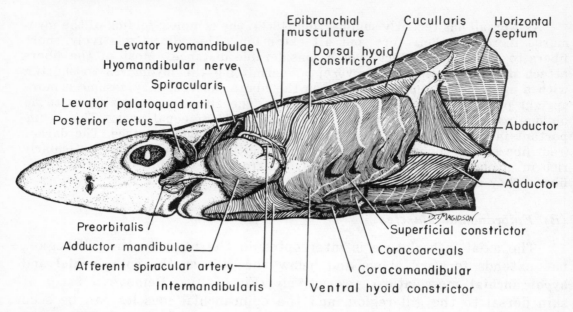

Figure 6–3. Lateroventral view of the muscles of the pectoral appendage and head of *Squalus*.

terior end of the rectus cervicis, and then dissect anteriorly, gradually separating the branchial region from the hypobranchial muscles. Be careful not to unduly injure blood vessels. This will expose a deep group of muscles that arise from the coracoid, from the wall of the pericardial cavity beneath the posterior part of the rectus cervicis, and from the dorsal surface of the rectus cervicis itself. These muscles insert onto the ventral ends of the branchial arches and are called the **coracobranchials**.

All the hypobranchial muscles support the floor of the pharynx and help to open the mouth and expand the gill pouches.

(D) Extrinsic Muscles of the Eye

The segments anterior to the epibranchial muscles disappear during embryonic development except for the most anterior three, and these give rise to the extrinsic muscles of the eye. These muscles extend from the wall of the orbit to the surface of the eyeball, and are responsible for the movements of the eyeball. They will be considered when the eye is studied (p. 165).

Appendicular Muscles

(A) Muscles of the Pectoral Fin

The appendicular group of somatic muscles is small and simple in fishes, for the paired fins are concerned with increasing stability and

maneuverability, not with propulsive thrusts. Skin the base of the pectoral fin on its dorsal, ventral, and anterior surfaces, if this has not been
done. It can then be seen that the appendicular muscles consist of a
single dorsal and a single ventral mass, each of which arises from the
the pectoral girdle, and inserts by slips onto either the dorsal or ventral
surface of the pterygiophores of the fin as far distal as the radial cartilages. The dorsal mass (**abductor** or **extensor**) elevates the fin and pulls
it posteriorly. The ventral mass (**adductor** or **flexor**), which also spreads
onto the anterior surface of the fin, depresses the fin and pulls it forward.

(B) Muscles of the Pelvic Fin

Similarly, skin the base of the pelvic fin. Study a female first, in
which the conditions are uncomplicated by the presence of copulatory
structures. The dorsal mass (**abductor**) arises from the surface of the
posterior trunk myomeres, from the iliac process, and from the metapterygium. It inserts by slips onto the radial cartilages. Most of the
ventral mass (**adductor**) is divided into proximal and distal portions. The
former arises from the pubo-ischiadic bar and inserts on the metapterygium;
the latter arises from the metapterygium and inserts on the radials.

The appendicular muscles of the male are fundamentally the same,
but portions of both the dorsal and ventral mass extend onto the clasper
as distinct little muscles. Also, the adductor cannot be seen until one
reflects a long muscular sac (the siphon, Fig. 11-3) that lies between it
and the skin. Do not remove the siphon completely, as it will be studied
in connection with the reproductive system.

Branchiomeric Muscles

The basic pattern of the branchiomeric, or branchial musculature, is illustrated in Figure 6-4. As can be seen, most of the embryonic muscle plate of
a given arch forms an **interbranchial** and **superficial constrictor**, but parts of
the muscle plate separate as a **levator**, **interarcual**, and **adductor**. The interbranch-
ials, superficial constrictors, interarcuals, and adductors pull the parts of a
branchial arch together and compress the branchial pouches, thus aiding in expelling water. The branchial pouches are expanded and take in water through
the elastic recoil of the branchial skeleton aided by the contraction of the
levators and the coracobranchials (hypobranchial group) already described. As
would be expected, this basic pattern is modified considerably in the region
of the mandibular and hyoid arches since the muscles of these arches are concerned primarily with the movement and support of the jaws.

To see the branchiomeric muscles, it is necessary to remove the skin
from the side of the gill region, jaws, spiracle, and from the back and
underside of the eye. Cut close to the skin and do not remove any muscular
tissue. In the spiracular region, remove the skin, but do not otherwise

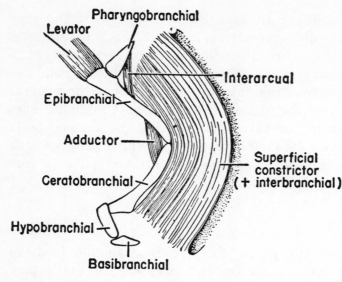

Figure 6–4. A representative visceral arch of a selachian, and its branchial muscles. Anterior is toward the left. (From Romer, The Vertebrate Body.)

injure the anterior surface of the spiracular valve (p. 32). After the skin is removed, carefully clean the area by cutting and picking away connective tissue. Be particularly meticulous in the vicinity of the jaws, spiracle, and eye.

Study the muscles, beginning with those of the mandibular arch. The large mass of muscle at the angle of the jaws is the **adductor mandibulae** (Fig. 6–3). It arises from the posterior part of the palatoquadrate (note a large process in this region on a skeleton) and inserts on Meckel's cartilage. Dorsal to this, and in front of the spiracle, is another muscular mass that arises from the side of the otic capsule and inserts on the palatoquadrate. Careful dissection reveals that this mass is divided into an anterior **levator palatoquadrati** and a posterior **spiracularis**. The latter is on the anterior wall of the spiracular valve. Lift the eye, and a **preorbitalis** will be seen ventral to it. This muscle arises from the underside of the chondrocranium and inserts by a tendon on Meckel's cartilage. The last of the mandibular muscles is the ventral **intermandibularis** which was cut through during the dissection of the hypobranchial muscles. It originates from Meckel's cartilage and the fascia over the adductor mandibulae, and extends diagonally posteriorly and laterally to insert on a midventral raphe.

The intermandibularis overlies a deeper transverse sheet of muscle, the **interhyoideus**. Separate the two and note that the fibers of the interhyoideus are more nearly transverse and originate from the ceratohyal. Posteriorly, the interhyoideus is continuous with the **ventral hyoid constrictor**. Dorsally, the hyoid musculature is represented by a **dorsal hyoid constrictor**, which lies above and posterior to the angle of the jaw, and a more anterior **levator hyomandibulae**. The latter lies posterior to the spiracle and dorsal to the hyomandibular. It arises from the surface of the epibranchial musculature and otic capsule, and inserts primarily on the hyomandibular. A few fibers, however, extend onto the palatoquadrate.

The musculature of the remaining visceral arches (branchial arches), except for the last, is much the same. Note the four **superficial constrictors** that lie above and below the last four gill slits. The fibers of the superficial constrictors attach to connective tissue raphes and incline toward the gill slits. Open all the branchial pouches by cutting through the superficial constrictors along a line parallel to a raphe. Open them as wide as you can, without cutting through the margins of the internal gill slits (the openings communicating with the pharynx). The interbranchial septa, thus mobilized, support the gill lamellae. Remove the skin and gill lamellae from the anterior face of one septum, and it can be seen that the septum is composed largely of circularly arranged muscle fibers. This muscle, which is a part of the constrictor musculature, is appropriately called the **interbranchial**. There are also only four of these, for the last branchial arch lacks an interbranchial septum and associated muscles. Cut completely across the interbranchial septum you have started to dissect in the frontal plane, the corresponding branchial arch, and into the pharynx. Examine the cut surface medial to the branchial arch, and you will see a cross section of the **branchial adductor**, a short muscle that extends from the epibranchial to the ceratobranchial (Figs. 6–4 and 9–9). There are five of these, one for each branchial arch.

The **levators** of the five branchial arches have united to form a triangular muscle, the **cucullaris** (partly homologous to the trapezius of higher vertebrates, and sometimes called by that name), which lies dorsal to the superficial constrictors. Separate the cucullaris from the superficial constrictors. Observe that it arises from the surface of the epibranchial musculature, and passes posteriorly to insert on the pectoral girdle and last branchial arch. Completely separate the cucullaris and the levator hyomandibulae from the epibranchial muscles, and push the branchial area ventrally. The large cavity exposed is the anterior cardinal sinus, a part of the venous system. Clean out the coagulated blood, if necessary, and examine the tops of the branchial arches. Some of the epibranchial muscles may be cut away if the arches cannot be seen clearly. Small muscles will be seen extending from the epibranchials to the pharyngobranchials, and also between the pharyngobranchials. These are the **lateral** and **dorsal interarcuals** respectively.

PRIMITIVE TETRAPODS

Many changes occur in the muscular system during the evolution from fish through tetrapods. These are correlated primarily with changes in the mode of support and locomotion, with the development of head movements independent of those of the trunk, with changes in the method of respiration, and with the development of a mobile tongue. As regards locomotion, the primitive dorsal and ventral appendicular muscle masses differentiate into a number of separate muscles, for the appendages become the organs of propulsive thrust,

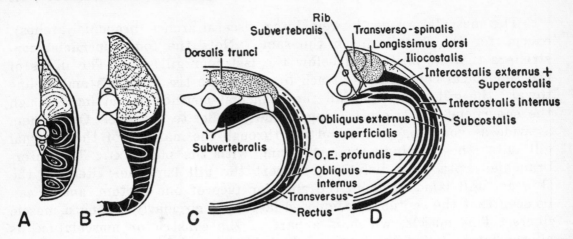

Figure 6–5. Diagrammatic cross sections to show the divisions of the trunk musculature in *A*, a shark tail; *B*, the shark trunk; *C*, a urodele; *D*, a lizard. The epaxial muscles are stippled; the hypaxial are black. In *D*, the dorsal labels of the hypaxial muscles pertain to a region in which ribs are present; the more ventral labels are those of the corresponding abdominal muscles. (From Romer, The Vertebrate Body. Mainly after Nishi.)

and their movements more complex. The appendicular muscles, however, retain their primitive dorsal and ventral groupings (Table 2). Concomitantly, the trunk and tail muscles become reduced and lose their segmentation, for lateral undulations play a less important role in locomotion. This is especially true for the hypaxial muscles. The epaxial muscles are retained, and differentiate further in connection with the movement of the vertebral column and head. In tetrapods that raise themselves off the ground, the epaxial musculature plays an important role in bracing and tying the elements of the vertebral girder together and in supporting the body against gravity. Some of the hypaxial musculature that persists forms thin layers on the ventrolateral portion of the body wall, some becomes associated with the ventral surface of the vertebral column (Fig. 6–5), and, especially in the pectoral region, some attaches onto the limb girdles.

With the shift from gill to pulmonary respiration, the more posterior visceral arches and most of their muscles are greatly reduced. But some of the posterior musculature, the cucullaris, becomes associated with the pectoral girdle, and it is retained and elaborated in this connection. The anterior visceral arches contribute to the jaws, auditory ossicles, and hyoid apparatus. Their muscles have a comparable history to a large extent, exept for the hyoid musculature. Most of the hyoid musculature moves to a superficial position, and forms, in mammals, the **platysma** and **facial muscles**. The hyoid apparatus and newly evolved tongue are moved instead by the hypobranchial muscles.

Necturus exemplifies the primitive tetrapod condition very well, for the changes just described are in an early stage. It must be remembered, however, that *Necturus* is a permanent larva that retains external gills, and hence retains also more of the branchial apparatus and muscles than would be found in a true adult tetrapod.

Axial Muscles

(A) *Muscles of the Trunk*

Confine your study of the muscles of the *Necturus* to one side of the body, preferably the side that has not been cut open to inject the circulatory system. Remove, from the middle of the trunk, a strip of skin that is about two inches long and extends from the middorsal to the midventral line. Clean off any extraneous connective tissue, and note that the musculature still consists of **myomeres**, separated by **myosepta**, and divided into **epaxial** and **hypaxial** portions. The myomeres, however, are not as complexly folded as they are in fishes, nor are the myosepta as obvious.

Most of the epaxial portions of the myomeres, including those in the epibranchial region, form a dorsal, longitudinal bundle of muscle called the **dorsalis trunci** (Figs. 6-5, *C* and 6-6). Although the epaxial musculature is fairly uncomplicated in amphibians, the deeper portions of it are already beginning to become specialized. Make a deep transverse cut through the dorsalis trunci, take a firm grip with a pair of forceps on a bundle of muscle fibers beside the middorsal line, and pull the fibers caudad. This will expose the tops of the neural arches, and a series of short **interspinalis** muscles. Each interspinalis arises on the edge of a posterior zygapophysis of one vertebra, and inserts on the neural arch of the next posterior vertebra.

Examine the hypaxial musculature and note that most of its superficial fibers incline posteriorly and ventrally. These constitute the **external oblique** muscle. The fibers next to the midventral line, however, extend longitudinally, and form an incipient **rectus abdominis**. Since *Necturus* is

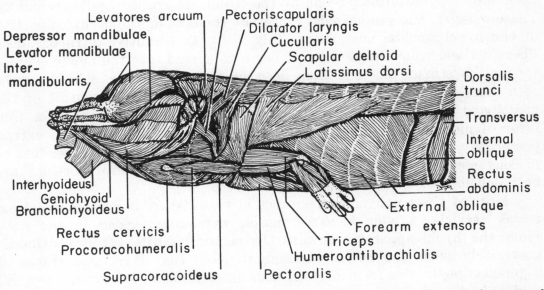

Figure 6-6. Lateral view of the trunk, pectoral, and head musculature of *Necturus*.

an incompletely metamorphosed species, the rectus abdominis is not as completely differentiated as it is in adult amphibians. Carefully cut into a myomere, reflecting the fibers of the external oblique as you do so. Presently, you will see a second layer, the **internal oblique**, whose fibers extend anteriorly and ventrally at right angles to those of the external oblique. Uncover the internal oblique in several myomeres. Its ventral fibers form a part of the rectus. Now cut into the internal oblique in one segment and reflect its fibers. You will soon see a third layer, the **transversus,** whose fibers are circular and lie nearly in the transverse plane. Cut through this muscle, and you reach the parietal peritoneum and body cavity. Another hypaxial muscle is the **subvertebralis,** which lies on the sides of the centra ventral to transverse processes (Fig. 6-5, C). It extends from vertebra to vertebra, attaching to the centra and transverse processes. It may be seen now by making a small incision through the body wall slightly to one side of the midventral line, and pushing the viscera aside; or observation may be postponed and it may be noted when the abdominal viscera are studied. Several other hypaxial muscles attach onto the girdles, and will be seen later.

(B) *Hypobranchial Muscles*

To get at the hypobranchial muscles, remove the skin from the entire underside of the head and neck as far posteriorly as the pectoral region. Carefully cut through the thin transverse sheet of branchiomeric muscles that lies under the head, and turn it to the sides. Cut slightly to one side of the midventral line. A pair of midventral, longitudinal muscles will be seen extending posteriorly from the symphysis of the jaw. Their caudal end inserts on the posterior portion of the hyoid apparatus (basibranchial 2). These muscles, the **geniohyoids**, represent the bulk of the prehyoid portion of the hypobranchial musculature (Fig. 6-7). In primitive tetrapods, deep fibers derived from the geniohyoid begin to spread into the newly evolved tongue. *Necturus* has one such deep layer, **genioglossus**, which can be seen most clearly after cutting through the posterior attachment of the geniohyoid and pulling it forward. The genioglossus appears as a thin layer of more or less longitudinal fibers that arise on the chin, extend posteriorly in the floor of the mouth, and insert on a transverse fold of mucous membrane at the base of the tongue. Its lateral portions are best developed.

Most of the posthyoid hypobranchial musculature is represented by a **rectus cervicis**, a wide band of muscle extending caudad in the neck from the hyoid apparatus toward the pectoral girdle. It is continuous posteriorly with the rectus abdominis. Note that it retains traces of segmentation in the form of tendinous inscriptions. When the pectoral region is dissected, slips of the rectus cervicis that are called **omoarcuals** will be seen to attach onto the procoracoid process of the girdle. A **pec-**

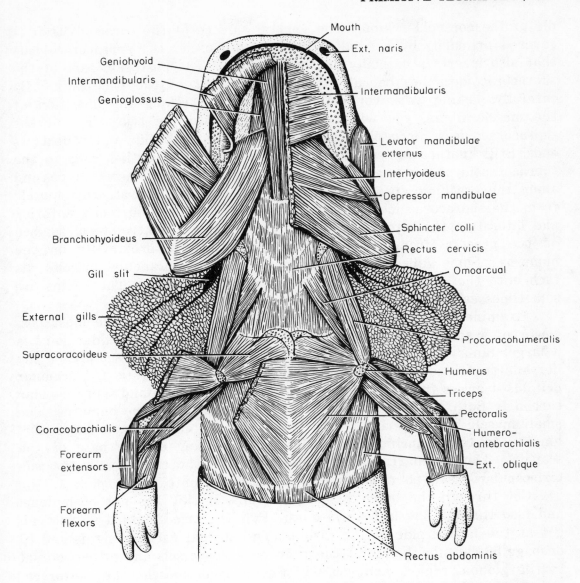

Figure 6–7. Ventral view of the head and pectoral muscles of *Necturus*.

toriscapularis. which is a derivative of the rectus cervicis, will also be seen at that time.

Appendicular Muscles

(A) *Pectoral Muscles*

Cut off the external gills on one side and skin the pectoral appendage and the lateral and ventral surfaces of the trunk in its vicinity. A large fan-shaped muscle, the **pectoralis**, will be seen on the ventral surface (Fig. 6-7). It arises from the linea alba and hypaxial musculature, and inserts on the proximal end of the humerus. It lies posterior to the middle of the coracoid region. The ventral surface of the coracoid itself is covered primarily by a smaller fan-shaped muscle, the **supracoracoideus** that also inserts proximally on the humerus. The posterior part of the supracoracoideus lies beneath the anterior part of the pectoralis. By carefully picking away connective tissue and watching the direction of the muscle fibers, you will be able to separate the two. To see the supracoracoideus more clearly, cut across the pectoralis and reflect its ends. Still another fan-shaped muscle, the **coracoradialis** lies deep to the supracoracoideus, but wait until you find its tendon of insertion passing along the brachium (p. 125) before you attempt to separate this muscle from the supracoracoideus. A **procoracohumeralis** will be found anterior and lateral to the supracoracoideus. It lies on the ventrolateral surface of the process, from which it arises, and inserts proximally on the humerus. Some slips of the rectus cervicis (a hypobranchial muscle) attach onto the medial surface of the procoracoid process. These slips are sometimes called the **omoarcuals**.

Examine the lateral surface of the girdle (Fig. 6-6). The most posterior of a series of muscles that converge toward the shoulder joint is a large, fan-shaped **latissimus dorsi**. This arises from the surface of the dorsalis trunci, sometimes by several slips. Next anterior is the **scapular deltoid**. It arises from the lateral surface of the scapula and suprascapular cartilage. Both of these insert on the proximal end of the humerus. The small muscle just anterior to the scapula is the **cucullaris**, a muscle belonging to the branchiomeric series, but inserting on the base of the scapula. The still narrower muscle anterior to the cucullaris is the **pectoriscapularis**, a hypobranchial muscle attaching onto the scapula.

Cut through the belly of the pectoralis, if this has not been done, and also through the latissimus dorsi. Pull the arm forward so you can get at the medial side of the girdle. (The sternal cartilages referred to on page 56 can now be seen.) Clean away blood vessels and nerves dorsal to the glenoid region, and you will find a small muscle, the **subcoracoscapularis**, arising from the posterior edge of the scapula, and inserting on the humerus. It passes between two muscles on the arm (triceps and coracobrachialis). Reflect the origin of the latissimus dorsi and examine the medial side of the dorsal end of the scapula. A band of muscle will be seen extending from the hypaxial musculature forward to insert on the scapula. This muscle, the **thoraciscapularis** (or **serratus**), belongs to the hypaxial group. Another, and narrower, band of muscle may also be seen extending forward from the front of the dorsal end of the scapula. A better view of it will be obtained by cutting and reflecting the cucullaris. This muscle, which is also a hypaxial derivative, is called the **levator scapulae**. Its anterior end attaches to the occipital region of the skull. It is this muscle which becomes the opercularis in terrestrial urodeles (p. 59).

The upper arm is covered for the most part by three muscles. A **humeroantibrachialis**, arising from the humerus and inserting on the proximal end of the radius, is located on the ventrolateral surface. A **cora-cobrachialis** (sometimes divided into **brevis** and **longus**) is located on the ventromedial surface. It arises from the coracoid and inserts on the shaft and distal end of the humerus. The tendon of the **coracoradialis**, described earlier, can be found between the coracobrachialis and humero-antibrachialis. After you find it, trace it medially to the belly of the muscle. A **triceps** covers the dorsal surface. Although the triceps arises from the girdle and humerus by three heads, only two of these are readily distinguishable — a medial **coracoid head** located on the medial side of the arm dorsal to the coracobrachialis, and a dorsolateral **humeral head** separated from the medial coracoid head by the subcoracoscapularis. All the heads converge and unite, forming a common tendon that passes over the elbow to insert on the proximal end of the ulna.

Clean the fascia from the surface of the forearm, and note that the muscles are aggregated into two groups — the **extensors** and the **flexors**. The extensors lie on the dorsal surface of the forearm, and arise from the lateral surface of the distal end of the humerus. The flexors have the opposite relationships. They lie on the ventral surface, and arise from the medial surface of the distal end of the humerus.

By studying Table 2, you should learn which of these muscles represent dorsal appendicular muscles and which ventral. In general, the ventral appendicular muscles protract and adduct the limb and flex its distal segments, while the dorsal muscles retract, abduct, and extend. Exceptions occur at the anterior end of the dorsal series where some of the muscles (the deltoid group) protract, and at the posterior end of the ventral series where some (the posterior fibers of the pectoralis) retract. In addition to moving the limbs, the appendicular muscles also serve as braces that can fix the bones at a joint and support the body. In general, the protractors are innervated more anteriorly than the retractors so that, as a wave of activity passes caudad along the spinal cord, the limb is first moved forward and then backward. In this way the limb of ancestral tetrapods probably aided locomotion as a "swimming wave" of contraction passed caudad along the myomeres.

(B) *Pelvic Muscles*

Skin one of the pelvic appendages and the adjacent trunk and tail. Also skin the cloacal region and remove, on one side, the large **cloacal gland** that lies between the skin and muscles. Clean the area, and note the ilium on the lateral surface dorsal to the base of the appendage. The hypaxial musculature attaches to it. Other parts of the girdle are covered by muscle.

Study the ventral surface of the girdle and thigh first. One of the most distinct muscles, and hence one that will serve as a good landmark, is the **pubotibialis** — a narrow band on the ventral surface of the thigh that arises from the lateral edge of the pubic cartilage and inserts on the proximal end of the tibia (Fig. 6-8). The ventral surface of the girdle, and of the thigh medial to the pubotibialis, is covered by a large triangular mass of muscle. The anterior two thirds of this represents a **puboischiofemoralis externus**; the posterior third, a **puboischiotibialis**. The line of separation between them can best be picked up near the distal end of the thigh deep to the pubotibialis, and then followed toward the mid-ventral line of the body. Note that the front of the puboischiotibialis overlaps the back of the puboischiofemoralis externus. The puboischiofemoralis externus, in turn, overlaps the attachment of the rectus abdominis on the front of the pubic cartilage. These two muscles arise from the ventral surface of the girdle. The puboischiofemoralis externus inserts on the femur; the puboischiotibialis, on the proximal end of the tibia. An **ischioflexorius** lies along the medioventral surface of the thigh posterior to the puboischiotibialis, and at first appears to be a part of this muscle. The two can be separated most easily at the proximal end of the shank, and from the dorsal side. The ischioflexorius inserts on the fascia of the distal end of the shank. Try to follow the muscles medially. It arises from the posterior end of the ischium. It may be noticed that the muscle is divided into distal and proximal bellies separated by a short tendon.

Dissect in the area lateral to the cloaca, and you will find two longitudinal muscles. The narrower one beside the cloaca is the **ischiocaudalis**. It inserts on the caudal vertebrae and arises from the posterior end of the ischium. The more lateral muscle, the **caudopuboischiotibialis** (or **caudocruralis**), has a similar insertion, but arises from the puboischiotibialis. Both these muscles are hypaxial muscles that attach in the pelvic region. They serve to flex the tail. Dissect dorsal to the caudopuboischiotibialis and you will find a third longitudinal muscle. This is the **caudofemoralis**, an appendicular muscle arising from the caudal vertebrae and inserting on the proximal end of the femur.

Cut across the puboischiotibialis and reflect its ends. The small triangular-shaped muscle lying posterior to the puboischiofemoralis externus is the **ischiofemoralis**. It arises from the ischium and inserts on the proximal end of the femur. The slender muscle, arising from the underside of the femur near the insertion of the puboischiofemoralis externus and continuing across the underside of the knee joint to the fibula, is the **popliteus**. It is really a shank, rather than a thigh, muscle, (An **adductor femoris**, another deep muscle found anterior to the origin of the puboischiofemoralis externus in some urodeles, is absent as such in *Necturus*.)

The large muscle lying dorsal to the pubotibialis along the preaxial edge of the thigh is the **puboischiofemoralis internus**. Its origin is from the internal surface of the pubic cartilage and ischium. To get at it, one must cut through the pubic attachment of the rectus abdominis, and dissect deeply. The muscle inserts along most of the femur.

The dorsal surface of the thigh is occupied by a pair of muscles that arise from the base of the ilium, extend over the knee as a tendon, and

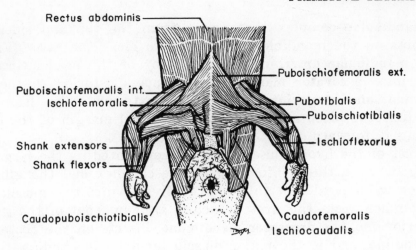

Rectus abdominis

Puboischiofemoralis ext.

Puboischiofemoralis int.
Ischiofemoralis

Pubotibialis
Puboischiotibialis

Shank extensors

Ischioflexorius

Shank flexors

Caudopuboischiotibialis

Caudofemoralis
Ischiocaudalis

Figure 6–8. Ventral view of the pelvic muscles of *Necturus*.

insert on the tibia. The anterior one is the **iliotibialis**; the posterior one is the **ilioextensorius**. Often these two appear as one muscle, for the separation between them may be obscure. Posterior to these, along the postaxial surface of the thigh, is a very slender **iliofibularis**. It, too, arises from the base of the ilium, but it separates from the others and passes to insert on the fibula. Dissect away blood vessels and nerves posterior to the origin of the iliofibularis, and you will find a small, deep, triangular muscle that arises from the base of the ilium and inserts on the posterior edge of the femur. This is the **iliofemoralis**.

As with the forearm muscles, the muscles of the shank fall into extensor and flexor groups. The **extensors** lie on the dorsal surface of the shank; the **flexors**, on the ventral. The extensors arise for the most part from the dorsal surface of the distal end of the femur; the flexors, from the ventral.

Branchiomeric Muscles

Skin the underside of the head, if this has not already been done, and the top and one side of the head. Be particularly careful above and below the stumps of the external gills. Note again the transverse sheet of muscle on the underside of the head that was cut and reflected during the study of the hyobranchial muscles. Approximately the anterior half of this represents an **intermandibularis**, a mandibular arch muscle; the posterior half an **interhyoideus**, a hyoid muscle (Fig. 6-7). Both insert on a median raphe, but they are distinct in their origins. The intermandibularis arises from the mandible; the interhyoideus, from the hyoid arch medial to the angle of the jaw and from the surface of a large muscle posterior to the angle of the jaw. The latter portion of the interhyoideus is sometimes called the **sphincter colli**.

The rest of the mandibular arch musculature is represented by a levator mandibulae complex that arises from the top and sides of the skull, inserts on the mandible, and serves to close the jaws (Fig. 6-6). A levator mandibulae anterior lies on the top of the head lateral to the middorsal line; a levator mandibulae externus, posterior to the eye; a levator mandibulae posterior, deep to the externus. The last may be found by dissecting deep along the posteroventral margin of the externus and pushing it forward.

The rest of the hyoid musculature is represented by the large muscle mass lying between the levator mandibulae complex and the gills. Close examination will reveal that this mass is divided into two muscles which arise in common from the first branchial arch in front of the gills. The more anterior one, the depressor mandibulae, inserts on the posterior end of the mandible, and is the major muscle involved in opening the jaws. The larger and more posterior one, the branchiohyoideus, continues forward beneath the interhyoideus and intermandibularis to insert along the ventral portion of the hyoid arch (the ceratohyal).

The remaining branchiomeric muscles belong to the branchial arches, although some have acquired other attachments. A series of levatores arcuum will be found dorsal to the bases of the external gills. Posterior to these, and crossing the origin of the cucullaris, you will see a dilatator laryngis. The dilatator laryngis looks like a levator, but note that it curves in toward the base of the neck (on its way to the larynx) rather than attaching onto the gills. The cucullaris, observed during the dissection of the shoulder region, completes the dorsal branchial muscles. The ventral branchial muscles can be found by dissecting deep between the branchiohyoid and rectus cervicis. Pull the branchiohyoid forward, and the rectus cervicis posteriorly. A series of subarcuals will be seen — a longitudinal one extending from the ceratohyal to the first branchial arch; another longitudinal one extending from the first and second to the third branchial arch; and an oblique one medial to this and extending between the first and second branchial arches. The transverse fibers posterior to the subarcuals, and deep to the rectus cervicis, constitute the transversi ventrales (Fig. 10-9, p. 286). These fibers extend from the third branchial arch on one side to that on the other. The remaining ventral muscles are a series of three depressores arcuum that extend from the ventral portion of the three branchial arches to the bases of the three external gills. They are small, and may have been removed in skinning.

MAMMALS

The changes that were seen beginning in the muscles of primitive tetrapods continue in the evolution toward mammals. (Review the general remarks on tetrapod muscle evolution on page 119.) But many of the groups of muscles have lost their clear definition in adult mammals, for some of the muscles have migrated from their original positions. This makes it more difficult to study the muscles by natural groups, but it can still be done, if a few concessions are made to topography.

Skinning and Integumentary Muscles

Lay your specimen of the cat or rabbit on its belly, and make a middorsal incision through the skin that extends from the back of the head to the base of the tail. Make additional incisions from this cut around the neck; around the tail, anus, and external genitals; down the lateral surface of each leg, and around the wrists and ankles. Skin is to be left for the time being on the head, tail and perineum, and feet.

Beginning on the back, gradually separate the skin from the underlying muscles by tearing through the superficial fascia with a pair of blunt forceps. As you separate the skin from the trunk, notice the fine, parallel, brown lines that adhere to its undersurface. They represent the **panniculus carnosus**, the largest of several integumentary muscles that move the skin. The panniculus carnosus arises from the surface of certain appendicular muscles of the shoulder (latissimus dorsi and pectoralis), and from a midventral band of connective tissue (**linea alba**), fans out over most of the trunk, and inserts on the underside of the skin. It should be removed with the skin, except for that portion attached to the shoulder muscles posterior to the armpit. Several smaller integumentary muscles, derived from the posterior musculature, become associated with the posterior part of the panniculus carnosus. They may not be noticed.

Much of the top and sides of the neck is covered by another integumentary muscle, the **platysma**, which is derived from the hyoid musculature. As may be noticed later when the head is skinned, a number of smaller dermal muscles are located over the face where they are associated with the lips, nose, eyes, ears, etc. They are collectively known as the **facial muscles**, or **muscles of expression**. Like the platysma, they are derived from the hyoid musculature.

As you continue to skin your specimen, you will come upon narrow tough cords passing to the skin. These are **cutaneous blood vessels** and **nerves**, and must be cut. Note that they tend to be segmentally arranged along the trunk. If your specimen is a pregnant or lactating female, the **mammary glands** will appear as a pair of large, longitudinal, glandular masses along the ventral side of the abdomen and thorax. They should be removed with the skin.

After the specimen is skinned, clean away the excess fat and superficial fascia on the side that is to be studied, but do not clean an area thoroughly until it is being studied. If your specimen is a male, be particularly careful in removing the wad of fat in the groin, for it contains on each side the proximal part of the **cremasteric pouch** — a part of the scrotum containing blood vessels and the sperm duct extending between the abdomen and scrotal skin. First find this pouch. It is large in the rabbit, but rather small in the cat. Clean away connective tissue deep in the groin, or inguinal region, so as to find the actual boundary between the thigh and abdomen. In the rabbit, a shiny, white **inguinal ligament** will be seen in this region, extending from the pubis to the ilium.

Posterior Trunk Muscles

All the axial muscles cannot be studied at the same time, for many of them are located deep to the shoulder muscles. Those located on the trunk between the pectoral and pelvic appendages will be examined now, and the more anterior ones will be considered after the appendages have been studied (p. 152).

(A) *Hypaxial Muscles*

Continue to clean off the surface of the trunk between the pectoral and pelvic appendages. The wide sheet of tough, white fascia covering the lumbar region on the back is the **thoracolumbar** or **lumbodorsal fascia**. The wide sheet of muscle that runs anteriorly and ventrally from the anterior part of this fascia and disappears in the armpit, is the latissimus dorsi (an appendicular muscle) (Fig. 6-10, p. 133). The large triangular muscle that covers the underside of the chest is the pectoralis (another appendicular muscle). The borders of the latissimus dorsi and pectoralis appear to run together posterior to the armpit. Separate the two in this region by removing the panniculus carnosus, and carefully trace their edges forward. Lift up the posteroventral edge of the latissimus dorsi, the posterolateral edge of the pectoralis, and remove the fat and loose connective tissue from beneath them. The hypaxial trunk muscles can now be studied.

As in other tetrapods (Fig. 6-5, C, D) the abdominal wall is composed of three layers of muscle, plus a paired longitudinal muscle along the midventral line. All serve to compress the abdomen. The **external oblique** forms the outermost layer (Fig. 6-9). This muscle arises by slips from the surface of a number of posterior ribs, and from the thoracolumbar fascia. Part of its origin lies beneath the posterior edge of the latissimus dorsi. Its fibers then extend obliquely posteriorly and ventrally to insert by an aponeurosis along the length of the linea alba. In the rabbit, some of the fibers insert on the inguinal ligament. Using a sharp scalpel, or a razor blade, make a cut 2 or 3 inches long through the external oblique at right angles to the direction of its fibers. Do not cut deeply, as you cut watch for a deeper layer of muscle having a different fiber direction. When you reach the deeper layer, reflect part of the external oblique.

The **internal oblique** lies beneath the external oblique. It is most

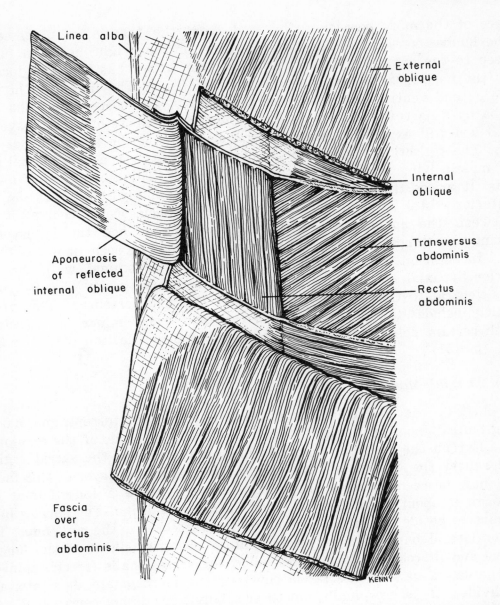

Linea alba

External oblique

Internal oblique

Transversus abdominis

Rectus abdominis

Aponeurosis of reflected internal oblique

Fascia over rectus abdominis

KENNY

Figure 6–9. Ventral view of the abdominal muscle layers on the left side of a cat. Part of the external oblique and internal oblique have been reflected.

apparent high up on the side of the abdomen near its main origin from the thoracolumbar fascia. Its fibers extend ventrally and slightly anteriorly at right angles to the fibers of the external oblique, and soon lead into a wide aponeurosis that inserts along the linea alba. Only the dorsal half of the muscle is fleshy. The ventral half is represented by an aponeurosis, but this may not be apparent at first, for one can see the third muscle layer, the transversus abdominis, through this aponeurosis.

To get at the **transversus abdominis**, make a longitudinal cut through the fleshy portion of the internal oblique, and reflect a part of this muscle. The transversus abdominis arises primarily from the medial surface of the more posterior ribs, and from the transverse processes of the lumbar vertebrae. The latter portion of the origin lies in a furrow deep to the epaxial muscles. Note that this furrow has the same location as the horizontal skeletogenous septum of fishes. The fibers of the muscle extend ventrally, and slightly posteriorly, to insert along the linea alba by a narrow aponeurosis. Part some of the fibers of the transversus, and you will expose the **parietal peritoneum** lining the abdominal cavity.

The reflection of the internal oblique also exposes a longitudinal band of muscle lying lateral to the midventral line. This is the **rectus abdominis**. It arises from the pubis and passes cranially to insert on the anterior costal cartilages and sternum. For much of its course, it lies between the aponeurosis of the internal oblique and the transversus. Transverse tendinous inscriptions can sometimes be seen in it.

In addition to the muscular layers of the abdominal wall, the posterior hypaxial musculature includes a subvertebral group (Fig. 6–5) that lies ventral to the lumbar and posterior thoracic vertebrae. This group, which includes the **quadratus lumborum** and **psoas minor**, is associated with certain pelvic muscles and is described in connection with the hind leg (p. 151).

(B) Epaxial Muscles

Lift up the thoracolumbar fascia with a pair of forceps, and make a longitudinal incision through it about 1/4 inch to one side of the middorsal line. Extend the incision from the latissimus dorsi to the sacral region, and reflect the superficial layer of the fascia. A deeper layer of this fascia will now be seen encasing the epaxial muscles. Make a longitudinal cut through it about one-half inch from the middorsal line. The narrow band of muscle beside the neural spines of the vertebrae is the **multifidus**. The wider lateral band of muscle represents a posterior union of the **longissimus and iliocostalis**. It may be called the **sacrospinalis** (**erector spinae**). In the cat, a part of the fascia covering the sacrospinalis dips into, and subdivides, the sacrospinalis, but these subdivisions do not correspond with the longissimus and iliocostalis of the anterior part of the trunk. Ignore them until the anterior part of the epaxial musculature is studied.

Pectoral Muscles

Most of the muscles of the pectoral region are, of course, appendicular muscles, but a number of axial and several branchiomeric muscles have become associated with the girdle and appendage.

(A) Pectoralis Group

The large triangular mass of muscle covering the chest is the **pectoralis**. It arises from the sternum, and passes to insert primarily along the humerus. Its major actions are to pull the humerus toward the chest (adduction) and posteriorly (retraction), but it also helps in transferring body weight from the trunk to the pectoral girdle and appendage (p. 139). Since its subdivisions differ in the cat and rabbit, they are described separately. Clean the surface of the muscle enough so that you can see the direction of its fibers.

Cat

Most authorities recognize four subdivisions to the pectoralis in the cat. A band of parallel fibers about 1/2 inch wide arises from the anterior end of the sternum, and passes laterally and slightly posteriorly across the rest of the muscle to insert on the fascia at the proximal end of the forearm. This is the **pectoantibrachialis** (Fig. 6-10). The muscle on the anterior surface of the shoulder anterior to the pectoantibrachialis is the clavodeltoid. It will be considered later, but at this time it should be mobilized for part of the pectoralis is located deep to it. Lift up the clavodeltoid and clean away connective tissue from beneath it. The clavicle is imbedded on the underside of the medial portion of this muscle.

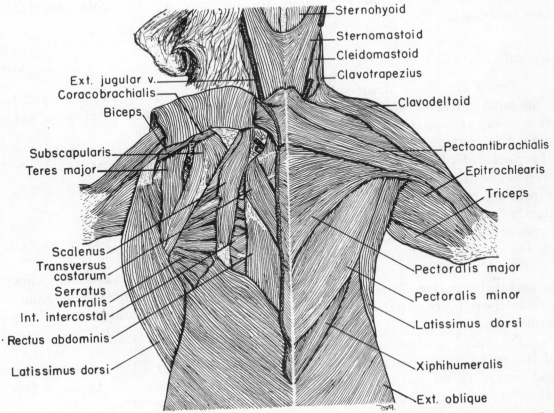

Figure 6–10. Ventral view of the muscles in the pectoral region of a cat. Superficial muscles are shown on the right side of the drawing. The pectoralis has been reflected on the left side in order to show certain deeper muscles.

Cut through the connective tissue that binds the clavicle to the manubrium, so you can push the clavodeltoid forward.

That portion of the pectoralis deep to the pectoantibrachialis and clavodeltoid is the **pectoralis major**. That portion of the pectoralis posterior to the major is the **pectoralis minor**. The line of separation between major and minor can best be detected on the lateral border of the pectoralis posterior to the middle of the pectoantibrachialis. At this point it can be seen that certain fibers of the pectoralis pass out along the brachium posterior to, and nearly parallel with, the pectoantibrachialis. These are the most posterior fibers of the pectoralis major. Trace them medially, separating them from the deeper fibers of the pectoralis minor. It will be noted that much of the pectoralis minor lies deep to the pectoralis major, and that the fibers of the minor extend forward at a more acute angle than those of the major. Despite its name, the pectoralis minor of the cat is larger than the major.

The fourth subdivision of the pectoralis, the **xiphihumeralis**, arises from the xiphisternum posterior to the pectoralis minor. It consists of a band of more or less parallel fibers that pass laterally and anteriorly, and soon go deep to the pectoralis minor. The muscle inserts on the proximal end of the humerus dorsal to the insertion of other parts of the pectoralis.

Rabbit

There are three main subdivisions to the pectoralis in the rabbit. The most anterior and superficial part is the **pectoralis primus**—a narrow band arising from the manubrium, and passing laterally to insert near the center of the shaft of the humerus. The muscle anterior to the pectoralis primus, and partly covering its insertion, is the clavodeltoid. It will be discussed later, but at this time mobilize the muscle by removing connective tissue from beneath it. The clavicle will be found partly imbedded in the medial end of the clavodeltoid. Remove loose connective tissue from around the clavicle, and note that it is connected to the manubrium by denser connective tissue.

Most of the rest of the pectoralis that one sees in a ventral view is **pectoralis major**. This portion arises from the length of the sternum, passes deep to the primus, and inserts along the proximal half of the humerus.

The third subdivision, **pectoralis minor**, lies deep to the major. A part of it can be seen emerging from beneath the anterior edge of the major and inserting on the clavicle. Follow it posteriorly, cutting through the major as you go. The muscle arises from the anterior third of the sternum. Its more superficial fibers insert on the clavicle, but its deeper fibers continue dorsal to the clavicle, pass over the front of the shoulder, and insert along the spine of the scapula. You can get a glimpse of this

portion of the muscle by dissecting along the dorsal margin of the clavodeltoid. More of the muscle will be seen later. This unusual insertion
of a part of the pectoralis is probably correlated with the extreme ex-
cursions of the foreleg and scapula when the animal is hopping. The
posterior part of the origin of the pectoralis minor is hard to separate
from the deeper fibers of the pectoralis major, but the insertions of the
two muscles are quite distinct.

(B) Trapezius and Sternocleidomastoid Group

The muscles belonging to the trapezius and sternocleidomastoid
group are branchiomeric muscles that have become associated with the
pectoral girdle, for they evolved from the cucullaris of lower vertebrates.
The various subdivisions of the trapezius act upon the girdle, primarily
the scapula. They pull it toward the middorsal line (abduction), anteriorly
(protraction) and posteriorly (retraction). In the cat, the anteromost
part of the trapezius (clavotrapezius) inserts on the clavicle, and together with the clavodeltoid protracts the arm. The sternocleidomastoid
complex of muscles acts primarily upon the head, turning it to the side
and flexing it. If the head is fixed, parts of the complex may act on

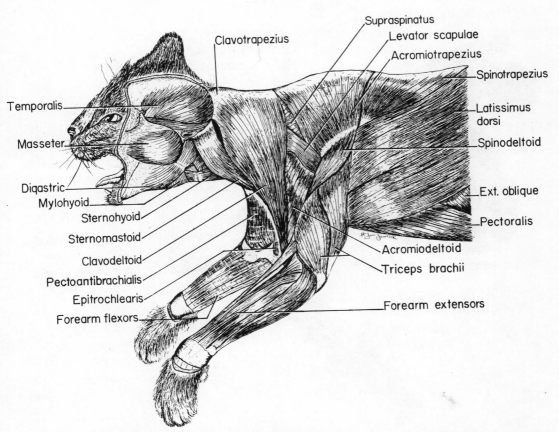

Figure 6–11. Lateral view of the pectoral and neck muscles of the cat.

the clavicle. Clean off connective tissue from the ventral, lateral, and dorsal surfaces of the neck, and from the dorsal part of the shoulder. You may have to remove more skin from the back of the head. Do not injure the large external jugular vein located superficially on the ventro-lateral surface of the neck.

Cat

The band of muscle that extends from the manubrium anteriorly and dorsally toward the mastoid region of the skull is the **sternomastoid**. It passes deep to the external jugular. Part of the muscle arises from the manubrium and part on a midventral raphe of connective tissue at the base of the neck. Cut through the latter portion of its origin, thereby freeing the muscle from its mate of the opposite side. Its insertion is primarily on the mastoid region of the skull. A narrower band of muscle will be seen dorsal to the posterior half of the sternomastoid. This is the **cleidomastoid**. It arises from the clavicle, and inserts on the mastoid region. The front of the cleidomastoid passes deep to the sternomastoid.

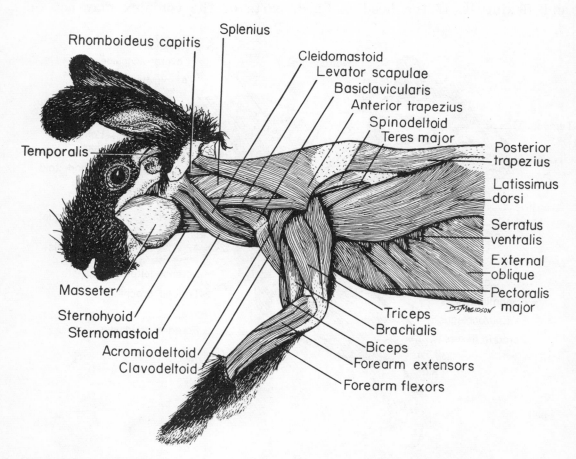

Figure 6–12. Lateral view of the pectoral and neck muscles of the rabbit.

The wide sheet of muscle located on the back and side of the neck dorsal and lateral to the cleidomastoid is the **clavotrapezius**. It arises from the back of the skull (nuchal crest) and the middorsal line of the neck, and inserts on the clavicle. The **clavodeltoid** mentioned earlier continues from the insertion of the clavotrapezius to the arm. These two muscles have united to some extent, and are sometimes described as one —the **cephalobrachial**.

The sheet of muscle covering the dorsal and lateral surfaces of the scapular region posterior to the clavotrapezius is the **acromiotrapezius**. Its anterior part arises from the middorsal line of the neck and front of the thorax; its posterior part from a tendon that crosses the midline and is continuous with the acromiotrapezius of the opposite side. It inserts on the metacromion and adjacent parts of the scapular spine.

The triangular sheet of muscle on the back, posterior to the acromiotrapezius and overlapping the front of the latissimus dorsi, is the **spinotrapezius**. It arises from the neural spines of the thoracic vertebrae, and passes forward to insert on the fascial covering of the scapular muscles.

Rabbit

A deeper portion of the platysma, known as the **depressor conchae posterior,** may have been left on the neck. If so, it extends from the manubrium forward and laterally to the base of the ear, passing superficial to the external jugular vein. Remove this muscle. This band of fibers beneath it, which arises from the manubrium and extends forward and dorsally deed to the external jugular, is the **sternomastoid** (Fig. 6–12). It inserts on the mastoid region of the skull. Lateral to its posterior portion are two other bands of muscle. The more superficial one, which arises from the clavicle and inserts on the mastoid, is the **cleidomastoid**. The deeper band is the **basiclavicularis**. The basiclavicularis arises from the clavicle lateral to the origin of the cleidomastoid. From here it passes deep to the cleidomastoid and sternomastoid to its insertion on the basioccipital region of the skull.

The wide sheet of muscle passing to the shoulder from the back of the neck is the **anterior trapezius**. Its origin is from the back of the skull and middorsal line of the neck. It inserts on the metacromion and spine of the scapula. The sheet of muscle posterior to it, and overlapping the front of the latissimus dorsi, is the **posterior trapezius**. It arises from the thoracolumbar fascia and from the neural spines of the thoracic vertebrae, and inserts on the spine of the scapula.

(C) *Remaining Superficial Muscles of the Shoulder*

The band of muscle that lies on the side of the neck between the clavotrapezius and acromiotrapezius (cat), or between the sternocleidomastoid complex and trapezius (rabbit) is the **levator scapulae**. It arises primarily from the occipital region of the skull, and inserts on the metacromion process of the scapula ventral to the insertions of the acromiotrapezius (cat) or anterior trapezius (rabbit). In the cat much of it lies deep to the clavotrapezius. The levator scapulae is a hypaxial muscle that pulls the scapula forward (protraction).

The deltoid lies posterior and ventral to the insertion of the levator scapulae ventralis. In both the cat and rabbit, the deltoid is subdivided into three parts. The **clavodeltoid** has been observed arising from the clavicle. It inserts on the shaft of the humerus in the rabbit, but passes to the ulna along with the brachialis (see the following section) in the cat. The **acromiodeltoid** lies posterior to the clavodeltoid. It arises from the acromion deep to the levator scapulae, and inserts on the proximal portion of the humeral shaft. The **spinodeltoid** lies along the lower border of the scapula. It arises from the scapular spine, and, in the rabbit, from the surface of the muscle covering the lower half of the scapula (infraspinatus). It inserts on the proximal end of the humerus. The muscle is thinner in the rabbit, and passes ventral to the large metacromion. But if the metacromion is broken and raised, the essential relationships will be seen to be the same as in the cat. The clavodeltoid protracts the arm, but the posterior parts of the complex are retractors and abductors of the humerus.

The **latissimus dorsi** has been observed on the side of the trunk posterior to the arm. It arises from the thoracolumbar fascia, and from the neural spines of the last thoracic vertebrae. From here it passes forward and ventrally to insert on the proximal end of the humerus in common with the teres major. In the cat, a small part of the muscle often forms a tendon that inserts with the pectoralis. The latissimus dorsi retracts the humerus.

(D) *Deeper Muscles of the Shoulder*

Cut across the center of the latissimus dorsi at right angles to its fibers, and also across all the subdivisions of the trapezius muscle. Reflect the ends of these muscles, and clean out the fat and loose connective tissue from beneath them so as to expose the deeper muscles of the shoulder. In the rabbit, the pectoralis minor can be seen sweeping over the front of the shoulder and scapula to insert on the scapular spine. It should be cut near its insertion and reflected.

The supraspinous fossa of the scapula of both cat and rabbit is occupied by the **supraspinatus**. The muscle inserts on the greater tubercle of the humerus, and protracts the humerus. The infraspinous fossa is occupied by the **infraspinatus**. Most of the muscle is covered by the spinodeltoid in the rabbit. The infraspinatus inserts on the greater tubercle of the humerus, and rotates this bone outward.

A **teres major** arises from the axillary border of scapula posterior and ventral to the infraspinatus. It passes forward to insert on the proximal end of the humerus in common with the latissimus dorsi. Its action is to rotate the humerus inward and to retract it.

Dissect deeply between the anterior part of the infraspinatus and the long head of the triceps (the large muscle on the posterior surface of the brachium). You will eventually come upon a very small triangular muscle arising by a tendon from the anterior part of the axillary border of the scapula, and inserting on the greater tubercle of the humerus. This is the **teres minor**. It helps the infraspinatus rotate the humerus outward.

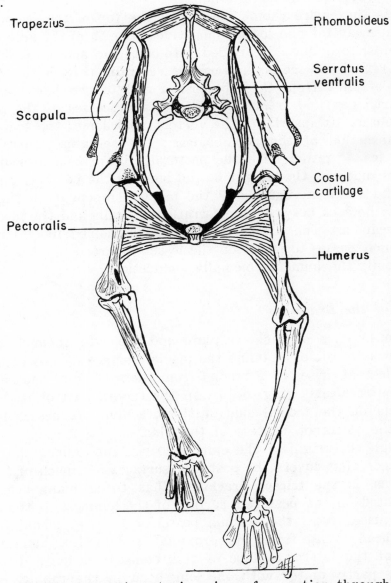

Figure 6–13. Diagrammatic anterior view of a section through the thorax at the level of the pectoral girdle to show muscular connections between the trunk and appendage.

Examine the scapula region from a dorsal view. The large muscle that arises from the tops of the posterior cervical and anterior thoracic vertebrae, and inserts along the vertebral border of the scapula, is the **rhomboideus**. It is divided into a number of loosely associated bundles which give a coarse texture to the muscle. In the rabbit these bundles are grouped into an anterior **rhomboideus minor** and a posterior **rhomboideus major**. In both animals, the most cranial bundle extends more anteriorly than the others to its origin from the back of the skull, and is called the **rhomboideus capitis**. The rhomboideus is a hypaxial muscle that pulls the scapula toward the vertebrae, helps to hold it in place, and assists in its protaction and retraction.

Cut across the entire rhomboideus, pull the top of the scapula laterally, and clean away fat and loose connective tissue from the area exposed. The large fan-shaped muscle (divided into anterior and posterior parts in the rabbit) that you see is the **serratus ventralis** (Fig. 6-13). It arises by a number of slips from the ribs just dorsal to the junction of ribs and costal cartilages, and from the transverse processes of the posterior cervical vertebrae. It inserts on the vertebral border of the scapula ventral to the insertion of the rhomboideus. The serratus ventralis is a hypaxial muscles. Together with the pectoralis, it forms a muscular sling that transfers much of the weight of the body to the pectoral girdle and appendage. The serratus ventralis is the major component in the sling.

Clean out the area between the serratus ventralis and the medial surface of the scapula. The subscapular fossa is occupied by a large **subscapularis**, which passes to insert on the lesser tuberosity of the humerus. This muscle pulls the humerus medially (adduction).

(E) Muscles of the Brachium

Clean the muscles of the brachium, and separate them from each other as much as possible. Examine the medial surface of the upper arm. The flat muscle that arises by a tendon from the surface of the latissimus dorsi is the **epitrochlearis** (Fig. 6-10). In the rabbit, part of its origin is from the fascia on the medial side of the brachium. It passes laterally to insert on the olecranon process of the ulna.

Cut through the origin of the epitrochlearis, and reflect the muscle. The large muscle that covers the posterior surface, and much of the sides of the humerus, is the **triceps brachii**. It has three main heads. The **long head**, located on the posterior surface of the humerus, is the largest. Note that it arises from the scapula posterior to the glenoid fossa. A large **lateral head** arises from the proximal end of the humerus, and covers much of the lateral surface of this bone. It is quite distinct in the cat, but is partly united with the long head in the rabbit. A small **medial head** can be found on the medial surface of the humerus deep to several nerves and blood vessels. It arises from most of the shaft of the humerus. In the cat, a still smaller **fourth head** arises from the distal portion of the lateral surface of the humerus. To see it, cut through, and reflect, the distal end of the lateral head. All parts of the triceps insert in common on the olecranon process. The function of the triceps and epitrochlearis is to extend the forearm.

The anterolateral surface of the humerus is covered by the **brachialis**. It arises from the shaft of the humerus and inserts on the proximal end of the ulna (cat), or on the ulna and radius in common with the biceps (rabbit, Fig. 6-12).

The **biceps brachii** lies on the anteromedial surface of the humerus. To see it clearly, cut through and reflect the pectoralis near its insertion on the humerus. (The mass of nerves passing to the arm dorsal to the pectoralis belong to the brachial plexus.) The biceps arises by a tendon from the edge of the glenoid fossa. This tendon lies in the bicipital groove between the greater and lesser tubercles. The muscle forms a prominent belly beneath the humerus, and inserts by a tendon on the radial tuberosity (cat), or on both the radius and ulna (rabbit). Biceps and brachialis are the major flexors of the forearm. The biceps also assists in supination of the forearm.

Clean off connective tissue from the medial side of the shoulder joint, and you will find a short band of muscle arising from the coracoid process and inserting on the proximal end of the humerus. This is the **coracobrachialis**. (Fig. 6-10). It helps to pull the humerus toward the body (adduction). In some cats and rabbits, a long portion of the coracobrachialis passes to the distal end of the humeral shaft.

(F) *Muscles of the Forearm*

The forearm is covered by a tough layer of deep fascia which must be removed to see the muscles. As in *Necturus,* the muscles fall into two groups: **extensors** located on the anterolateral surface of the forearm, and arising for the most part from the ectepicondylar region of the humerus; and **flexors** located on the posteromedial surface of the forearm, and arising for the most part from the entepicondylar region. The general function of the extensors is to extend the forearm, hand, and digits, and to supinate the forearm and hand. The flexors have an antagonistic action, namely, flexion and pronation.

Pelvic Muscles of the Cat

The muscles of the pelvic region of the cat and rabbit differ sufficiently in details to warrant separate description for the most part. Clean off the fat and superficial fascia from the surfaces of the pelvic region and thigh as a preliminary to the dissection. Be sure to remove the large wads of fat lateral to the base of the tail, and from the depression (**popliteal fossa**) posterior to the knee. Start to separate the more obvious muscles.

(A) *Lateral Thigh and Adjacent Muscles*

The most anterior muscle on the thigh of the cat is the **sartorius** (Fig. 6-14). It appears as a band extending from the crest and ventral border of the ilium (its origin) to the patella and medial side of the tibia (its insertion). Most of the muscle lies on the medial surface of the thigh, but part of it can be seen laterally. The muscle adducts and rotates the femur and extends the shank.

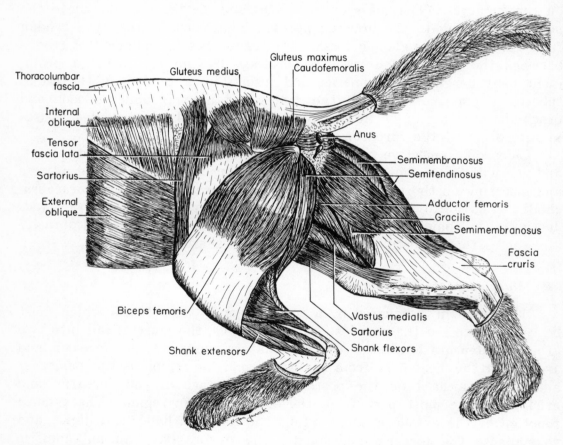

Figure 6-14. Superficial pelvic and thigh muscles of the cat. Lateral muscles can be seen on the left leg and medial muscles on the right leg.

Posterior to it on the lateral surface, you will see a tough, white fascia — the **fascia lata**. A triangular muscle mass (**tensor fasciae latae**) arises from the ventral border of the ilium and from the surface of adjacent muscles. It passes to insert in the dorsal part of the fascia lata, and it acts to tighten this fascia and extends to the shank.

The very broad muscle posterior to the fascia lata is the **biceps femoris**. It arises by a single head in the cat from the tuberosity of the ischium, and has a wide insertion that extends from the patella to the middle of the shaft of the tibia. The muscle forms the lateral wall of the popliteal fossa. Its action is to flex the shank and abduct the thigh.

The stout muscle arising from the ischial tuberosity, posterior to the origin of the biceps femoris, is the **semitendinosus**. It inserts on the medial side of the distal end of the tibia, and flexes the shank. The semitendinosus forms the posteromedial border of the popliteal fossa.

The band of muscle anterior and dorsal to the origin of the biceps femoris is the **caudofemoralis**. It arises from the second and third caudal vertebrae, passes beneath the anterior border of the biceps femoris, and forms a thin tendon that inserts on the patella. Lift up the biceps to follow it. The caudofemoralis abducts the thigh and extends the shank. This muscle appears to have evolved from a part of the gluteus maximus, and for this reason is sometimes called the gluteus longus.

Cut across the biceps near its origin, being careful not to cut a very slender band of muscle that lies beneath it. This band, which will be seen on reflecting the biceps, is the **tenuissimus** (Fig. 6-15, p. 145). It arises from the second caudal vertebra, inserts on the tibia with the biceps femoris, and assists this muscle in abduction of the thigh and flexion of the shank. The tenuissimus is closely associated with the biceps femoris, and has probably evolved from it.

(B) Superficial Muscles on the Lateral Surface of the Pelvic Girdle

The caudofemoralis crosses the lateral surface of the pelvic girdle, but three other muscles can be seen in this region as well. If you have cleaned away the fat from the base of the tail, you will see a small muscle lying just dorsal to the origins of the biceps and semitendinosus, and posterior to the origin of the caudofemoralis. This is the **obturator internus**. It arises from the dorsal surface of the ischium near the symphysis, passes over the lateral edge of the ischium anterior to the tuberosity, and inserts into the trochanteric fossa of the femur. Its insertion will be seen later. This muscle abducts the thigh.

Two muscles form the triangular mass anterior to the proximal portion of the caudofemoralis. The fibers immediately anterior to the caudofemoralis constitute the **gluteus maximus**. This muscle arises from the last sacral and first caudal vertebrae. It passes ventrally, and inserts on the fascia lata and greater trochanter. The more diagonal fibers anterior to it form the **gluteus medius**, a larger muscle in the cat. The gluteus medius arises primarily from the crest and lateral surface of the ilium, passes posteriorly and ventrally, and inserts on the greater trochanter. Some of the gluteus medius lies beneath the maximus, and the two should be carefully separated. Part of the tensor fasciae latae arises from the anteroventral border of the gluteus maximus. Both glutei abduct the thigh.

(C) Quadriceps Femoris Complex

Cut across the belly of the sartorius, and across the fascia lata at right angles to its fibers. Make any further cuts necessary to reflect the fascia lata and its tensor fasciae latae. Clean the area exposed. The large muscle on the anterolateral surface of the thigh, which was covered by the fascia lata, is the **vastus lateralis** (Fig. 6-15). It arises from the greater trochanter and shaft of the femur. The narrower muscle on the front of the thigh, medial to the vastus lateralis, is the **rectus femoris**. It arises from the ilium anterior to the acetabulum. The muscle on the medial surface of the thigh, posterior to the rectus femoris, is the **vastus medialis** (Fig. 6-14). It arises from the shaft of the femur. Both the rectus femoris and the vastus medialis lie beneath the sartorius. Pull the rectus femoris and vastus lateralis apart. The deep muscle observed between them is the **vastus intermedius**. It, too, arises from the shaft of the femur. All these muscles converge and insert onto the patella and its ligament. The patella ligament attaches on the tuberosity of the tibia. This whole complex is often referred to as the **quadriceps femoris**, and it is the major extensor of the shank.

(D) Posteromedial Thigh Muscles

The medial surface of the thigh, posterior to the quadriceps femoris, is largely covered by the **gracilis** (Fig. 6-14), a wide, flat muscle that arises from the ischial and pubic symphyses. It inserts by an aponeurosis onto the tibia and fascia of the shank. The gracilis adducts and retracts the leg. Cut across its belly, reflect its ends, and clean the area exposed. Note again the semitendinosus observed previously from the lateral surface. The large muscle aterior to it, which arises from the posterior edge of the ischium, is the **semimembranosus**. It inserts on the medial epicondyle of the femur and adjacent part of the tibia, and acts primarily to retract the thigh. The large muscle anterior to it is the **adductor femoris**. Its origin is from the ischial and pubic symphyses; its insertion is on most of the length of the femoral shaft, and its action is adduction and retraction of the thigh.

Two small, triangular-shaped muscles lie anterior to the adductor femoris. The one just in front of the adductor femoris is the **adductor longus**, and the one anterior to the adductor longus is the **pectineus**. The pectineus also lies just posterior to the point at which the prominent femoral artery and vein emerge from the body wall. Both abductor longus and pectineus arise from the front of the pubis, and insert near the proximal end of the femur. Both are adductors of the thigh.

(E) Deeper Muscles of the Pelvic Girdle

The deeper muscles of the pelvis of the cat are difficult to dissect.

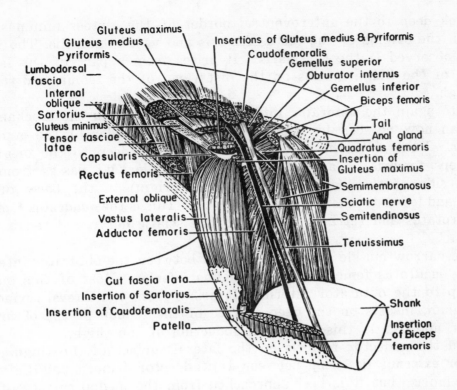

Figure 6–15. Lateral view of the deep muscles of the pelvis and thigh of a cat. The sartorius, tensor fasciae latae, biceps femoris, gluteus medius, gluteus maximus, caudofemoralis, and piriformis have been largely cut away to expose the deeper muscles. However, the origins and insertions of these muscles are shown as points of reference.

To expose these muscles, separate the tensor fasciae latae from the lower border of the gluteus medius, and cut and reflect the caudofemoralis, gluteus maximus, and gluteus medius. Cut the first two near their origin, the gluteus medius near its insertion. Carefully clean away connective tissue and fat from the area exposed. The wide, but thin, triangular muscle that lies deep to the gluteus medius is the **piriformis** (Fig. 6–15). It arises from the last two sacral and first caudal vertebrae, and inserts on the greater trochanter. Its action is abduction of the thigh. Since it is thin, it may have been cut and reflected with the gluteus medius. The prominent **sciatic nerve** passes deep to the piriformis.

Cut the piriformis near its insertion, and reflect the muscle. Two muscles lie beneath the piriformis — an anterior **gluteus minimus** and a posterior **gemellus superior**. These two are more or less united and are difficult to separate. The former can be recognized by the conspicuous tendinous fibers that form its anteroventral border; the latter is crossed by the sciatic nerve, and extends as far caudad as the obturator internus. The gemellus superior arises from the dorsal border of the ilium and ischium; the gluteus minimus from the lateral surface of the ilium. Both insert on the greater trochanter and rotate and abduct the thigh.

Look deep to the anteroventral border of the gluteus minimus, and between the origins of the rectus femoris and vastus lateralis. The short muscle observed is the **capsularis**. It arises from a part of the lateral surface of the ilium, inserts on the distal end of the femur, and rotates the thigh.

Note again the **obturator internus**. Its insertion in the trochanteric fossa can now be seen by lifting up the caudal border of the gemellus superior. Ventral to the obturator internus, and deep to the origin of the biceps femoris, is a thick rectangular muscle that passes from the ventral side of the ischial tuberosity (its origin) to the bases of the greater and lesser trochanters (its insertion). This is the **quadratus femoris**, and it rotates and retracts the thigh. The sciatic nerve and tenuissimus cross it.

The narrow muscle which can be seen between the obturator internus and the quadratus femoris is the **gemellus inferior**. Most of this muscle lies deep to the obturator internus. It arises from the lateral surface of the ischium, inserts on the deep surface of the tendon of the obturator internus, and assists this muscle in abduction of the thigh.

Find the adductor femoris on the lateral surface of the thigh. The **obturator externus** lies deep between the adductor femoris and the quadratus femoris, but it is best approached from the medial surface of the thigh by cutting and reflecting the adductor longus and pectineus. Note that the obturator externus arises from the ventromedial border of the obturator foramen, and inserts deep in the trochanteric fossa. Its action is rotation of the thigh.

(F) Iliopsoas Complex and Adjacent Muscles

(G) Muscles of the Shank

The iliopsoas group of muscles and the general features of the shank muscles of the cat and the rabbit are much the same. They are described together in Subdivisions F and G, page 151.

Pelvic Muscles of the Rabbit

The actions of the pelvic muscles in the rabbit are essentially the same as those of the comparable muscles in the cat, described on pages 142 to 146, so they will not be repeated here.

Clean off the superficial fascia from the surface of the pelvic region and thigh as a preliminary to the dissection. Be sure to remove the wad of fat from the depression (**popliteal fossa**) posterior to the knee.

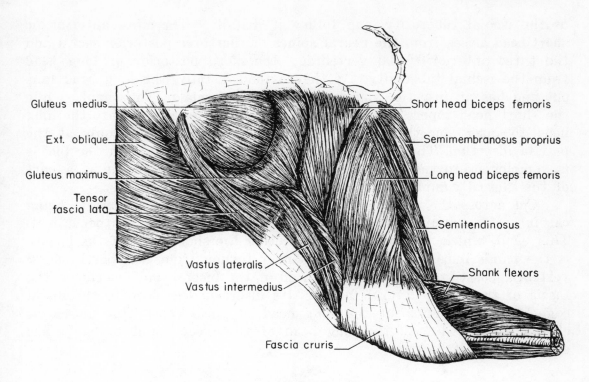

Figure 6–16. Lateral view of the pelvic and thigh muscles of the rabbit. Part of the fascia lata has been removed.

(A) *Lateral Thigh Muscles*

Note that the anterolateral surface of the thigh is covered by a sheet of tough, white fascia — the **fascia lata**. A triangular muscle mass lies on the anterodorsal portion of the thigh between the fascia lata and the fascial covering of the sacral region. Most of this muscle is the **tensor fasciae latae** (Fig. 6–16), but other muscles have fused with it, as will be seen presently. The tensor fasciae latae arises from the ventral border of the ilium and from the surfaces of adjacent muscles. It inserts in the fascia lata.

Make whatever cuts are necessary to reflect the fascia lata and the fascia covering the sacral region. Certain muscle fibers attach into the fascia, and they must be cut. The lateral surface of the thigh posterior to the fascia lata is covered by the large **biceps femoris**. This muscle also forms the lateral wall of the popliteal fossa. The thick muscular band posterior to the biceps, and forming the posteromedial border of the popliteal fossa, is the semitendinosus. It will be described later, but separate it from the biceps at this time. Note that a thin head of the semitendinosus crosses the posterodorsal portion of the biceps to arise from fascia which has probably been removed. Reflect this head, and remove any remaining fascia covering the biceps. The biceps femoris arises by two heads. Find the separation between them anterior to the region of the ischial tuberosity, and follow it distally. The more anterior or **short head** arises from the neural spines of the three posterior sacral and the three anterior caudal vertebrae; the more posterior or **long head** from the ischial tuberosity. As they extend distally, the long head fans out and has an insertion on the patella and fascia of the shank, while the short head tapers to a narrow tendon which passes deep to the long head to insert on the patella. The two heads are united only at the patella insertion. The muscle usually called the short head of the biceps is homologous to the caudofemoralis of the cat. This muscle is a part of the gluteal complex, and may be called the gluteus longus.

Cut across the middle of the short head of the biceps femoris, being careful not to injure a very slender band of muscle that lies beneath it. This band, which will be seen on reflecting the short head of the biceps, is the **tenuissimus**. It arises by a slender tendon from the fourth sacral vertebra, passes deep to the long head of the biceps, and inserts on the fascia of the shank (crural fascia). Its insertion lies deep to the distal edge of the insertion of the long head of the biceps. This muscle is closely associated with the long head of the biceps, and probably has evolved from it.

(B) Superficial Muscles on the Lateral Surface of the Pelvic Girdle

The triangular muscle mass that lies on the dorsal surface of the sacrum between the short head of the biceps and the tensor fasciae latae is part of the gluteus complex. The posterior and the ventral portions of this mass constitute the V-shaped **gluteus maximus**. The limbs of the V represent the origin of the muscle; the apex represents its insertion which is on the third trochanter (a small process of the femur distal to the greater trochanter). The posterior portion the muscle lies anterior to, and partially beneath, the short head of the biceps. This portion arises from the sacral fascia and from the neural spines of the sacral vertebrae. The anteroventral portion of the muscle arises from the sacral fascia and ventral border of the ilium. It is partially fused with the tensor fasciae latae.

The **gluteus medius** lies between, and is partly covered by the two limbs of the gluteus maximus. It arises primarily from the crest of the ilium and inserts on the greater trochanter.

(C) Quadriceps Femoris Complex

The large muscle on the anterolateral surface of the thigh, which was covered by the fascia lata, is the **vastus lateralis**. It arises from the greater trochanter and adjacent parts of the femur. A narrow band of muscle will be seen between the vastus lateralis and biceps femoris. This is the **vastus intermedius**, and it arises from the greater trochanter and shaft of the femur. The long band of muscle covering the front of the thigh, anterior to the vastus lateralis and deep to the fascia lata, is the **superficial head** of the **rectus femoris**. It arises in common with the tensor fasciae latae, to which it is intimately fused, from the ventral border of the ilium. Note also that its distal end is loosely united with the vastus lateralis, and with muscles on the medial side of the thigh. Separate the latter fusion, and cut across the belly of the superficial head of the rectus femoris. The **deep head** of the **rectus femoris** lies on the anteromedial surface of the thigh deep to the superficial head, and medial to the vastus lateralis (Fig. 6-17). It arises from the ilium anterior to the acetabulum. The muscle on the medial side of the thigh, posterior to the deep head of the rectus femoris, is the **vastus medialis**. It arises from the shaft of the femur. All these muscles converge and insert in common on the patella and its ligament, which, in turn, extends to the tuberosity of the tibia. Together these muscles form the great **quadriceps femoris**.

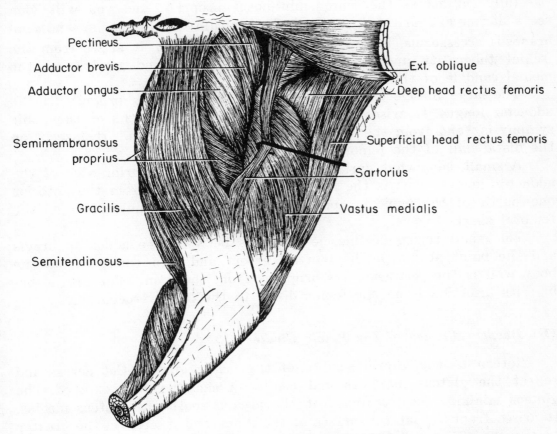

Figure 6-17. Medial veiw of the thigh muscles of the rabbit.

(D) *Posteromedial Thigh Muscles*

Examine the medial surface of the thigh posterior to the quadriceps femoris. The narrow band of muscle that covers the femoral artery and vein is the **sartorius** (Fig. 6-17). It arises from the inguinal ligament, and passes distally to insert on the medial condyle of the tibia. Its distal portion is fused with the **gracilis** — a wide, thin muscle covering most of the medial surface of the thigh posterior to the sartorius. The gracilis arises from the pubic symphysis and adjacent fascia. It inserts by an aponeurosis onto the tibia and fascia of the shank.

Cut across these two muscles and reflect their ends. Note again the **semitendinosus** seen previously from the lateral aspect. Its main origin is from the ischial tuberosity, but, as noted above, it has a small head that arises from the fascia overlying the biceps femoris. It inserts on the tibia and fascia of the shank deep to the insertion of the gracilis.

The large muscle anterior to the semitendinosus is the main part of the semimembranosus (**semimembranosus proprius**). It arises from the posterior and ventral surfaces of the ischial tuberosity, and inserts on the medial epicondyle of the femur and on the proximal end of the tibia. Carefully cut across the semimembranosus proprius, and you will soon see a darker cylindrical muscle imbedded within it. This is the **semimembranosus accessorius**. Follow it through the proprius. It arises from the ischial tuberosity and inserts by a prominent round tendon on the medial condyle of the tibia.

The large muscle anterior to the semimembranosus proprius is the **adductor longus**. It arises from the posterior three-fourths of the pubic symphysis, and from the ventral portion of the ischium. It inserts on the distal half of the femoral shaft.

A small, somewhat darker muscle lies upon the anterior edge of the adductor longus. This is the **adductor brevis**. It arises from the anterior one-fourth of the pubic symphysis and inserts near the middle of the femoral shaft.

The small triangular muscle that lies between the adductor brevis and the point at which the femoral artery and vein emerge from the body wall is the **pectineus**. Its origin is from the front edge of the pubis; its insertion is on the femur distal to the lesser trochanter.

(E) Deeper Muscles of the Pelvic Girdle

Return to the lateral surface of the pelvic girdle. Cut across and reflect the gluteus maximus and medius. Clean the area exposed. The **gluteus minimus** occupies most of the area beneath the gluteus medius. It arises from the lateral surface of the ilium, and inserts on the greater trochanter. (Figure 6–15, showing the deeper pelvic muscles of the cat, will also be helpful in this dissection.)

The **piriformis** lies beside, and partly covers, the posterior border of the gluteus minimus. It arises from the second and third sacral vertebrae, and inserts on the greater trochanter. The insertions of the gluteus minimus and piriformis have fused to some extent.

The **sciatic nerve** emerges from beneath the posterior border of the piriformis and soon crosses a large, white tendon complex inserting in the trochanteric fossa. Careful dissection will reveal that three muscles contribute to this complex. The most anterior one, which is fairly distinct, is the **gemellus superior**. It arises from the dorsal border of the ischium. The **obturator internus** lies posterior to, and is partly covered by, the gemellus superior. It arises on the inside of the pelvis from the borders of the obturator fenestra, and passes over the dorsal rim of the ischium to its insertion. Only the latter part of the muscle is apparent in this dissection. The third muscle in the group, the **gemellus inferior**, is more or less fused with the obturator internus. It is represented by the fibers which originate from the ischial tuberosity and side of the ischium anterior to the tuberosity.

Find the adductor longus on the lateral surface. The rectangular muscle located between its dorsal border and the ventral border of the gemellus inferior is the **quadratus femoris**. It arises from the ventral surface of the ischial tuberosity and inserts on the third trochanter and distal border of the trochanteric fossa.

The **obturator externus** is located deep between the quadratus femoris and adductor longis. However, it is best approached from the medial surface of the thigh by cutting and reflecting the adductor brevis and pectineus. It arises on the lateral surface of the pelvis from the borders of the obturator foramen, and inserts deep in the trochanteric fossa.

(F) *Iliopsoas Complex and Adjacent Muscles* (*Cat and Rabbit*)

Note the thick bundle of muscle that emerges from the body wall medial to the origin of the rectus femoris. This is the iliopsoas complex. It may be studied now, or it may be postponed until the abdominal viscera have been dissected. If the former choice is made, trace the bundle forward by cutting through the muscle layers of the abdominal wall. The complex lies in a retroperitoneal position. The main part of the bundle represents the **psoas major** (Fig. 10-12, p. 294). This portion arises primarily from the centra of the last two or three thoracic vertebrae and from the centra of all the lumbar vertebrae. It inserts on the lesser trochanter, and protracts and rotates the thigh.

Lateral and slightly dorsal to its extreme posterior portion you will see a group of fibers arising from the ventral border of the ilium plus (in the rabbit) adjacent vertebrae. These fibers represent the **iliacus**. They insert and act in common with the psoas major. Psoas major and iliacus are more intimately united in the cat than in the rabbit, and together may be called the **iliopsoas**.

The thin muscle medial to the psoas major is the **psoas minor**. It arises (cat) from the centra of the posterior thoracic and anterior lumbar vertebrae, or (rabbit) from the centra of the posterior lumbar vertebrae. It inserts on the girdle near the origin of the pectineus. The psoas minor is one of the subvertebral hypaxial muscles referred to on page 132, rather than an appendicular muscle. Its action is flexion of the back.

Lift up the lateral border of the psoas major near its middle. The thin muscle that lies on the ventral surface of the transverse processes is the **quadratus lumborum**, another subvertebral hypaxial muscle. It arises primarily from the centra and transverse processes of the lumbar and last several thoracic vertebrae. It extends further forward in the rabbit, and some of its fibers also spring from the ribs. It inserts on the ilium anterior to the origin of the iliacus. Its action is lateral flexion of the vertebral column.

(G) Muscles of the Shank (Cat and Rabbit)

Remove the tough fascia (**fascia cruris**) covering the shank, and separate the more obvious muscles. As in other vertebrates, the muscles fall into extensor and flexor groups. The extensors lie on the anterolateral surface of the shank; the flexors on the posteromedial surface. Although the muscles of the distal parts of the appendage are not being considered, one of the flexors on the shank is worthy of note, for it often is used in experimental studies on muscular contraction and it is even referred to in English literature. This is the **gastrocnemius** — the most superficial of the flexors, and the one which forms the greater part of the calf of the leg. It arises by two heads from the lateral and medial condyles of the femur and adjacent parts of the tibia. It inserts by a large tendon (the **calcaneus tendon of Achilles**) on the calcaneus (fibulare). Deeper muscles generally unite with the gastrocnemius, or its tendon, and together they bring about plantar flexion of the foot. This action is sometimes referred to as extension in mammalian anatomy, but if this term is used it should not be forgotten that the gastrocnemius is, from an evolutionary point of view, one of the ventral limb muscles or flexors.

Anterior Trunk Muscles

The posterior trunk muscles were described before the appendages were dissected, and other trunk muscles were seen during the dissection of the shoulder. Now that the appendages have been examined, it is possible to resume the study of the anterior trunk muscles.

(A) Hypaxial Muscles

All the trunk muscles that become associated with the pectoral girdle belong to the hypaxial group. They are the **levator scapulae**, **rhomboideus**, **rhomboideus capitis**, and **serratus ventralis**. Find these again.

Lay your specimen of the cat or rabbit on its back, reflect the pectoralis, and examine the muscles on the ventrolateral portion of the thoracic wall. The **rectus abdominis** will be seen passing forward to its insertion on the sternum and anterior costal cartilages (Fig. 6-10, p. 133). The thoracic wall is composed of three layers of muscle comparable to those of the abdominal wall. Observe that the outermost layer, the **external intercostals**, consists of fibers that pass from one rib posteriorly and ventrally to the next posterior rib. This layer does not extend all the way to the midventral line. Cut through, and reflect, an external intercostal, and you will find an **internal intercostal**. Its fibers extend anteriorly and ventrally. The third layer, **transversus thoracis**, is incomplete and found only near the midventral line. To see it, lift up the rectus abdominis, and cut through and reflect the ventral portion of an internal intercostal. The transversus thoracis arises from the dorsal surface of the sternum, and is inserted by a number of slips into the costal cartilages. A better view of the muscle will be had when the thorax is opened (p. 253).

In addition to these layers, other muscles are associated with the thoracic wall. The diagonal muscle that arises near the middle of the sternum, and crosses the anterior end of the rectus abdominis to insert on the first rib, is the **transversus costarum** (Fig. 6-10). Dorsal to it is a fan-shaped muscle which extends between the cervical vertebrae and the ribs. This is the **scalenus**. It arises from the transverse processes of all the cervical vertebrae (cat), or the last four (rabbit), and has multiple insertions on various ribs. In the cat, one portion of its insertion extends as far caudad as the ninth rib. In the rabbit, much of the insertion passes between the anterior and posterior halves of the serratus ventralis.

Turn your specimen on its side, reflect the latissimus dorsi, and pull the top of the scapula away from the trunk. Medial to the scapula and serratus ventralis, you will see a number of short muscular slips that arise from the thoracolumbar fascia and insert on the dorsal portion of the ribs. These constitute the **serratus dorsalis** (Fig. 6-18). They can be seen better by making a longitudinal incision through the thoracolumbar fascia and reflecting it.

All these muscles of the thorax, together with the muscular diaphragm and the abdominal muscles, are concerned with respiration. The muscular movements of respiration are very complex, but the major movements during inspiration are a contraction of the diaphragm, and a forward movement of the ribs through the action of such muscles as the external intercostals, scalenus, and the more anterior portions of the serratus dorsalis. During expiration, these muscles relax; the diaphragm is pushed anteriorly by the contraction of abdominal muscles and the pressure of the abdominal viscera and the ribs move posteriorly by their elastic recoil and by the action of such muscles as the internal intercostals, transversus thoracis, and the more posterior parts of the serratus dorsalis.

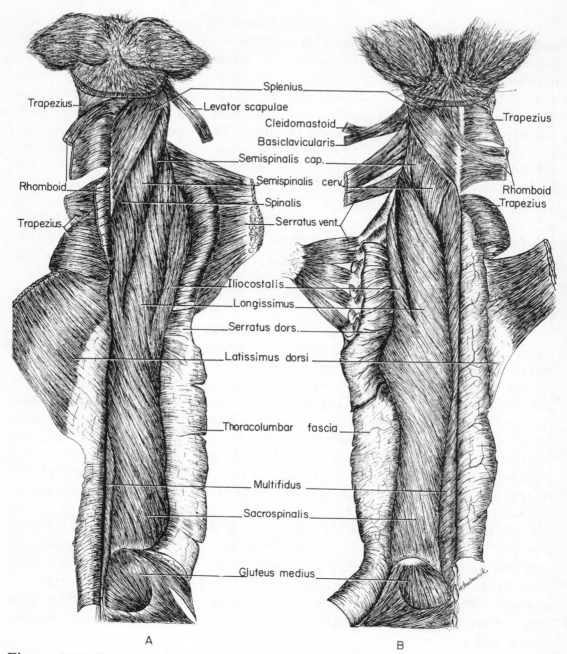

Trapezius

Rhomboid

Trapezius

Splenius

Levator scapulae

Cleidomastoid

Basiclavicularis

Semispinalis cap.

Semispinalis cerv.

Spinalis

Serratus vent.

Iliocostalis

Longissimus

Serratus dors.

Latissimus dorsi

Thoracolumbar fascia

Multifidus

Sacrospinalis

Gluteus medius

Trapezius

Rhomboid
Trapezius

A

B

Figure 6–18. Dorsal views of the epaxial musculature of the cat, *A*, and rabbit, *B*. The overlying muscles have been reflected, but their attachments are shown.

The subvertebral portion of the anterior hypaxial musculature is represented primarily by the **longus colli**. This muscle appears as a band in the neck lying ventral and medial to the origin of the scalenus. It arises from the centra of the first six thoracic vertebrae, and as it passes forward it receives other slips of origin from the transverse processes and centra of the cervical vertebrae. Portions of it insert on the centra and transverse processes of each of the cervical vertebrae. Its action is flexion and lateral flexion of the neck.

(B) Epaxial Muscles

The most superficial of the anterior epaxial muscles is the **splenius** (Fig. 6-18). It is a thin but broad triangular muscular sheet that covers the back of the neck deep to the trapezius and the anterior portions of the rhomboideus and serratus dorsalis. The splenius arises from the mid-dorsal line of the neck, and passes forward and laterally to insert on the occipital region of the skull (nuchal crest) and transverse process of the atlas. Each splenius individually acts as a lateral flexor of the head; together they elevate the head (extension).

Much of the thoracolumbar fascia was reflected during the study of the posterior epaxial muscles and the serratus dorsalis. Complete the reflection of this fascia. The epaxial mass should now be well exposed. Find the multifidus and sacrospinalis posteriorly (p. 132) and trace them forward.

In the rabbit, the **multifidus** continues forward as a narrow band beside the neural spines to the base of the neck. The **sacrospinalis** lies lateral to the multifidus. Near the posterior end of the thorax, the sacrospinalis splits into two longitudinal bundles, the more medial and larger of which is the **longissimus**. The anterior portion of the longissimus inserts on the transverse processes of the posterior cervical vertebrae. The more lateral and smaller bundle is the **iliocostalis**. It lies deep to the fleshy portion of the serratus dorsalis, and inserts on the ribs.

In the cat, as the **multifidus** continues forward next to the neural spines, it soon disappears beneath more superficial muscles. It actually continues in a deep position to a point near the middle of the thorax. As the lateral **sacrospinalis** continues forward, it appears to split into three longitudinal divisions. The most medial, the **spinalis**, consists of diagonal fibers that pass to insert on the neural spines. This portion continues forward to the base of the neck. The middle and longest division, **longissimus**, consists of longitudinal fibers that extend as far forward as the transverse processes of the posterior cervical vertebrae. The most lateral division, **iliocostalis**, lies deep to the fleshy portion of the serratus dorsalis, and inserts on the ribs.

In both cat and rabbit, two longitudinal muscles will be found deep to the splenius. These muscles arise from the vertebrae between the front of the longissimus dorsi and the front of the multifidus (in the rabbit) or the spinalis (in the cat). The muscles extend forward to the neck and head. The more dorsal, more medial, and wider of the two is the **semispinalis cervicis**; the more ventral, more lateral, and narrower the **semispinalis capitis**. It is not always possible to find a clear division between these two muscles. Still deeper epaxial muscles are not described. All the epaxial muscles are concerned with the extension and lateral flexion of the back, neck, and head.

As shown in Figure 6–5, the epaxial muscles of reptiles fall into three groups — transversospinalis, longissimus, and iliocostalis. The epaxial group becomes further complicated in mammals, but the reptilian subdivisions can still be recognized. The transversospinalis system of reptiles is represented by several deep muscles (interspinalis, intertransversarii, occipitals) and by the multifidus spinae and semispinalis; the longissimus by the longissimus and splenius; and the iliocostalis by the iliocostalis. In addition some mammals have a spinalis which lies medial to the longissimus. Its classification is one of the major problems.

In some mammals, including the rabbit, the situation appears to be rather simple (Fig. 6–18). The multifidus extends as far craniad as the base of the neck, and is replaced on the neck by the semispinalis. The longissimus and iliocostalis lie lateral to these. They are united posteriorly but are distinct in the thoracic region. A spinalis is absent. This might well be the primitive mammalian condition.

In the cat, the situation is more complex. The multifidus extends forward and gradually disappears. It is absent near the anterior end of the thorax. A semispinalis resumes in the neck region. Laterally there is a longissimus and iliocostalis united in the lumbar region, but distinct in the thoracic. A spinalis is present. It is united posteriorly and laterally with the longissimus, but it occupies the area on the anterior end of the thorax that was occupied by the anterior part of the multifidus in the rabbit. The spinalis is often considered to be a medial extension of the longissimus complex, but it appears more logical to consider it to be the modified anterior portion of the primitive multifidus which has secondarily become associated with the longissimus. Its essential relationships are those of the anterior part of the primitive multifidus.

Hypobranchial Muscles

The hypobranchial muscles are an axial group located on the ventral side of the neck and throat. All are concerned with the movements of the larynx, hyoid apparatus, and tongue. The group is utilized in swallowing. Complete the skinning of this region in your cat or rabbit as far forward as the chin. Clean away the loose connective tissue and fat. You may turn back, but do not destroy, prominent glands, ducts, blood vessels, and nerves in this region.

(A) Posthyoid Muscles

Find the sternomastoid muscles (pp. 136, 137), and push them laterally. The thin, midventral band of muscle that covers the windpipe (trachea), and extends from the anterior end of the sternum to the hyoid, is the **sternohyoid** (Fig. 6–19). Actually there is a pair of sternohyoids, but they are generally fused. Their origin is the sternum, their insertion, the hyoid.

Carefully separate the sternohyoid from another band of muscle that lies dorsal and lateral to it. This band, the **sternothyroid**, has a similar origin, but passes forward to insert on the thyroid cartilage of the larynx. The larynx, or Adam's apple, lies posterior to the hyoid. Thus the

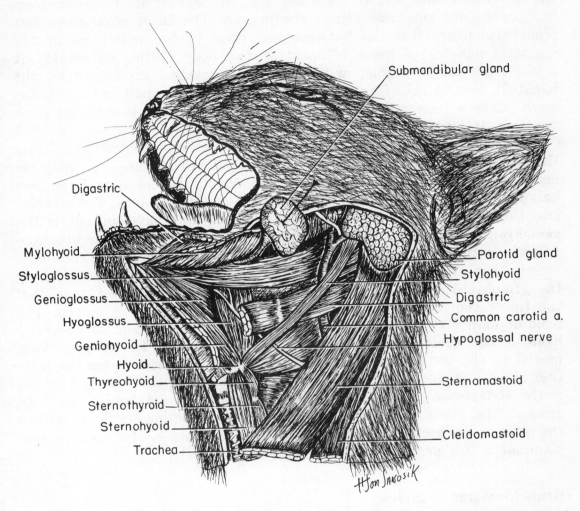

Submandibular gland

Digastric

Mylohyoid
Styloglossus
Genioglossus
Hyoglossus
Geniohyoid
Hyoid
Thyreohyoid
Sternothyroid
Sternohyoid
Trachea

Parotid gland
Stylohyoid
Digastric
Common carotid a.
Hypoglossal nerve
Sternomastoid
Cleidomastoid

HJon Janosik

Figure 6–19. Lateroventral view of the musculature in the anterior part of the neck and the floor of the mouth in the cat.

sternothyroid is not as long a muscle as the sternohyoid. A short band of muscle, the **thyreohyoid**, lies on the lateral surface of the larynx. It arises at the point of insertion of the sternothyroid, and passes forward to insert on the hyoid. Sternothyroid and thyreohyoid appear as one band unless their attachments on the larynx are carefully exposed.

(B) Prehyoid Muscles

Two branchial muscles must be studied and reflected in order to get at the prehyoid muscles. Note the stout band of muscle that is attached to the ventral border of the mandible. This is the **digastric**. It has a fleshy origin in the cat from the paraoccipital and mastoid processes of the skull. In the rabbit, its origin is by a tendon from just the paraoccipital process. In both animals it inserts along the ventral border of the lower jaw, and acts to open the jaw. Disconnect its insertion from the jaw on one side, and reflect the muscle. The sheet of more or less transverse fibers that lies between and deep to the insertions of the digastric muscles of opposite sides of the body is the **mylohyoid**. It arises from the mandible, and inserts on a median raphe and on the hyoid. It acts to raise the floor of the mouth and to pull the hyoid forward. Make a longitudinal incision through the muscle, and reflect it on one side. It is not very thick.

The longitudinal muscles that lie deep to the mylohyoid constitute the prehyoid portion of the hypobranchial musculature. It may be necessary to cut through the mandibular symphysis and spread the two halves of the lower jaw apart to see these muscles clearly. The midventral band of muscle (really a pair of muscles that have united) is the **geniohyoid** (Fig. 6-19). It arises from the front of the mandible and inserts on the hyoid. Cut across the geniohyoid and reflect its ends. The muscle that arises from the hyoid lateral, and deep, to the insertion of the geniohyoid is the **hyoglossus**. It passes forward into the tongue. Pull the tip of the tongue, and you will note that the muscle is moved. The band of muscle that arises from the chin deep to the origin of the geniohyoid is the **genioglossus**. It passes posteriorly into the tongue, lying medial to the hyoglossus. The band of muscle that arises from the mastoid process at the base of the skull and passes forward into the tongue is the **styloglossus**. It lies lateral to the anterior portion of the hyoglossus. The glossus muscles, together with intrinsic muscle fibers within the tongue (**musculi linguae**), form the substance of the tongue and manipulate this organ.

Branchiomeric Muscles

(A) *Mandibular Muscles*

The **mylohyoid** and the anteroventral half of the **digastric**, which were seen during the dissection of the hypobranchial muscles, are branchiomeric muscles of the mandibular arch. To see the other mandibular arch muscles in either the cat or rabbit, skin the top of the head and the cheek region on one side. The auricle should also be cut off. The platysma, facial muscles (both belonging to the hyoid group), and loose connective tissue must be removed, but be careful not to injure glands, nerves, and vessels in this region. Special care should be exercised in skinning and cleaning the cheek, for the duct of one of the salivary glands crosses the cheek just beneath the facial muscles. Find the zygomatic arch. The powerful muscle that lies ventral to the arch is the **masseter**. It arises from the arch, and inserts in the coronoid fossa and adjacent parts of the mandible (Figs. 6-11 and 6-12).

Another mandibular muscle, **temporalis**, lies dorsal to the zygomatic arch. In the cat it is a sizable muscle, and fills the large temporal fenestra. It arises primarily from the surface of the cranium, but some fibers spring from the top of the zygomatic arch. It passes deep to the zygomatic arch and inserts on the coronoid process of the mandible.

In the rabbit, the temporal fossa and muscle are very small. To see the muscle, cut through a powerful facial muscle that passes from the top of the skull posteriorly to the base of the pinna. The **temporalis** extends from a point above the base of the auricle to the back of the orbit (Fig. 6-12). Its insertion is by a tendon that passes down the posterior wall of the orbit to the coronoid process of the mandible.

Other mandibular muscles, are impractical to dissect at this stage. Two pterygoid muscles pass from the base of the skull to the medial side of the ramus of the mandible. A tensor tympani lies in the middle ear, where it attaches to the malleus, and a tensor veli palati extends from the base of the skull into the soft palate. The tensor tympani is described in more detail in connection with the ear (p. 179).

The masseter, temporalis and pterygoids close the jaws. The digastric is the major muscle involved in opening the jaws, but the mylohyoid and hypobranchial muscles can assist under certain circumstances. The mylohyoid also compresses the floor of the mouth and elevates the hyoid. The function of the tensor tympani and tensor veli palati is implied by their names.

Jaw mechanics are quite different in a cat and a rabbit (Fig. 6-20). The jaw joint of a cat, as is characteristic of a carnivore, is in line with the tooth row so that the upper and lower jaws come together in the manner of a pair of scissors. The posterior teeth mesh first. A large, powerful temporal muscle inserts on a well developed coronoid process. The masseter is also powerful, but not so large as the temporalis. In the rabbit, as is characteristic of gnawing and herbivorous mammals, the jaw joint is well above the tooth row so that all the teeth of the upper and lower jaws come together simultaneously. The masseter and the coronoid fossa in which it inserts are very large. The masseter is also very complex with respect to the layers of muscle fibers and their direction, for it not only shuts the lower jaw but pulls it back and forth in a grinding action. The temporalis, on the other hand, is not important in

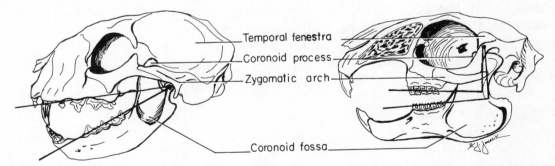

Temporal fenestra
Coronoid process
Zygomatic arch
Coronoid fossa

Figure 6-20. Diagrams to show the mechanics of the jaws of a cat, left, and rabbit, right.

this type of jaw action, and it, together with the temporal fenestra and coronoid process to which it attaches, is very small.

(B) Hyoid Muscles

The major hyoid arch muscles are the **platysma**, **facial muscles**, posterior half of the **digastric**, and the **stylohyoid**. The first three have been seen. The stylohyoid is a small ribbon of muscle that lies lateral (cat) or caudal (rabbit) to the posterior portion of the digastric (Fig. 6-19). In the cat, it arises from the stylohyal bone and inserts on the body of the hyoid. In the rabbit, it arises from the base of the skull and inserts onto the posterior horn of the hyoid. It acts on the hyoid.

Another hyoid arch muscle, which is not feasible to find at this time, is the stapedius. It lies within the middle ear, and acts on the stapes. The muscle is described more fully on page 179.

Considerable variation is seen in the digastric of different mammals. In many species, man among them, the anterior and posterior halves of the muscles form distinct bellies which are interconnected by a prominent, round tendon. This is also the situation during one of the embryonic stages of the rabbit, but, as development proceeds, the posterior belly is reduced to a tendon. In other mammals, the two halves of the digastric are united by a tendinous inscription. Such an inscription is present embryologically in the cat, but it disappears in the adult.

(C) Posterior Branchiomeric Muscles

Much of the posterior branchiomeric musculature is lost during the course of evolution. But some becomes associated with the pectoral girdle as the **trapezius** and **sternocleidomastoid** complexes. These were described in connection with the shoulder region. Some of the remaining posterior branchiomeric musculature forms the **intrinsic muscles of the larynx**, and some is contributed to the wall of the pharynx. Certain of the intrinsic muscles of the larynx can be seen on the ventral surface of the larynx deep to the anterior end of the sternohyoid.

7 / The Sense Organs

ALTHOUGH THE sense organs and nervous system are concerned with the integration of all parts of the body, they may appropriately be considered at this time, for the most conspicuous effector organs are the muscles described in the previous chapter. The sense organs, the central nervous system, and the basic pattern of distribution of the peripheral nerves will be the topic of this and the following chapter. If separate heads cannot be provided for the study of *Squalus*, and separate brains for the mammal, this unit of work should be postponed until the end of the course.

Irritability

Irritability, that is, the ability to receive sensations and respond to stimuli, is a basic property of protoplasm. All the aspects of irritability are combined in the individual cells of the more primitive organisms, but in higher animals there is a division of labor. Certain cells receive, others transmit, and still others respond to the stimuli. The receptive cells are the sense organs, or **receptors**; the transmitting cells are the neurons of the **nervous system**; and the responding cells are the **effectors** (muscles, glands, cilia, etc.).

Classification of Sense Organs

The stimuli causing the sensation of pain appear to be received by free nerve endings, but stimuli producing other sensations in vertebrates are received by very specific sense organs. These range from microscopic sensory corpuscles to such large and complex organs as the eye. This array of sense organs, and the sensory nerves which lead from them, may be divided into two groups — somatic and visceral. **Somatic sensory organs** lie in the "outer tube" of the body — the skin (exteroceptors of the physiologist) and the somatic muscles (proprioceptors of the physiologist). Occasionally proprioceptive organs, which are the organs of muscle sense, are found in branchial muscles. **Visceral sensory organs** (interoceptors) are associated with the "inner tube" of the body, i.e., the viscera. Only those sense organs that can be seen grossly will be considered.

THE LATERAL LINE SYSTEM

Fishes and larval amphibians have a unique sensory system, known as the lateral line system, by which currents and other water movements can be detected. Experiments on a variety of fishes have shown that the lateral line system enables them to localize objects at a distance, even in turbid water where visibility would be reduced, either by the disturbances produced by a moving object, or by reflected water waves set up by their own movements. The system has been called one of "distant touch." The actual receptive organs in the system are groups of mechanoreceptive cells termed **neuromasts**. Hairlike processes from each cell in a neuromast protrude into a cap of gelatinous material, the **cupula**, that is bent by movements of the water. A special group of somatic sensory neurons, called the **lateralis neurons**, lead from the neuromasts and enter the brain through the seventh, ninth, and tenth cranial nerves.

Experiments have shown that the neuromasts have a spontaneous activity, but the level of this activity can be altered by water movement along the flank. In a ray, perfusion of the lateral line canal from the head toward the tail increases the frequency of discharge of certain units and decreases that of others. Perfusion in the opposite direction has opposite effects.

The distribution of the neuromasts varies. Most lie in canals, one of which extends along the side of the trunk; the others ramify over the head. These **lateral line canals** are usually beneath the skin, but open to the surface by pores. In a few fishes, the canals take the form of open grooves. Other neuromasts are in isolated, but often linearly arranged, pits (**pit organs**). Still others lie in little swellings at the bottom of jelly-filled tubes (**ampullae of Lorenzini**). The function of the ampullae of Lorenzini has been in dispute. Experiments made by Sand (1938) indicate that they are thermal receptors. But some investigators doubt this, for all other receptors in the acousticolateral system are fundamentally mechanoreceptors. Murray (1957) has reinvestigated their function in the ray and finds that they are indeed mechanoreceptors sensitive to differences in pressure between the inside and outside of the ampullae. Natural stimuli that might bring about such pressure differences are contacts with objects and local changes in water pressure over the body surface as the fish swims in varying relation to nearby objects. Thus the function of the ampullae overlaps that of the cutaneous receptors on the one hand and lateral line on the other. Murray also finds that they are sensitive to thermal changes, but he believes this to be an incidental effect as is the case in some other sense organs sensitive to thermal changes.

As stated, the lateral line system is limited to fishes and larval amphibians. It is completely lost during the evolution of reptiles, and is never reacquired, even in such aquatic tetrapods as the Cetacea. Parts of the system have been noted during the study of the external features of *Petromyzon* (p. 14), *Squalus* (p. 30), and *Necturus* (p. 34). The system will be considered in detail only in *Squalus*.

Pit Organs

Pit organs are said to occur on various parts of the head of the dogfish, but they are rather difficult to distinguish grossly from other lateral line organs.

Ampullae of Lorenzini

Examine your specimen of a large head of *Squalus*. The pores on the snout through which a jellylike substance extrudes when the area is squeezed are the openings of the ampullae of Lorenzini. Note their distribution. They are very abundant on the ventral surface of the snout, and form two, large, V-shaped groups on the dorsal surface of the snout (Fig. 7-1). Other smaller groups are located posterior to the jaws. Make a deep V-shaped cut through the skin in one of the dorsal snout patches. The apex of the V should be directed posteriorly. Pull the skin flap and underlying tissue forward, and examine its underside. Note that each pore leads into a long, jelly-filled tubule which, in this region, extends anteriorly to a round swelling, the ampulla proper. Small nerve twigs attach onto each ampulla.

Lateral Line Canals

Make a fairly long transverse cut through the skin posterior and medial to the spiracle. Examine the cut surface. The hole that you see in the deeper layers of the skin is a cross section of the **lateral line**

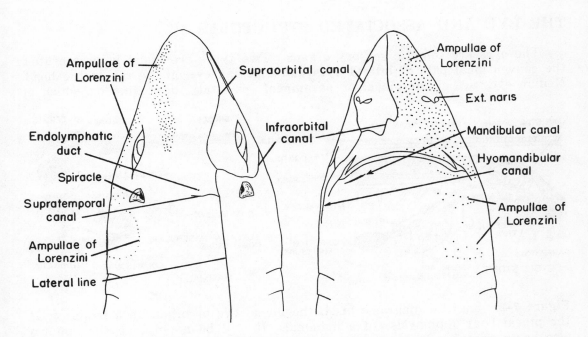

Figure 7-1. Dorsal and ventral views of the lateral line canals, and ampullae of Lorenzini, of the dogfish, *Squalus suckleyi*. (From Daniel, The Elasmobranch Fishes, University of California Press.)

canal proper. This canal extends the length of the trunk and tail. Trace it forward by carefully slicing off the superficial layers of the skin with a sharp scalpel. The object is to make a horizontal section of the canal, taking the top half off and leaving the bottom half on the specimen. Keep your eye on the canal as you expose it to be sure you are cutting at the right level. Pores leading into the canal can be seen on the underside of the skin you remove. Farther forward on the head, the lateral line canal leads into others (Fig. 7-1), which should be uncovered in the same manner.

A **supratemporal canal** crosses the top of the head posterior to the endolymphatic pores. An **infraorbital canal** passes ventrally posterior to the eye, zigzags beneath the eye, and then extends forward to the tip of the snout in a somewhat meandering fashion. It passes medial to the nostril. A **hyomandibular canal** extends posteriorly from the bottom of the zigzag of the infraorbital canal. A **supraorbital canal** passes forward dorsal to the eye and onto the snout. It then turns on itself, and extends posteriorly to connect with the infraorbital canal. The latter portion of the supraorbital canal passes just dorsal to the nostril. A short **mandibular canal** is not connected with the others. It appears as a row of pores overlying the mandible posterior to the labial pocket. Make a cut through the skin at right angles to the row of pores, examine the cut surface, and verify that a canal is present.

THE EYE AND ASSOCIATED STRUCTURES

The eyes are somatic sensory organs. Two types are found in vertebrates, the conventional pair of **lateral eyes** and a **median eye** on the top of the head. Median eyes are either **pineal** or **parapineal** (parietal), depending on which of

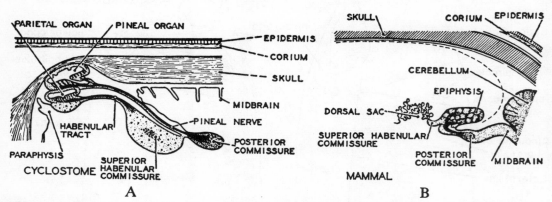

Figure 7-2. Sagittal diagrams of *A*, the pineal eye or organ of a lamprey; *B*, the pineal body (epiphysis) of a mammal. It will be noted that the lamprey also has a parapineal, or parietal, organ, but that it does not reach the surface. A parietal organ connects with the superior habenular commissure, a pineal organ with the posterior commissure. (From Rand, The Chordates, Blakiston Company. After Studnicka.)

two outgrowths from the diencephalic region of the brain reaches the surface (Fig. 7–2, *A*). Such eyes were present in all primitive groups of fishes and in early tetrapods. A few vertebrates, including the lamprey, certain lizards, and *Sphenodon*, retain median eyes, presumably as light-sensitive organs concerned in the adjustment of the organism to the diurnal cycle, but they have been lost, or reduced to vestiges, in most contemporary species (Fig. 7–2, *B*). Recall the pineal eye of the lamprey (p. 15). Vestiges of this type of eye will be seen during the dissection of the brain.

It is with the lateral eyes that you will be concerned at this time. Although the eyeball differs in its method of accommodation and in its adaptive details, its basic anatomy is much the same in all vertebrates. Evolutionary tendencies, however, can be seen in associated structures. For example, the surface of the fish eye is bathed in water which keeps it moist and clean. Tetrapods have evolved movable eyelids of various types and tear glands and ducts that protect, cleanse, and moisten the eye surface.

Fishes

Study the eye of *Squalus* on the same side of the head on which you dissected the lateral line. *Squalus*, like other cartilaginous fishes, has an upper and lower **eyelid** formed of skin folds, but they are immovable. (Most fishes lack eyelids altogether.) Note that the epidermis on the inner surface of the eyelids reflects onto and over the surface of the eyeball. This layer is the **conjunctiva**.

Remove the upper eyelid, and free the eye from the lower lid by cutting through the conjunctiva. The **eyeball** (**bulbus oculi**) lies in a socket, the **orbit**, on the side of the chondrocranium. Cut away the cartilage that forms the roof of the orbit (supraorbital crest, antorbital process, postorbital process), but do not cut into the otic capsule. A mass of gelatinous connective tissue surrounds, and helps to support, the eye. It must be picked away.

A group of ribbon-shaped muscles passes from the medial wall of the orbit to the eyeball. These are the extrinsic muscles of the eye, and they are responsible for the various movements of the whole eyeball. As explained on page 116, these muscles belong to the axial subdivision of the somatic muscles. Although they are derived from three myotomes (Table 2), they fall into an oblique and a rectus group grossly. Push the eyeball posteriorly and note the two muscles that arise from the anteromedial corner of the orbit and insert on the eye. The one that inserts on the dorsal surface is the **superior oblique**; the one that inserts on the ventral surface the **inferior oblique**. The muscles that arise from the posteromedial corner are all recti. Three can be seen in a dorsal view. The one that inserts on the top of the eyeball adjacent to the insertion of the superior oblique is the **superior rectus**; the one that lies medial to the eye is the **medial rectus**; the one that lies posterior to the eye is the **lateral rectus**. Lift up the eye and look at its ventral surface. An **inferior rectus** passes to insert on the ventral surface of the eyeball beside the insertion of the inferior oblique.

Most of the other strands that are seen passing to the muscles and through the orbit are nerves; they will be considered later, but find the **optic nerve** at this time. It is the large nerve passing from the eyeball posterior to the inferior oblique.

Cut across the extrinsic muscles of the eye near their insertions, trying not to injure the nerves going to them. The stalk of cartilage that will be seen passing to the back of the eyeball between the four rectus muscles is the **optic pedicle**. It is shaped like a golf tee, and helps to support the eyeball much as a golf tee does a golf ball. Disconnect it from the eyeball and also free the small nerve that crosses the medial surface of the eye. Cut the optic nerve, and remove the eye. Some **conjuctiva** will cling to the front of the eye.

The outermost layer of the eyeball proper is a tough **fibrous tunic**, the anterior portion of which is modified to form the transparent **cornea**, through which you can see an opening, the **pupil**, surrounded by the pigmented **iris**. Conjunctiva and cornea are fused.

Submerge the eyeball in a dish of water and cut off its dorsal third. Note the large, round, **crystalline lens**. Try not to pull it away from

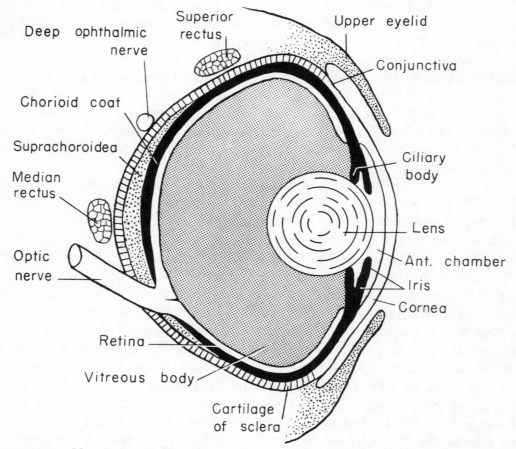

Figure 7-3. Vertical section through the right eyeball of *Squalus* at the level of the optic nerve. Viewed from behind.

surrounding tissues. The three layers, or coats, of which the eyeball is composed can be seen at the back of the eye. As mentioned, the outermost layer is the **fibrous tunic**. Its front portion is modified as the **cornea**, and the rest of it constitutes the **sclera** (Fig. 7-3.). Much of the dogfish's sclera is cartilaginous, and this provides a great deal of support for the eyeball. The pigmented layer internal to the sclera is the **chorioid coat (vascular tunic)**. The **iris** is the modified anterior portion of the chorioid, and the **pupil** is a hole through the iris. The whitish layer internal to the chorioid is the **retina**. It is an incomplete layer that disappears near the lens. The point at which the optic nerve connects with the retina is the **blind spot**.

Carefully pull the lens away from the iris. A ring of black material will probably adhere to it. This black material is in the form of small radiate folds, and it represents the **ciliary body** — a modification of the chorioid layer located near the base of the iris. A gelatinous material (the **zonule**), which will not be seen, extends between the ciliary body and the lens. The lens is supported by the ciliary body, zonule, a middorsal suspensory ligament (which can not be seen grossly), and the **vitreous body**. The vitreous body is the gelatinous material that fills the large cavity (**chamber of the vitreous body**) between the lens and the retina. Other cavities, which are filled with a watery **aqueous humor**, are located in front of the lens. The very small cavity that lies between the lens and the iris is the **posterior champer**; the larger one between the iris and the cornea, the **anterior chamber**.

Cut across the back of the eyeball near where the optic pedicle attached. The relatively thick layer of material between the sclera and chorioid is known as the **suprachoroidea**. It is a vascular connective tissue that is found only in species with an optic pedicel.

Cut into the lens and observe that it is composed of layers of modified epithelial cells arranged concentrically like the skins of an onion.

The sclera is the supporting layer of the eyeball, and the presence of cartilage within it is not surprising, since the sclera develops in part from the optic capsule of the embryonic chondrocranium (p. 46). **Sclerotic bones**, when they are present, ossify in this cartilage. The chorioid layer performs several functions. It is vascular, and helps to nourish the light-sensitive, avascular retina. Its pigment, along with pigment of the retina, prevents internal reflections of light that would otherwise blur the image. In addition the chorioid of many vertebrates, including the dogfish, is so constructed that it can reflect some light back onto the retina. Such a reflecting device, known as a **tapetum lucidum**, is found in vertebrates that live under conditions of low light intensity — certain fishes and nocturnal tetrapods. The elasmobranch tapetum depends on **guanine crystals** within certain chorioid cells. It is among the most remarkable of vertebrate tapeta in that adjacent pigment cells have the ability to extend or retract their pigment across the guanine layer, thus adapting the eye to light or dark conditions (Fig. 7-4). Further adjustments to light and dark are, of course, made by the iris. The dogfish's iris is unusual in that the sphincter

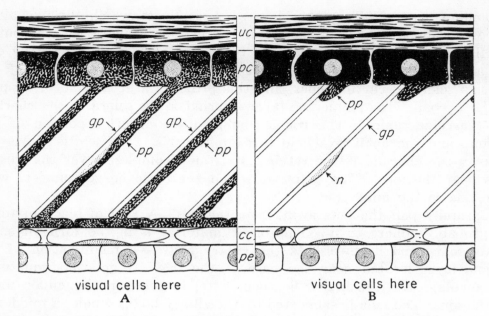

Figure 7–4. Diagrams of section through the chorioid layer of the dogfish *Mustelus mustelus*, to show the tapetum lucidum adapted for light, *A*, and dark, *B*, conditions. In the light-adapted eye, processes of the pigment cells cover the guanine plates; in the dark-adapted eye they are withdrawn and allow the guanine to reflect light back to the visual cells of the retina. Abbreviations: *cc*, chorioid capillaries; *gp*, guanine plates; *n*, nucleus of guanine cell; *pc*, migratory pigment cells of the chorioid; *pe*, pigmented layer of the retina (in this case devoid of pigment); *pp*, pigment processes; *uc*, unmodified pigment cells of the chorioid. (From Walls, The Vertebrate Eye, Cranbrook Institute of Science. After Franz.)

muscle fibers within it are independent effectors which contract upon direct stimulation by light. The radial muscle fibers in the iris, which dilate the pupil, are innervated by neurons in the oculomotor nerve.

Since the refractive index of the cornea and humors of the eye is nearly the same as that of water, they do not take part in the refraction of light rays. Nearly all the refraction in fishes occurs as the rays pass through the thick, spherical lens. In elasmobranchs, the lens is held in such a position that distant objects are in focus on the retina. In bright light, acuity can be increased, regardless of the distance of the object, by the contractions of the iris until the pupil is a pin point. Additional accommodation for near objects is accomplished by the forward movement of the lens through the contraction of a small **protractor lentis** muscle that is located ventrally in the ciliary body. The pressure of the aqueous humor pushes the lens back to its resting position when this muscle relaxes.

Mammals

We may pass directly to the mammal eye, for that of the aquatic *Necturus* contributes little to an understanding of the evolution of this organ. The basic structure of the mammalian eyeball is much the same as in *Squalus*. Details differ, however. Accommodation is accomplished primarily by changing the shape of a biconvex lens. When the eye is at rest, distant objects are in focus. To accommodate for near objects, the curvature of the front of the lens is increased through the action of muscles in the ciliary body. The cat has a tapetum lucidum, but one is absent in the rabbit and man. That of the cat is dependent on a compact layer of endothelial cells in the deeper layers of the chorioid, rather than on refractive granules of guanine.

Only the accessory structures of the eye need be dissected. Examine the eye of your cat or rabbit on the side of the specimen that was used for the study of the muscles. Movable **upper** and **lower eyelids** (**palpebrae**) are present. The slitlike opening between them is called the **palpebral fissure**. The corners of the eye, where the lids unite, are the **canthi**. Cut through the lateral (posterior) canthus and pull the lids apart. A third lid, known as the **nictitating membrane**, can now be seen clearly. It is attached at the medial (anterior) canthus, but its posterior edge can spread over the surface of the eye if the eyeball is retracted slightly. In man, the nictitating membrane is reduced to a vestigial semilunar fold (**plica semilunaris**) that can be seen covering the medial corner of the eye. Examine the edge of each lid 3 or 4 millimeters from the medial canthus with a hand lens. If you are fortunate, you will see on each a minute opening (**lacrimal punctum**) that leads into a lacrimal duct. If you cannot find them in your specimen, look for one on the human eye by pulling down the lower lid and examining its edge near the most medial eyelash.

Cut off the upper and lower lids, leaving a bit of skin around the medial canthus. As you remove the lids, note that the **conjunctiva** on the underside of the lids reflects over the **cornea**. If the cornea is not too opaque, the **pupil** and **iris** can be seen. A facial muscle, the **orbicularis oculi**, encircles the eyelids, and will be cut off with them. It closes the lids.

Free the eyeball and associated glands from the bony rims of the orbit by picking away connective tissue. Do not dissect deeply in the region of the medial canthus, and try not to destroy a loop of connective tissue attached to the anterodorsal wall of the orbit. One of the ocular muscles passes through the latter. Using bone scissors, cut away the zygomatic arch beneath the eye, the postorbital processes, and the crest of bone above the orbit (supraorbital arch). Push the eye anteriorly. The dark glandular mass that lies on the posterodorsal surface of the eyeball is the **lacrimal gland**. It often adheres to the adjacent portion of the orbital roof in the rabbit. Its secretions, the tears, enter near the lateral canthus, bathe the surface of the cornea, and pass into the lacrimal duct through the **lacrimal puncta**. The lacrimal duct enters the nose through a canal in the **lacrimal bone**. Find the nasolacrimal canal on a skull (p. 81). You may be able to find the lacrimal duct by dissecting ventral and anterior to the medial canthus. First find the anterior border of the orbit and the position of the nasolacrimal canal. Do not injure a muscle that is attached to the orbital wall near the lacrimal bone.

Other glands associated with the eye differ sufficiently in the cat and rabbit to warrant separate description. The nictitating membrane of the cat contains a small **harderian gland**, which may be found by removing and looking on the medial surface of the base of the nictitating membrane. It supplements the secretions of the lacrimal gland. Push the eyeball dorsally, and carefully remove fat and connective tissue from the floor of the orbit. A small **infraorbital salivary gland** lies on the floor of the orbit in line with, and just posterior to, the upper tooth row.

Push the eye of the rabbit dorsally and carefully dissect away connective tissue from beneath it. You may have to cut away more of the zygomatic arch and some of the muscles of mastication. The gland that forms a band along the ventrolateral border of the eyeball is an accessory portion of the lacrimal gland, for its secretions enter near the lateral canthus. It is called the **inferior lobe of the lacrimal gland**. An **infraorbital salivary gland** lies ventral to the anterior part of the inferior lobe of the lacrimal gland. These two are sometimes hard to separate. A large **harderian gland** lies beneath the eyeball medial to the inferior lobe of the lacrimal gland. It is divided into a posteroventral red lobe and an anterodorsal white lobe, but this color differentiation is not always apparent in an embalmed specimen. The secretions of the harderian gland enter on the inner surface of the nictitating membrane, and supplement those of the lacrimal complex.

Remove the glands and connective tissue from around the eye of your cat or rabbit, and expose the extrinsic muscles of the eyeball. The pattern of the muscles is much the same as in fishes, except for two additional muscles. A **levator palpebrae superioris** arises from the medial wall of the orbit dorsal to the optic foramen, and inserts on the upper eyelid which it raises. Its lateral end may be found unattached, for its insertion was probably cut in removing the eyelids. Two oblique muscles pass to the anterior wall of the eyeball. The **superior oblique** arises from the wall of the orbit slightly anterior to the optic foramen and goes through a connective tissue pulley (the **trochlea**), which is attached to the anterodorsal wall of the orbit, before inserting on the eye. The **inferior oblique** arises from the maxillary (cat) or lacrimal (rabbit) bone. Four **recti** (**superior**, **inferior**, **medial**, and **lateral**) arise from the margins of the optic foramen and pass to the posterior portion of the eyeball. A **retractor bulbi**, which can be divided into four parts, passes to the eye deep to the recti. The derivation of levator palpebrae superioris and retractor bulbi is shown in Table 2.

THE NOSE

In fishes, the nose typically consists of a pair of sacs, each of which connects to the surface by a pair of openings (**external nares**), through which water carrying odoriferous particles circulates. In sarcopterygian fishes and tetrapods, each **olfactory sac**, or passage, has but one external naris, but each opens into the mouth through an internal naris (or **choana**). Thus the nose serves as an air passage as well as retaining its original olfactory function. In the evolution through tetrapods, the olfactory and respiratory roles of the nasal passages become segregated to some extent, the olfactory epithelium becoming restricted for the most part to the dorsal parts of the passages. But in many tetrapods a part of the olfactory epithelium remains in the ventral part of the passage where it forms the **vomeronasal organ of Jacobson**. This organ is apparently concerned with smelling food within the mouth, and it sometimes has a separate connection with the mouth (p. 81). In mammals the respiratory passages are prolonged through the evolution of a bony secondary palate (p. 75) and a fleshy soft palate.

In order for a substance to stimulate the olfactory epithelium it must be in solution. This is no problem for the fish, but it is for a terrestrial animal. Tetrapods have met the problem by the evolution of glands whose secretions keep the epithelium moist. The secretions of these glands, and the mucosa of the nasal passages as a whole, also condition the air that passes to the lungs by moistening, cleansing, and (in birds and mammals) warming it. In birds and mammals the mucosal surface is increased through the evolution of **conchae** or **turbinal bones** (p. 79).

The receptive olfactory cells of the nasal organ are of a unique type in that the cells not only receive the stimuli but also have long processes that conduct the impulses back to the brain. This type of cell, which is believed to be very primitive, is called a **neurosensory cell**. Although related to the visceral sensation of taste, olfaction is considered to be somatic sensory.

Fishes

Study the nose of *Squalus*. Note that each **external naris** is divided into two openings (Fig. 7-5, *A*) by a superficial flap of skin and a deeper ridge. The lateral opening, which also faces anteriorly to a slight extent, is the incurrent aperture through which a current of water enters the olfactory sac; the lateral one, the excurrent aperture. A thin, flaplike valve along its posterior margin prevents water from entering the excurrent aperture. Remove the skin and other tissue from around the **olfactory sac** on the side of the head used for the dissection of the eye. The nasal capsule and antorbital process of the chondrocranium also must be cut away. It will then be seen that the olfactory organ is a round sac having no connection with the mouth. An **olfactory tract** of the brain extends to the posteromedial surface of the sac and there expands into an **olfactory bulb**. Cut open the sac and note how its internal surface is increased by many septa-like folds (**olfactory lamellae**) which in turn bear minute secondary folds (Fig. 7-5, *B*).

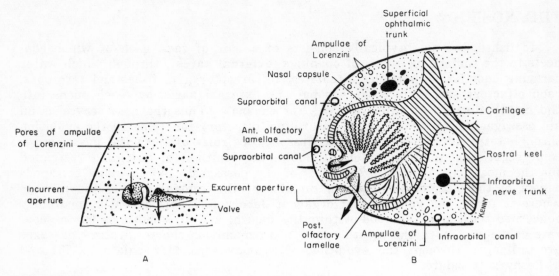

Figure 7–5. *A*, ventral view of the right external naris of *Squalus*; anterior is toward the top of the page. *B*, diagram of the left olfactory sac as seen in a cross section taken just posterior to the external naris. Viewed from behind. Arrows indicate the direction of the water current.

Primitive Tetrapods

Note the pair of widely separated **external nares** on your specimen of *Necturus*. Remove the skin between the external naris and the eye on the side of the head used for the dissection of the muscles. The long, tubular **olfactory passage** will be seen after picking away surrounding connective tissue. Open the mouth, cutting through the angle of the jaw on the side being studied, and find the slitlike **internal naris**. It lies lateral to the most posterior, and shortest, row of teeth (pterygoid teeth). Cut open the olfactory passage and find the **olfactory lamellae** within it. What is their function? Jacobson's organ is not developed in *Necturus*, but is present in metamorphosed amphibians.

Mammals

The nose of mammals is to be studied on sagittal sections of the head cut in such a way that one half shows the nasal septum, and the other the inside of the nasal passage. The nose should be studied from demonstrated preparations, unless this unit of work has been postponed to the end of the course, or unless heads from specimens of a previous year's class have been saved for the purpose. This work should also be correlated with the description of the sagittal section of the skull on page 79. The following account is based on the cat, but is applicable to many other mammals.

The **external nares**, which are close together in mammals, lead into paired **nasal cavities**. The nasal cavities occupy the area of the skull anterior to the cribriform plate of the mesethmoid bone and dorsal to the bony, secondary palate. On the larger section, it can be seen that they are completely separated from each other by a **nasal septum**. The ventral portion of the septum is formed by the **vomer bone**, the postero-dorsal portion by the **perpendicular plate of the mesethmoid**, and the rest by **cartilage**.

On the smaller section, it can be seen that each nasal cavity is filled to a large extent by three folded **conchae** or **turbinal bones**. (Fig. 9–14, p. 249). These are, of course, covered with the nasal mucosa. The **inferior concha** (**maxilloturbinal**) is represented by a simple fold that extends from the dorsal edge of the external naris posteriorly and ventrally to about the middle of the secondary palate. The nasolacrimal duct from the eye enters lateral to the inferior concha. It is best seen on a skull in which this concha has been removed. The **superior concha** (**nasoturbinal**) is represented by a single longitudinal fold in the dorsal part of the nasal cavity lying deep to a median, perpendicular plate of the nasal bone. The area between and posterior to these two conchae is filled by the highly folded **middle concha** (**ethmoturbinal**). The spaces among the folds of the middle concha are referred to as the **ethmoid cells**.

Air passages lead from the external naris between the conchae. The most prominent of these is the **inferior meatus**, which lies between the inferior and middle conchae on the one hand and the secondary palate on the other. It opens by the **internal naris** (**choana**) into the nasopharynx. The choanae are located at the posterior border of the secondary palate. The nasopharynx is separated from the mouth cavity by the fleshy soft palate. A **superior meatus** lies ventral to the superior concha and leads to the **frontal sinus** (an air space in the frontal bone), and the posterior ethmoid cells. The **middle meatus**, which is a distinct passage in some mammals, is represented in the cat by the labyrinth of ethmoid cells. Most of these cells ultimately lead into the inferior meatus. Some also connect with a **sphenoid sinus** in the anteroventral part of the sphenoid bone. Most of the olfactory epithelium is limited to the more posterior ethmoid cells. **Olfactory nerves** may be seen, with a hand lens, passing through the holes of the cribriform plate to the olfactory bulb of the brain.

Paired **vomeronasal organs of Jacobson** are present in the cat. The entrance to one can be seen on the roof of the mouth just posterior to the first incisor tooth. From here a **nasopalatine duct** leads through an incisive foramen (p. 81) to the organ. Jacobson's organ can be found by carefully dissecting away the anteroventral portion of the nasal septum. It appears as a cul-de-sac with a cartilaginous wall lying on the secondary palate and against the nasal septum. It extends between 1/4 and 1/2 inch caudad to the incisor teeth.

THE EAR

The ear of vertebrates is closely related to the lateral line system. Indeed, the inner, receptive part of the ear is often regarded as a deeply set portion of this system. Among the features supporting this homology are the common mode of embryonic formation, similarity of the receptive "hair cells" of the ear to neuromasts, and the close relationship of the statoacoustic nerve to nerves from the lateral line.

The inner ear consists of a series of thin-walled canals and sacs filled with a fluid known as the **endolymph**. These canals and sacs are collectively called the **membranous labyrinth**, and they are imbedded within a series of parallel canals and chambers within the otic capsule known as the **osseous labyrinth**. The membranous labyrinth and osseous labyrinth are separated from each other by spaces filled with fluid and crisscrossed by minute strands of connective tissue. This fluid is the **perilymph**.

Only an internal ear is present in fishes (Fig. 7-6). As in higher vertebrates, it is an organ of equilibrium, and in many teleosts the lower part of it (sacculus and lagena) has also been shown to be sensitive to sound waves that reach it by passing from the water through the adjacent tissues, or by way of special sound concentrating mechanisms such as the swim bladder and weberian ossicles. Evidence that elasmobranchs can also hear is twofold. Firstly, they show behavior responses to low frequency sound vibrations in the water even if the lateral line and major cutaneous nerves are cut, but they lose these responses when the statoacoustic nerve is cut. Secondly, nerve impulses have been recorded from the nerve twigs coming from the sacculus and utriculus of a ray while it was being subjected to sound stimulations up to 120 cycles per second. It has not been possible to record similar impulses from the lagena twig. In terrestrial vertebrates, a part of the inner ear is definitely specialized for the reception of sound waves.

Whereas sound waves pass easily from water into the tissues of a fish, which are mostly water, they do not pass easily from air into water. Tetrapods sensitive to air-borne sound vibrations have evolved a special mechanism that receives such vibrations, and increases their pressure amplitude sufficiently to overcome the inertia in the liquids of the inner ear. In most living amphibians and reptiles, sound waves impinge upon the ear drum, or **tympanum**, located on the body surface or at the base of a canal, the **external acoustic meatus**. They are transmitted across a **tympanic cavity** (homologous to the spiracular pouch) by the **stapes** (homologous to the hyomandibular). The foot plate of the stapes fits into a **fenestra vestibuli**, or **oval window**, on the side of the otic capsule, and a specialized part of the perilymph transmits vibrations from there to the receptive part of the membranous labyrinth. The difference in size between the large tympanum and small fenestra ovalis increases the pressure amplitude. Vibrations are finally released from the inner ear through a **fenestra cochlea**, or **round window**.

The orthodox view of ear evolution (Fig. 7-6) assumes that labyrinthodonts and cotylosaurs had an ear of this type. This has been retained in most of their descendants and has been further elaborated in mammals. The sound detecting portion of the mammalian inner ear is a long, spiral cochlea; two additional ossicles, **malleus** and **incus**, appear in the middle ear as a corollary of a new jaw joint and the inward movement of the articular and quadrate (p. 73); and an **auricle**, or **pinna**, develops about the entrance of the external auditory meatus. But some investigators question the presence of a tympanum in early mammal-like reptiles and their cotylosaur and labyrinthodont ancestors. They

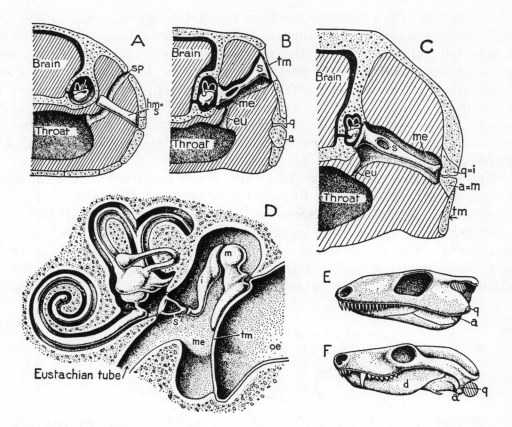

Figure 7-6. Diagrams to show the orthodox hypothesis of the evolution of the middle ear and auditory ossicles. The cross sections are through the otic region of *A*, a fish; *B*, a primitive amphibian; *C*, a primitive reptile; and *D*, a mammal. The lateral views are of *E*, a primitive amphibian; and *F*, a mammal-like reptile. They show in particular the shift of the eardrum (hatched) from the otic notch to a point behind the jaw articulation. Abbreviations: *a*, articular; *d*, dentary; *eu*, eustachian tube; *hm*, hyomandibular; *i*, incus; *m*, malleus; *me*, middle ear, or tympanic, cavity; *oe*, outer ear cavity (external acoustic meatus); *q*, quadrate; *s*, stapes; *sp*, spiracle; *tm*, tympanum, or eardrum. (From Romer, Man and the Vertebrates, University of Chicago Press.)

believe that these early tetrapods could detect vibrations only by bone conduction, especially through the lower jaw. According to this view, an ear with a tympanum has evolved independently in those amphibians that have it (frogs), in most living reptile groups, and in mammals. Animals such as *Necturus* have not lost an air-sensitive ear, but evolved from ancestors that never had one.

Fishes

Dissect the ear of *Squalus* on the side of the head used for the study of the other sense organs. First uncover the otic capsule by removing the skin and muscles from its dorsal, lateral, and posterior surfaces. The spiracle and adjacent parts of the mandibular and hyoid arch may also be cut away, but leave some skin in the vicinity of the paired endolymphatic pores (p. 32). Note the duct leading from one of the pores to the endolymphatic fossa on the top of the chondrocranium. It ultimately connects with the sacculus of the membranous labyrinth. This duct, usually called the **endolymphatic duct**, is not strictly homologous to the endolymphatic duct of higher vertebrates, for it represents a retention of the pathway along which the ear invaginated from the surface during embryonic development. The endolymphatic duct of higher vertebrates is a secondary evagination from the inner ear toward the surface, and not the original invagination pathway.

To expose the **membranous labyrinth** of the inner ear, you must carefully shave away the surrounding cartilage of the otic capsule beginning on the dorsal and lateral surfaces and gradually working ventrally. You can generally see the various canals and chambers through the cartilage shortly before you reach them. Use special care as you dissect the cartilage from around the parts of the membranous labyrinth, and try not to break them. You will first expose the three **semicircular canals — ante-**

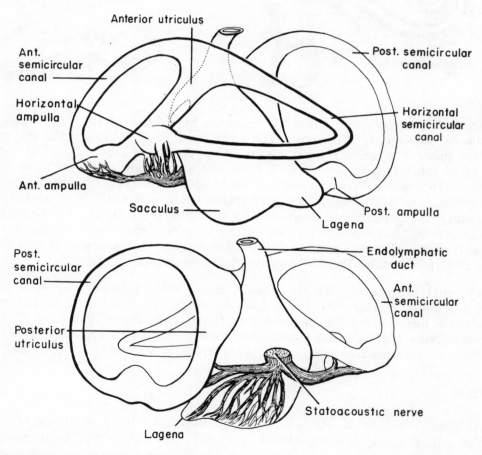

Figure 7–7. Left ear (membranous labyrinth) of the selachian, *Heptanchus maculatus.* Lateral view above, medial view below. (Redrawn from Daniel, The Elasmobranch Fishes, University of California Press.)

rior **vertical**, **posterior vertical**, and **horizontal**. As you continue to dissect, you will come upon the **sacculus** lying in a large cavity medial to the horizontal canal. The sacculus is generally partly collapsed, so that the dorsolateral wall of the cavity in which it lies can be dissected away without much danger. Continue to remove cartilage from around the sacculus and semicircular canals, tracing the latter to their points of attachment on chambers called utriculi. Each end of each canal attaches to one of two utriculi. The anterior and horizontal canals attach onto the **anterior utriculus**; the posterior canal onto the **posterior utriculus**. The utriculi connect with the sacculus by inconspicuous openings. The ventral end of each canal bears a round swelling, the **ampulla**, containing a sensory patch known as a **crista**. Branches of the **statoacoustic (vestibulocochlear) nerve** can be seen leaving each ampulla, but the cristae may not be apparent. Dissect away as much of the cartilage as possible from the ventral side of the sacculus and observe the short extension or bulge from its posteroventral portion. This is the **lagena**. The sacculus and lagena contain large sensory patches called **maculae**. These sensory patches are overlaid by a mass of calcareous concretions and sand grains that form an **otolith**. Remove as much cartilage as possible from the medial side of the membranous labyrinth, cut the nerves coming from the sensory patches, and gently lift out the ear and float it in a dish of water. It is well to leave a bit of cartilage on the medial and ventral surfaces. Review the parts of the ear mentioned and compare your dissection with Figure 7-7.

The basic structure of the membranous labyrinth can be seen better in the cartilaginous fishes than in any other vertebrates, for the labyrinth is relatively large and can be freed more easily from cartilage than from bone. The elasmobranch ear is atypical, however, in having two utriculi. Usually all semicircular canals connect into one. Also the connection of the endolymphatic duct to the surface is lost in the adults of other vertebrates.

Primitive Tetrapods

The structure of the ear of a primitive tetrapod should be studied on demonstration dissections of a bullfrog, since the urodele ear lacks a **tympanum** and middle ear cavity, and is not sensitive to air-borne vibrations. The large external tympanum is easily seen. Remove this on one side, and the **tympanic cavity** or **middle ear cavity** will be exposed. Open the mouth and note that the middle ear cavity communicates with the back of the mouth cavity by the **auditory** or **eustachian tube**. A long, rodlike **stapes** crosses the middle ear cavity.

Expose the **otic capsule**, which contains the inner ear, by removing the overlying skin and muscle. Also remove the musculature lying posterior to the middle ear. Cut through the back of the middle ear cavity, and trace the stapes to the otic capsule. Its inner end is associated with the knoblike, specialized **operculum** (homologous to the urodele structure of the same name, page 59). Remove the operculum and the inner end of the stapes, and you will see the **oval window** into which they fit. Vibrations enter the inner ear at this point, and are released through a fenestra, which, in living amphibians, opens into the cranial cavity dorsal to the acoustic nerve. It may be found by opening the cranial cavity and removing the brain. This fenestra is analogous to the **round window**, which opens into the middle ear cavity of certain reptiles and mammals and is usually given this name.

Mammals

The external ear can be seen easily on your specimen of a cat or rabbit. It consists of the external ear flap, the **auricle** or **pinna**, a canal with a cartilaginous wall leading inward to the skull (**external acoustic meatus**), and the **tympanum**. The last can be found on the side of the head on which the pinna was removed by cutting away as much of the external acoustic meatus as possible and shining a light into the remainder. Note that the tympanum is set at an angle, its anterior portion extending more medially than its posterior portion. The opaque line seen through the membrane is the handle of the malleus.

The rest of the ear is difficult to dissect, and should be observed on demonstration preparations. These can be prepared from the sagittal sections of the head used for the study of the nose. The following account is based on the cat, but is applicable to many other mammals. Remove the muscles and other tissue from around the tympanic bulla except at its anteromedial corner. The middle ear cavity lies within the bulla and opens into the nasopharynx by the **auditory**, or **eustachian**, **tube**. The opening of the tube appears as a slit in the lateral wall of the nasopharynx (Fig. 9–14). Careful dissection between this slit and the anteromedial corner of the bulla will reveal the tube, part of whose wall is cartilaginous and part bony.

Using bone scissors, break away the posteromedial portion of the bulla, and also the mastoid and paraoccipital processes and adjacent parts of the nuchal crest. This exposes the posteromedial chamber of the **tympanic**, or **middle ear**, **cavity**. Note that it is largely separated from a smaller anterolateral chamber by a more or less vertical plate of bone. A hole through the dorsolateral portion of this plate passes between the two chambers of the middle ear cavity. The **fenestra cochlea**, or **round window**, can be seen through this hole. Carefully break away all of this plate of bone and open up the anterolateral chamber. The handle of the **malleus** can be seen on the inside of the tympanum, and it will be noted that the fenestra cochlea is situated on a round promontory of bone. A finger-like process of cartilage extends from the posterolateral wall of the tympanic cavity between these two structures. The tiny nerve that runs along it and leaves its tip is the **chorda tympani**, a branch of the facial nerve going to the taste buds of the tongue and certain salivary glands. Break away bone from the anteromedial corner of the bulla and find the entrance of the auditory tube. The other auditory ossicles (**incus** and **stapes**), and the **fenestra vestibuli** or **oval window** in which the stapes fits, lie dorsal to the fenestra cochlea. To see them, one must remove a piece of bone posterior and dorsal to the external acoustic meatus without injuring the plate of bone supporting the tympanic membrane. You will also notice two small muscles passing to certain of the ossicles. A **stapedius** arises from the medial wall of the tympanic cavity caudad to the fenestra cochlea and inserts on the stapes. A **tensor tympani** arises from the medial wall anterior to the fenestra vestibuli and inserts on the malleus. The groups to which these muscles belong are indicated in Table 2. They adjust the auditory ossicles and tympanum to the intensity of the sound waves. Their contraction, for example, reduces the amplitude of the vibration of the ossicles and protects the delicate structures of the inner ear from violent movements resulting from loud noises. In this respect, they are analogous to the muscles in the iris of the eye, which adjust the eye to light intensity.

The inner ear lies within the periotic portion of the temporal bone. Portions of it may be noted by removing the brain and chipping away pieces of the periotic, but it cannot be dissected satisfactorily by this method. The **internal acoustic meatus** for the vestibulocochlear and facial nerves lies in the cranial cavity on the posteromedial surface of the periotic.

CHAPTER
8 / The Nervous System

AS EXPLAINED in the introduction to the chapter on sense organs, the nervous system is concerned with the integration, or coordination, of the various parts of the body, so that all will function in harmony with each other and with the external environment. But it should be pointed out that the nervous system is not the only system involved in body coordination. The endocrine glands also play an important role. Nervous coordination is implemented by nerve impulses that travel along the nerve cells or neurons. It is characterized by being rapid and specific. Endocrine coordination, on the other hand, is by way of the secretions of hormones that are carried by the circulatory system. It tends to be slower and less specific. Since the endocrine glands are widely scattered, they will not be discussed in one place but as they are observed. The nervous system will be considered at this time, as it is the last of the group of organ systems that are concerned directly with the general function of body support and movement.

Divisions of the Nervous System

Grossly, the nervous system can be divided into a central and a peripheral portion (Fig. 8–1). The **central nervous system** consists of the **brain** and **spinal cord**. Both cord and brain are hollow, for they contain a **central canal**, or **neurocoele**, which expands to form **ventricles** in certain regions of the brain. You will recall that a single, dorsal, tubular nerve cord is one of the diagnostic chordate characteristics. The distribution of the neurons within the central nervous system is such that we can speak of **gray** and **white matter**. The gray matter contains the cell bodies of neurons and unmyelinated fibers; the white, myelinated fibers. Most of the gray matter is centrally located. In the cord, it forms a continuous column that has the appearance of an H, or a butterfly, in cross section. In the brainstem, the gray matter tends to break up into distinct aggregations referred to as **nuclei**, but the nuclei have a relationship to each other similar to the relationship between the parts of the gray matter in the cord. In higher vertebrates, some gray matter migrates to the surface parts of the brain, where it forms a gray **cortex**.

The **peripheral nervous** system consists of the nerves containing the sensory and motor neurons that supply the receptors and effectors of the body. It is often subdivided into the cranial and spinal nerves on the one hand, and the autonomic nervous system on the other. **Spinal nerves** are segmentally arranged, and each connects to the central nervous system by a **dorsal** and **ventral root** (Fig. 8–1). The dorsal root bears a **ganglion** containing the cell bodies of sensory neurons. More distally, the spinal nerve breaks up into branches, or rami, going to various parts of the body — a **dorsal ramus** to the epaxial region, a **ventral ramus** to the hypaxial region, and often one or more **communicating rami** with visceral connections. Although most of the cranial nerves may have had this pattern at one stage in their evolution, their segmentation and division into dorsal and ventral roots is not as apparent in living vertebrates as that of the spinal nerves.

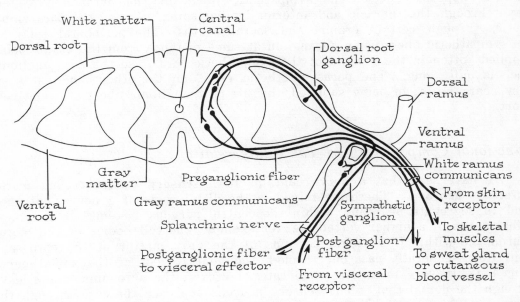

Figure 8–1. Diagram of a cross section through the spinal cord and a pair of spinal nerves to show the organization and functional components of the nervous system. Two internuncial neurons are shown in the gray matter between the sensory and motor neurons. (From Villee, Walker, and Smith, General Zoology.)

The **autonomic** portion of the peripheral nervous system is usually considered to consist of the motor neurons going to the visceral organs, glands, and smooth muscle generally. These leave the cord or brain in certain of the spinal and cranial nerves. Some of the fibers of the autonomic system remain in the spinal and cranial nerves, but some leave to travel in special branches to the organs in question. Thus the autonomic system, while clearly definable functionally, is not completely separated from the rest of the peripheral nervous system morphologically. Sometimes the autonomic nerves to the viscera, and the sensory fibers from the viscera, are referred to as the **visceral nervous system**. This is opposed to a **somatic nervous system** supplying somatic muscles and sense organs. But this, too, is a functional subdivision that is not completely valid morphologically.

A unique feature of the autonomic system is that there is a peripheral relay (Fig. 8-1). A **preganglionic neuron**, having its cell body in the gray matter of the cord, passes out through the ventral root of a spinal nerve, and goes to a peripheral ganglion. There it synapses with a **postganglionic fiber** that continues to the visceral organ. In contrast, only one neuron is involved in the innervation of the branchiomeric and somatic musculature.

Another unique feature of the autonomic nervous system is its subdivision, in the higher vertebrates at least, into **sympathetic** and **parasympathetic** portions. A given visceral organ receives both types of fibers. One activates the organ; the other inhibits it. Sympathetic innervation increases the activity of the heart, slows down digestive processes, and has other effects that help the body adjust to conditions of stress. Parasympathetic innervation has the opposite effect. The sympathetic fibers of the autonomic system leave through the thoracic and anterior lumbar spinal nerves; the parasympathetic, through certain cranial and sacral nerves. The peripheral relay of the sympathetic fibers is in a ganglion at some distance from the organ being supplied (often in the ganglia of the sympathetic cord), so the postganglionic fiber is quite long. The parasympathetic relay, on the other hand, is in, or very near, the organ being supplied, thus its postganglionic fibers are relatively short.

Functional Components and Their Interrelations

We have seen that a nerve contains both **sensory** (**afferent**) and **motor** (**efferent**) **neurons**. These are mixed in the distal parts of a nerve, but they tend to segregate near and within the central nervous system. In a typical spinal nerve of a higher vertebrate, for example, the cell bodies of the motor neurons lie within the ventral portions of the gray matter of the cord, and their fibers generally pass out through the ventral root of the spinal nerve (Fig. 8-1). In very primitive vertebrates, however, the autonomic fibers leave through the dorsal root. The sensory neurons approach the cord through the dorsal root of the spinal nerve, and their cell bodies are located in the dorsal root ganglion. The sensory neurons ultimately enter the dorsal portions of the gray matter, but they may travel for some distance in the white matter of the cord before doing so.

Besides distinguishing between sensory and motor neurons, it is possible to distinguish between sensory neurons coming from somatic receptors and those coming from visceral receptors. The same is true for motor neurons. Thus there are four major types of neurons within the peripheral nerves — **somatic sensory**, **visceral sensory**, **visceral motor**, and **somatic motor**. These four types of neurons also correlate with areas of the gray matter in that they begin, or end, in definite regions, or columns, as shown in Figure 8-1. Somatic sensory neurons end in the most dorsal part of the gray matter, which constitutes a **somatic sensory column**; visceral sensory neurons, just ventral to them in a **visceral sensory column**. The cell bodies of the visceral motor neurons constitute a **visceral motor column** located in the lateral part of the gray matter; those of the somatic motor neurons constitute a **somatic motor column** located in the most ventral part of the gray matter. Certain of these four categories may be further subdivided into special and general types. For example, a distinction can be made between the visceral motor fibers to the viscera (autonomic system), and those to the branchiomeric musculature. The former are general visceral motor fibers, and the latter special visceral motor fibers.

As explained, somatic and visceral sensory neurons ultimately enter their respective columns of the gray matter. These neurons may continue through the gray matter to synapse with a motor neuron on their own side, or on the opposite side of the central nervous system. This is the mechanism for a simple two-neuron reflex. Or the sensory neurons may synapse with internuncial neurons whose cell bodies lie in the sensory columns of the gray matter. These **internuncial neurons** may pass through the gray matter to a motor neuron (also a reflex mechanism), or they may ascend in the white matter to the brain. Here they may form reflex connections, or be relayed to still other parts of the brain. Internuncial neurons that ascend to the brain are called **afferent internuncial neurons**, and at some point during their ascent they cross (decussate) to the side of the central nervous system opposite to the one on which they originated. Impulses that originate in the brain descend through the white matter to the motor neurons on

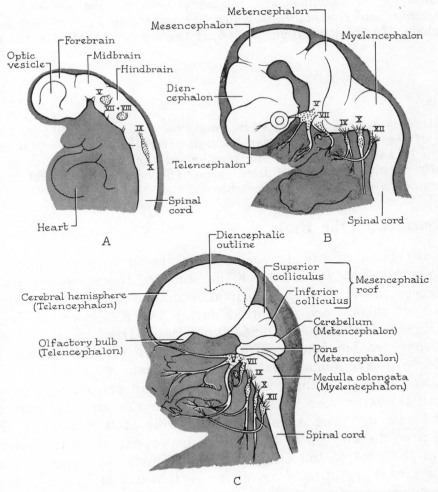

Figure 8-2. Diagrams of three stages in the development of the human brain to show the differentiation of the principal brain regions. (From Villee, Walker, and Smith, General Zoology. Modified after Patten.)

efferent internuncial neurons. These, too, decussate at some point during their descent.

Development of the Brain

As might be expected, the brain is the most complex part of the nervous system. An understanding of the various regions of which it is formed is best gained from a consideration of its development. As shown in Figure 8-2, the brain arises as an enlargement of the anterior end of the neural tube. Very soon it becomes divided by certain constrictions into three vesicles, or regions: an anterior forebrain or **prosencephalon**; a middle midbrain or **mesencephalon**; and a posterior hindbrain or **rhombencephalon**.

The mesencephalon does not divide further, but the other two regions do. The prosencephalon divides into an anterior **telencephalon** and a posterior **diencephalon**; the rhombencephalon into an anterior **metencephalon** and a posterior **myelencephalon**. Each of these five regions further differentiates. The telencephalon gives rise to the cerebral hemispheres and olfactory bulbs; the diencephalon to the thalamus and related structures; the mesencephalon to the optic lobes (superior colliculi) which are located in its roof (tectum); the metencephalon to the cerebellum and pons; the myelencephalon to the medulla oblongata.

FISHES

The nervous system of the dogfish should be studied carefully, for the basic structure of the nervous system can be seen exceptionally well. Not only can the system be easily exposed by the removal of cartilage, but it is in a morphologically primitive and generalized stage. The nervous system of *Squalus* is a good prototype for that of all vertebrates.

Dorsal Surface of the Brain

The dissection of the nervous system should be done on the large head of *Squalus* that was used for the study of the sense organs. The cranial nerves should be studied primarily on the intact side of the head, for certain of them were destroyed during the dissection of the sense organs. Remove the skin and underlying tissue from the dorsal surface of the chondrocranium and from around the eye. Be careful not to cut a large dorsal nerve (**superficial ophthalmic trunk**) that lies anterior to the orbit and lateral to the rostrum. Cut away the cartilaginous roof of the cranial cavity. As you do so, look into the anterior part of the cavity and you may see a delicate, threadlike stalk extending from a depressed area on the top of the brain (diencephalon) to the epiphyseal foramen in the roof of the cranial cavity. This is the **epiphysis**, a vestige of the pineal eye of more primitive vertebrates (Fig. 8-3). Also cut away the supraorbital crest, and as much of the lateral walls of the cranial cavity as is possible without injuring nerves. Much of the ear on the intact side will have to be cut away. Be particularly careful not

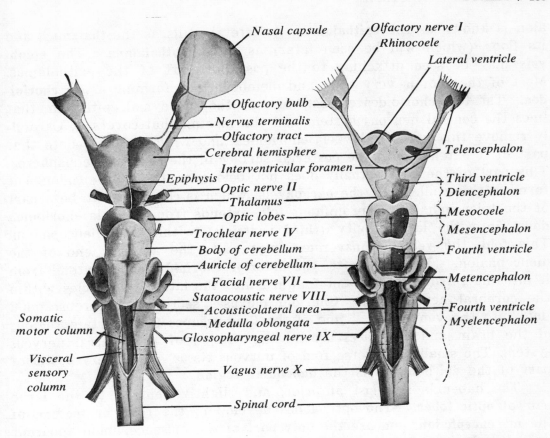

Figure 8–3. Dorsal views of the dogfish brain. The ventricles have been opened in the right figure. (From Ranson and Clark, Anatomy of the Nervous System.)

to break the small **trochlear nerve** that leaves the brain dorsally and passes to the superior oblique muscle.

The brain should now be well exposed. Its surface is covered with a delicate, vascular connective tissue, the **meninx primitiva**. Strands of connective tissue pass from the meninx to the connective tissue lining the cranial cavity (the **endochondrium**). In life, cerebrospinal fluid lies in the apparently empty perimeningeal space between the brain and the wall of the cranial cavity.

The paired **olfactory bulbs** form the anteromost part of the brain. They are the lateral enlargements in contact with the posteromedial walls of the olfactory sacs. An **olfactory tract** extends posteriorly from each olfactory bulb to the **cerebral hemispheres**, the large swellings at the anterior end of the median portion of the brain. The portions of the hemispheres that the tracts enter are the olfactory lobes, but they are difficult to distinguish grossly from other parts of the cerebrum. All of these parts of the brain belong to the telencephalon.

The **diencephalon** is the depressed area often with a dark roof situated posterior to the cerebral hemispheres. The roof of the diencephalon is known as the **epithalamus**, its lateral walls as the **thalamus**, and its floor (which will be seen later) as the **hypothalamus**. The **epiphysis** may be seen attaching to the posterior part of the epithalamus. Most of the roof is very thin, and membranelike, forming a **tela chorioidea**. The tela chorioidea consists of only the ependymal epithelium that lines the central nervous system and the meninx that covers it. Carefully remove this part of the tela chorioidea, and while doing so note that part of it extends onto the posterior surface of the cerebral hemispheres. This part of the roof actually does not belong to the diencephalon, but forms a thin-walled sac, the **paraphysis**, which is considered to be a part of the telencephalon. A prominent fold extends from the tela chorioidea down into the large cavity (**third ventricle**) within the diencephalon. This fold, the **velum transversum**, represents the anterior end of the diencephalon. Vascular tufts, the **anterior chorioid plexus**, extend from it into the third ventricle and forward into the lateral ventricles within the cerebral hemispheres. Chorioid plexuses secrete the cerebrospinal fluid into the ventricles. Some of the fluid escapes from certain regions of the brain and circulates around the outside of the central nervous system. The small transverse fold of nervous tissue at the very posterior part of the epithalamus is the **habenular region**.

The habenula lies just anterior, and slightly ventral, to the large pair of **optic lobes**. The optic lobes develop in the roof, or **tectum**, of the **mesencephalon**, and are the only part of the mesencephalon apparent in a dorsal view.

The **metencephalon** lies posterior to the mesencephalon, and consists dorsally of the **cerebellum**. The **body** of the cerebellum is the large, median, oval mass whose anterior end overhangs the optic lobes. Note that it is partially subdivided into four parts by a longitudinal and transverse groove. The pair of earlike flaps that lie on either side of the posterior part of the body of the cerebellum are the **auricles** of the cerebellum (Fig. 8-3). The ventral part of the metencephalon contributes to the medulla oblongata in fishes.

The **myelencephalon** lies posterior to the metencephalon and forms the greater part of the **medulla oblongata** — the elongated region of the brain that is continuous posteriorly with the spinal cord. Most of the roof of the medulla is a thin tela chorioidea from which the **posterior chorioid plexus** dips into the fourth ventricle. Carefully remove it and also its forward extension from the auricles of the cerebellum. The large cavity now exposed that lies within the medulla is the **fourth ventricle**. Lift up the posterior end of the cerebellum and it can be seen that the fourth ventricle extends into the auricles. Also observe at this time that the auricles are continuous with each other beneath the body of the cerebellum. Owing to its shape, the ventral portion of the fourth ventricle is often called the **fossa rhomboidea**.

The columns of gray matter that continue from the cord into the brain can be seen in the ventral and lateral walls of the fourth ventricle. The pair of midventral, longitudinal folds that lie on the floor of the ventricle are the **somatic motor columns**. They contain the cell bodies of somatic motor neurons. There is a deep, longitudinal groove lateral to each somatic motor column. The lateral wall of either groove constitutes the **visceral motor column**, a column containing the cell bodies of visceral motor neurons. A longitudinal row of small bumps lies dorsal to the visceral motor column. This is the **visceral sensory column** (Fig. 8-3), and it receives and relays impulses from the visceral sensory neurons. The longitudinal groove between the visceral motor and visceral sensory columns is the **sulcus limitans**, a landmark separating the ventral motor from the dorsal sensory portion of the cord and brainstem. (The brainstem is the brain minus the cerebrum and cerebellum.) The dorsolateral rim of the fourth ventricle constitutes the **somatic sensory column**, and it receives and relays impulses from the somatic sensory neurons. The anterior part of the somatic sensory column is enlarged. This portion, known as the **acousticolateral area** (Fig. 8-3), receives the neurons from the ear and lateral line organs. It is continuous anteriorly with the auricles of the cerebellum. Although it cannot be seen in a gross dissection, these sensory and motor columns continue forward through the metencephalon and into the mesencephalon. Their anterior ends, however, become discontinuous, forming discrete patches, or nuclei, of gray matter.

Cranial and Occipital Nerves

The cranial and occipital nerves must be considered before the ventral and internal parts of the brain can be examined. More of the lateral wall of the cranium will have to be removed as the nerves are studied.

(A) Nervus Terminalis

Fishes are usually described as having 10 pairs of cranial nerves, and these are both named and numbered. However, an additional anterior nerve has been left out of this numbering system. The **nervus terminalis** (Fig. 8-4) is a tiny nerve that lies along the medial surface of the olfactory tract, and extends between the olfactory sac and olfactory lobe. It is best seen at the medial angle formed by the junction of the olfactory tract with the olfactory lobe, for it separates slightly from the olfactory tract in this region. Although the terminalis is found in all vertebrates, except cyclostomes and birds, its function is uncertain. The consensus is that it carries both somatic sensory fibers (cutaneous, not olfactory fibers) from the nasal area, and visceral motor fibers of the autonomic system. The latter are probably vasomotor. Some investigators have reported the presence of ganglia along the nerve.

(B) Olfactory Nerve

The **olfactory nerve** (I) carries olfactory impulses (somatic sensory) from the olfactory sac to the olfactory bulb. Since the sac and bulb are adjacent to each other in the dogfish, the olfactory nerve is not compact, but consists of a number of minute groups of neurons passing between these structures. The olfactory neurons are of the neurosensory type (p.171).

(C) Optic Nerve

The **optic nerve** (II) brings in optic impulses (somatic sensory) from the eye. Find it in the orbit and trace it medially. It is a thick nerve. Push the brain away from the cranial wall and note that the optic nerve attaches to the ventral surface of the diencephalon. Since the retina of the eye develops embryologically from an outgrowth of the brain, the optic nerve is in one sense a brain tract rather than a true peripheral nerve.

(D) Oculomotor Nerve

The **oculomotor nerve** (III) carries somatic motor impulses to most of the extrinsic ocular muscles, receives proprioceptive impulses from these muscles (somatic sensory), and carries autonomic fibers (visceral motor) to the eye. To see it, mobilize the eye on the intact side of the head in the manner described in connection with the dissection of the eye (p. 165), remove the gelatinous connective tissue lying in the orbit, and look on the ventral surface of the eyeball. The branch of the oculomotor going to the **inferior oblique** muscle will be apparent. Follow it posteriorly and medially. It passes ventral to the inferior rectus, and at the posterior margin of this muscle crosses a small, often whitish, blood vessel. This vessel, an artery, follows the posterior margin of the inferior rectus and enters the eyeball. The autonomic fibers of the oculomotor form a small **ciliary nerve** that travels along this vessel, but these fibers can seldom be seen grossly. The branch of the oculomotor to the **inferior rectus** lies between the inferior rectus and the small vessel. Turn your specimen over, and pick up the oculomotor nerve from the dorsal side. It extends dorsal to the origin of the superior rectus and enters the cranial cavity. Just before it enters, it gives off one branch to the **superior rectus** and another to the **medial rectus**. (Do not confuse the oculomotor with another nerve of the same size, the profundus, which crosses the base of the oculomotor and extends along the medial surface of the eyeball.) Push the brain away from the cranial

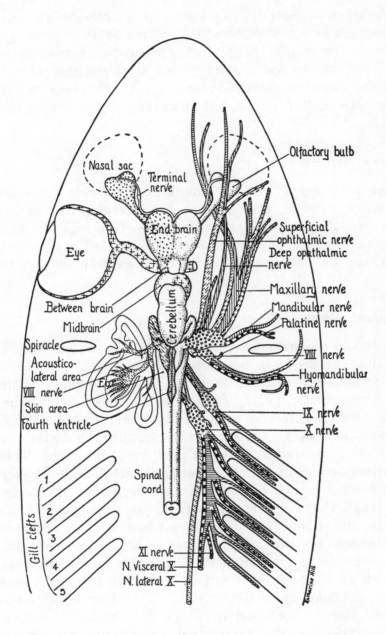

Figure 8–4. Diagram of the brain and most of the cranial nerves of the dog-fish showing functional components: olfactory, stipple; visual, crosses; acousti-colateral, broken oblique lines; visceral sensory, horizontal lines; general cutaneous, vertical lines; visceral motor, black and white rectangles; ganglia, circles. The branch of the vagus labeled "XI" goes to the cucullaris. (From Ranson and Clark, Anatomy of the Nervous System. After Herrick and Crosby.)

wall and note the attachment of the oculomotor on the ventral surface of the mesencephalon.

(E) Trochlear Nerve

The **trochlear nerve** (VI) has been noted crossing an optic lobe. Lift up the cerebellum and note where it attaches on the brain. The trochlear passes through the cranial wall, goes ventral to, or perforates, the large superficial ophthalmic trunk (sometimes it perforates this trunk), and extends to the **superior oblique muscle**. Like the oculomotor, it is primarily a somatic motor nerve, but it carries a few proprioceptive fibers (somatic sensory).

(F) Abducens Nerve

Skip the fifth nerve for a moment and consider the **abducens** (VI), which carries somatic motor fibers to the **lateral rectus** and returns proprioceptive (somatic sensory) impulses from this muscle. It can be seen on the ventral surface of the posterior rectus. Its attachment on the ventral surface of the medulla will be seen later when the brain is removed.

(G) Trigeminal Nerve

Return to the **trigeminal nerve** (V), which is the nerve of the mandibular arch and the general cutaneous sensory nerve of the head (Fig. 8–4). The trigeminal attaches more or less in common with the facial (VII) and statoacoustic (VIII) nerves on the dorsolateral surface of the medulla just posterior to the auricles of the cerebellum. It is difficult to separate these nerves at their attachments on the brain, but they can be sorted out to some extent by their peripheral branches. The trigeminal has four branches in fishes. A **superficial ophthalmic branch**, together with a comparable branch of the facial nerve, forms the large **superficial ophthalmic trunk** that has been noted passing through the dorsal region of the orbit and along the lateral surface of the rostrum. A **deep ophthalmic branch** (or **profundus nerve**) enters the orbit dorsal to the oculomotor, adheres to the connective tissue on the medial surface (back) of the eyeball, and leaves the front of the orbit through a small foramen to join the superficial ophthalmic trunk. The deep ophthalmic should be traced on the side of the head in which the eyeball has been left. Both ophthalmic branches of the trigeminal return somatic sensory impulses from general cutaneous sense organs (not lateral line organs) in the skin on the top and side of the head. In addition, the deep ophthalmic has several minute and inconspicuous branches to the eyeball. These are considered to be homologous to the long ciliary nerve of mammals, and for the most part return sensory fibers from parts of the eye other than the retina. In some selachians there is evidence that they also carry a few autonomic fibers to the eye (Norris and Hughes, 1920).

A **mandibular branch** of the trigeminal can be found by dissecting away the connective tissue on the posterior wall of the orbit. It is a fairly thick nerve that lies posterior to the lateral rectus. The mandibular carries visceral motor fibers to the branchial muscles of the first visceral arch, and it returns some somatic sensory fibers from general cutaneous sense organs in the skin overlying the lower jaw.

The **maxillary branch** of the trigeminal, together with the buccal branch of the facial, forms the large infraorbital trunk that extends anteriorly across the floor of the orbit. The **infraorbital trunk** is fully as wide as any of the ocular muscles, and is easily confused with a muscle. The infraorbital divides near the anterior border of the orbit, and is distributed to the skin overlying the upper jaw and the underside of the rostrum. The maxillary portion of this trunk returns somatic sensory fibers from general cutaneous sense organs in this region.

Cut away enough cartilage and connective tissue from the posteromedial corner of the orbit to be able to see where all of the branches of the trigeminal come together, and again note the attachment of the trigeminal to the medulla. The main part of the trigeminal bears a slight enlargement, the **semilunar** or **gasserian ganglion**, that contains the cell bodies of the sensory neurons. However, it is unlikely that you can distinguish this ganglion.

(H) Facial Nerve

The **facial nerve** (VII) is the nerve of the hyoid arch, spiracle, and the anterior lateral line organs. As mentioned above, its **superficial ophthalmic** and **buccal branches** contribute to the superficial ophthalmic and infraorbital trunks, respectively. They return somatic sensory impulses from the lateral line organs. The attachment of certain fibers from these trunks to the ampullae of Lorenzini can be noted. Cut away the skin adjacent to the posteroventral corner of the spiracle on the intact side of the head, and pick away the underlying connective tissue. The large nerve that will be seen is the **hyomandibular branch** of the facial (Fig. 6-3, p. 116). Follow it peripherally, noting that it is distributed to the hyoid muscles (visceral motor fibers), skin (somatic sensory fibers from lateral line organs), and lining of the mouth (visceral sensory fibers of both general and taste nature). Follow it medially to its union with the other branches, and its attachment on the brain, which, as stated, is more or less in common with that of the trigeminal and statoacoustic nerves. You will have to cut away some of the spiracle, surrounding muscles, and otic capsule as you go. About 1/4 inch from the brain, the hyomandibular bears a slight enlargement, the **geniculate ganglion**, that contains the cell bodies of sensory neurons. Another, and smaller, branch of the facial leaves from the anteroventral surface of this ganglion. This is the **palatine**, and it returns visceral sensory neurons from the mouth lining.

(*I*) *Statoacoustic Nerve*

A part of the **statoacoustic** (**vestibulocochlear**) **nerve** (VIII) may have been noted coming from the ampullae of the anterior vertical and horizontal semicircular canals during the dissection of the hyomandibular nerve. Continue to cut away the otic capsule and note another, and longer, part of this nerve coming from the ampulla of the posterior vertical semicircular canal, the sacculus, and parts of the utriculus. The statoacoustic contains only somatic sensory fibers from various parts of the inner ear.

(*J*) *Glossopharyngeal Nerve*

The **glossopharyngeal nerve** (IX) is the nerve of the third visceral arch and the first of the five definitive gill pouches. It can be seen crossing the floor of the otic capsule posterior to the sacculus. It passes ventral to the posterior branch of the statoacoustic, and at first may be confused with this nerve. Cut away this part of the statoacoustic and find the attachment of the glossopharyngeal on the side of the medulla. Trace the glossopharyngeal laterally. This will be facilitated if you open the first gill pouch by cutting through the skin and muscle dorsal and ventral to the first external gill slit. Cut all the way to, but not through, the internal gill slit (the opening between the gill pouch and pharynx). As the nerve leaves the otic capsule, it bears an oval-shaped swelling, the **petrosal ganglion**, that contains the cell bodies of sensory neurons. Several branches leave from the petrosal ganglion. A large **posttrematic** passes down the posterior face of the first gill pouch to carry visceral motor fibers to the branchial muscles, and return visceral sensory fibers from this region. A smaller **pretrematic** branch, which is entirely visceral sensory, passes down the anterior face of the first pouch. A still smaller **pharyngeal** branch follows the pretrematic a short distance, then curves around a tendon near the dorsal edge of the internal gill slit and is distributed to the wall of the pharynx. It, too, is entirely visceral sensory. There is finally a small **dorsal** branch distributed to lateral line organs, and often to the skin in the supratemporal region. Such a branch is present in *Squalus* but is impractical to find.

(*K*) *Vagus Nerve*

The vagus (X) is the nerve of the remaining visceral arches. Find its attachment on the dorsolateral surface of the posterior end of the medulla and follow it caudally out of the otic capsule. To see the rest of the nerve, you must cut open the remaining gill pouches in the manner described for the first. If you cut as far as you should, you will cut into a large blood space, the **anterior cardinal sinus**, lying dorsal to the internal gill slits. This sinus must also be opened by longitudinal incisions. A large branch of the vagus (**visceral branch**) lies beneath the connective tissue on the dorsomedial wall of the anterior cardinal sinus. It gives off four **branchial branches** that cross the floor of the sinus and are distributed to the remaining four visceral arches and pouches. Each branchial branch follows the pattern of the glossopharyngeal, having a sensory ganglion from which **posttrematic**, **pretrematic**, and **pharyngeal branches** arise. After giving off the branchial branches, the visceral branch continues as the **intestino-accessory branch**. It gives off a small branch, which probably will not be seen, to the cucullaris (Fig. 8-4), and then follows along the wall of the esophagus to the viscera. It also sends a branch to the pericardial cavity. The intestino-accessory branch contains visceral motor fibers of the autonomic system, visceral sensory fibers, and visceral motor fibers to the cucullaris.

Just before the visceral branch enters the anterior end of the anterior cardinal sinus, the vagus gives off a **lateral** (or **dorsal**) **branch**. This branch lies medial to the visceral branch and extends posteriorly between the epaxial and hypaxial musculature. It receives somatic sensory fibers from the lateral line canal proper, and in some elasmobranchs a few general cutaneous fibers from the skin in the gill region. Cell bodies of these neurons lie in a ganglion near the proximal end of this branch.

(L) Occipital Nerves

Two or three **occipital nerves** lie posterior to the vagus. Although the occipital nerves are closely related to cranial nerves, and are destined to contribute to the formation of a tetrapod cranial nerve (the hypoglossal), convention excludes them from the cranial series because the posterior limit of the cranium is not the same in all fishes. Sometimes these nerves emerge from the skull and sometimes just posterior to the skull. The occipital nerves are serially homologous with ventral roots of spinal nerves. The dorsal roots of these nerves are believed to have become incorporated in the vagus (p. 199).

In *Squalus,* the occipital nerves emerge from the back of the chondrocranium. Find the junction between cranium and vertebral column by cutting away the muscle and cartilage overlying this region. There will, of course, be an abrupt change in the width of the axial skeleton at this point. The nerve that emerges between the chondrocranium and first vertebra is the **first spinal nerve**. The small nerves that arise from the lateroventral surface of the neural tube between this point and the large root of the vagus are the occipital nerves. There are generally two such nerves, rarely three. At first they may appear to be accessory, posterior roots of the vagus, for they pass through the cartilage and appear to join the vagus. Actually they run with the vagus for only a short distance, and then leave to form the **hypobranchial nerve**. The hypobranchial nerve also receives contributions from the first two or three spinal nerves, and it is the nerve that carries somatic motor fibers to the hypobranchial muscles. It also contains some somatic sensory fibers of a proprioceptive and general cutaneous nature. The general cutaneous fibers enter the spinal cord through the spinal nerves.

The separation of the occipitals from the vagus can most easily be seen by finding the hypobranchial nerve and tracing it medially. Completely free the visceral branch of the vagus from the wall of the anterior cardinal sinus. The hypobranchial nerve emerges from the musculature, crosses the visceral branch of the vagus near the level of its last branchial branch, and curves ventrally near the point where the visceral branch passes onto the esophagus.

A summary of the distribution and functional components of the cranial and related occipital nerves is presented in Table 3. This appears to be a confusing array, but it is felt that, at one stage in evolution, most of these nerves were related to head segments, as spinal nerves still are to trunk segments. Subsequent evolution has led to such a great modification of the head that this relationship is now obscure. However, a clue to the relationship can be obtained by an analysis of the components in the cranial and occipital nerves and a comparison between them and primitive spinal nerves.

Primitive spinal nerves differ from those of higher vertebrates in failing to have a union of the dorsal and ventral roots, and in having visceral motor fibers traveling through the dorsal roots instead of the ventral. In view of this, the terminalis, trigeminal, facial, glossopharyngeal, and vagus are considered by some investigators to be serially homologous to primitive dorsal roots of spinal nerves. The profundus branch of the trigeminal also should be added to this series, for it is a separate nerve in agnathous vertebrates and in embryos of the dogfish. The oculomotor, trochlear, abducens, and occipitals are considered to be ventral roots that have secondarily acquired proprioceptive fibers in addition to their somatic motor fibers. The olfactory, optic, and statoacoustic are believed to be unique nerves that have developed in association with the special sense organs rather than with head segments.

Relating these dorsal and ventral nerves to each other, and to head segments, is still another problem. One clue to this has been found in the embryo of the dogfish (*Scyllium*), for in such a form there is a complete series of myotomes in the early stages of development (Fig. 6–1, p.110). Eight segments are found in the cranial region. The first three form the extrinsic eye muscles, one or more of those adjacent to the otic capsule degenerate, and the more posterior contribute to the epibranchial and hypobranchial muscles. Some of the segments posterior to the cranium also contribute to these muscles in certain vertebrates. The nerves that are believed to be related to these segments can best be expressed in tabular form (Table 4). It will be noted in this table that the terminalis has no segment. But the presence of such a nerve may indicate that a segment was present anterior to the ocular myotomes at a very

Table 3. Distribution and Components of the Cranial and Occipital Nerves of Selachians

The cranial and occipital nerves of selachians, together with their general distribution and functional components, are shown in this table. An X indicates the presence of the various components; and (X) that the component is found in some, but not all selachians.

Nerve	Branches	Distribution	Somatic Sensory — General (Cutaneous)	Somatic Sensory — Special including Lateral Line (L)	Somatic Sensory — Proprioceptive	Visceral Sensory — General	Visceral Sensory — Special (Taste)	Visceral Motor — General (Autonomic)	Visceral Motor — Special (Branchial)	Somatic Motor
Terminalis		Olfactory sac	X					X		
I. Olfactory		Olfactory epithelium		X						
II. Optic		Retina		X						
III. Oculomotor	4 Muscular Branches	Inferior oblique; inferior, superior, and anterior rectus			X					X
	Ciliary	Ciliary body of eye						X		
IV. Trochlear		Superior oblique			X					X
V. Trigeminal	Superficial Ophthalmic	Skin over top and side of head	X							
	Deep Ophthalmic (Profundus)	Skin over top and side of rostrum; sensory fibers from the eye and in some cases a few motor fibers to the eye	X					(X)		
	Mandibular (Post-trematic)	Mandibular muscles and skin over lower jaw	X						X	

Table 3. Distribution and Components of the Cranial and Occipital Nerves of Selachians (Continued)

The cranial and occipital nerves of selachians, together with their general distribution and functional components, are shown in this table. An X indicates the presence of the various components; and (X) that the component is found in some, but not all selachians.

Nerve	Branches	Distribution	Somatic Sensory			Visceral Sensory		Visceral Motor		Somatic Motor
			General (Cutaneous)	Special including Lateral Line (L)	Proprioceptive	General	Special (Taste)	General (Autonomic)	Special (Branchial)	
	Maxillary	Skin over upper jaw and underside of rostrum	X							
VI. Abducens		Posterior rectus			X					X
VII. Facial	Superficial Ophthalmic	Lateral line organs over top and side of head	(X)	X (L)						
	Buccal	Lateral line organs over upper jaw and underside of rostrum		X (L)						
	Hyomandibular (Post-trematic)	Hyoid muscles, mouth lining, and lateral line organs near lower jaw		X (L)		X	X		X	
	Palatine (Pharyngeal)	Mouth lining				X	X			
VIII. Stato-acoustic (Vestibulo-cochlear)		Sensory patches of inner ear		X						
IX. Glossopha-ryngeal	Dorsal	Supratemporal lateral line organs; adjacent skin	(X)	X (L)						

Table 3. Distribution and Components of the Cranial and Occipital Nerves of Selachians (Continued)

The cranial and occipital nerves of selachians, together with their general distribution and functional components, are shown in this table. An X indicates the presence of the various components; and (X) that the component is found in some, but not all selachians.

Nerve	Branches	Distribution	Somatic Sensory			Visceral Sensory		Visceral Motor		Somatic Motor
			General (Cutaneous)	Special including Lateral Line (L)	Proprioceptive	General	Special (Taste)	General (Autonomic)	Special (Branchial)	
	Pretrematic	Anterior wall of first typical gill pouch				X	X			
	Post-trematic	Posterior wall of first typical gill pouch				X	X		X	
	Pharyngeal	Pharyngeal lining				X	X			
X. Vagus	Lateral (Dorsal)	Most of lateral line canal; skin in dorsolateral gill region	(X)	X (L)						
	Visceral	4 branchial branches to remaining pouches and then continues as the intestino-accessory branch to the cucullaris, heart, and abdominal viscera. Each branchial branch has pretrematic, post-trematic, and pharyngeal branches. At least one post-trematic supplies ventral pit organs		X (L)		X	X	X	X	

Table 3. Distribution and Components of the Cranial and Occipital Nerves of Selachians (Continued)

The cranial and occipital nerves of selachians, together with their general distribution and functional components, are shown in this table. An X indicates the presence of the various components; and (X) that the component is found in some, but not all selachians.

| Nerve | Branches | Distribution | Somatic Sensory | | | Visceral Sensory | | | Visceral Motor | | Somatic Motor |
			General (Cutaneous)	Special including Lateral Line (L)	Proprioceptive	General	Special (Taste)	General	General (Autonomic)	Special (Branchial)	
Occipital Nerves (2–3 in Squalus)		Anterior epibranchial and hypobranchial musculature. (The first 2–3 spinal nerves also supply the hypobranchial musculature in Squalus)			X						X

early stage of vertebrate evolution. The other nerves can be related to myotomes, but it should be emphasized that such a scheme is hypothetical. It does, however, serve as a good working hypothesis, and helps to bring order to a seemingly chaotic pattern of nerves.

It is also apparent from this analysis that most of the dorsal nerves have shifted their orientation from the myotomes to the visceral arches, and thus have become essentially branchial nerves. The most generalized branchial nerve is believed to be the glossopharyngeal (Fig. 8–5). As was observed, the main branch of this nerve is its posttrematic, which extended down the posterior face of the first definitive gill slit, supplying visceral motor fibers to the musculature of the third visceral arch and visceral sensory fibers to this region. The nerve also has visceral sensory pretrematic and pharyngeal branches, and a somatic sensory dorsal branch. The vagus, facial, and to some extent the trigeminal represent modifications of this basic pattern, as can be seen by comparing their branches with those of the glossopharyngeal in Table 3.

Table 4. Segmentation of the Cranial Nerves

Myotome	Ventral Root	Dorsal Root	Visceral Arch
		Terminalis	
1	Oculomotor	Profundus	Trabeculae of chondrocranium
2	Trochlear	Trigeminal	Mandibular
3	Abducens	Facial	Hyoid
4		Glossopharyngeal	3rd
5	Occipitals, of which a variable number persist in the adult	Vagus	4th
6			5th
7			6th
8			7th

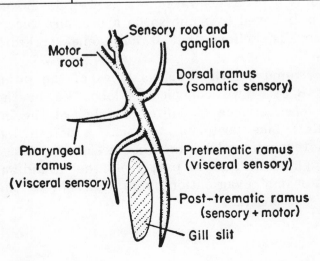

Sensory root and ganglion

Motor root

Dorsal ramus (somatic sensory)

Pharyngeal ramus (visceral sensory)

Pretrematic ramus (visceral sensory)

Post-trematic ramus (sensory + motor)

Gill slit

Figure 8–5. Diagram of a lateral view of a typical branchial nerve (dorsal root cranial nerve). This diagram is based on the glossopharyngeal nerve, but the vagus, facial, and, to some extent, the trigeminal represent modifications of this basic pattern. (From Romer, The Vertebrate Body.)

Ventral Surface of the Brain

It is now possible to return to the brain, remove it, and study its ventral surface. Before lifting the brain out, cut across the posterior end of the medulla, the olfactory tracts, and the roots of the cranial nerves. Try to leave stumps of the nerves attached to the brain. After you have cut across the trigeminal, facial, and statoacoustic nerves, push the brain to one side and note the small abducens nerve leaving. It, too, must be cut. Lift up the posterior end of the brain and carefully pull it anteriorly. You will soon see a part of the brain extending into a recess (the **sella turcica**) in the floor of the cranial cavity. It will probably be necessary to cut away some of the floor posterior to this recess to get the brain out intact.

After the brain has been removed, examine its ventral surface. Identify the attachments of the cranial nerves and the major regions of the brain. Several new structures can be seen in the region of the hypothalamus (floor of the diencephalon). The optic nerves attach at the anterior end of the hypothalamus, and their fibers decussate at this point, forming an X-shaped structure known as the **optic chiasma**. In all vertebrates except mammals with stereoscopic vision, the decussation is complete, for all the fibers of the left optic nerve pass to the right side of the brain, and vice versa. After passing through the optic chiasma, the optic fibers form a band, the **optic tract**, that leads posteriorly and dorsally to the optic lobes. The meninx primitiva will have to be removed to distinguish the tract. Most of the rest of the hypothalamus consists of a large, posteriorly projecting **infundibulum**. Just posterior to the optic chiasma, the infundibulum bears a pair of prominent lateral lobes known as the **inferior lobes** of the infundibulum (Fig. 8-6). The **hypophysis**, or **pituitary gland**, attaches between the inferior lobes and extends posteriorly. The infundibulum finally forms a thin-walled, dark **saccus vasculosus** that lies posterior to the inferior lobes and dorsal to the pituitary. The saccus vasculosus is well developed in deep sea fishes and hence is believed by some to be a pressure receptor; others, on cytological grounds, believe that it has some secretory function.

If the brain was carefully removed, the various lobes of the pituitary can be recognized (Fig. 8-6). Two conspicuous lobes lie in the midventral line — an **anterior lobe** between the inferior lobes of the infundibulum, and an **intermediate lobe** posterior to this. A pair of **superior lobes** lie between the intermediate lobe and the saccus vasculosus. The pituitary is an endocrine gland derived in part from the infundibulum, and in part from an embryonic evagination from the roof of the mouth known as the hypophyseal pouch.

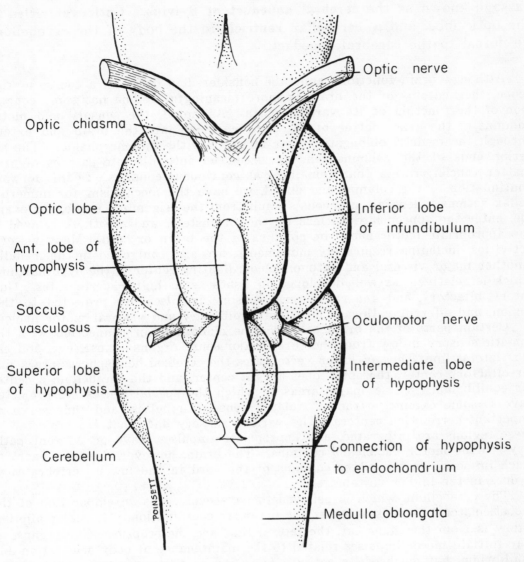

Optic nerve

Optic chiasma

Optic lobe

Inferior lobe
of infundibulum

Ant. lobe of
hypophysis

Saccus
vasculosus

Oculomotor nerve

Superior lobe
of hypophysis

Intermediate lobe
of hypophysis

Cerebellum

POINSETT

Connection of hypophysis
to endochondrium

Medulla oblongata

Figure 8–6. Ventral view of the hypophysis and adjacent parts of the brain of *Squalus.*

Ventricles of the Brain

The central cavity that characterizes the neural tube of chordates expands to form large chambers, or ventricles, in the brain (Fig. 8–3). The third and fourth ventricles have been observed. To see the others, cut off the roofs of the cerebral hemispheres, optic lobes, and body of the cerebellum in the horizontal plane. **Lateral ventricles**, or the **first and second ventricles**, lie in the paired cerebral hemispheres. Each connects with the **third ventricle** of the diencephalon by a narrow passage known as the **interventricular foramen of Monro**. The **third ventricle** connects with the fourth ventricle of the hindbrain by a narrow, ventral passage known as the **cerebral aqueduct of Sylvius**. **Optic ventricles** in the optic lobe, and a **cerebellar ventricle** in the body of the cerebellum, lie dorsal to the cerebral aqueduct.

Although brain function cannot be considered in detail in a course of this scope, the anatomy of the brain is more meaningful if one has some conception of the function of its various parts. As noted, the sensory and motor columns of the gray matter of the cord continue into the brain and extend through the medulla oblongata as far forward as the mesencephalon. The anterior ends of the columns tend to break up into discrete patches of gray matter (nuclei) rather than remaining as continuous columns. In this forward continuation of the columns and nuclei, we have the mechanism for numerous reflex actions. General cutaneous stimuli from the trigeminal nerve, for example, enter the appropriate nucleus of the brainstem, and are there relayed to the appropriate motor nuclei or columns of the brain or cord. Many visceral activities, including respiratory movements, are also controlled in the medulla. Another major visceral and autonomic coordinating center is the hypothalamus, which is relatively as well developed in fishes as in higher vertebrates. Gustatory, olfactory, and general visceral sensory impulses are projected to this region, and efferent pathways lead to the pituitary and visceral motor nuclei.

Certain parts of the brain have become elaborated in connection with the somatic sensory inflow from the three major sense organs — nose, eye, and ear plus lateral line. Thus in lower vertebrates the cerebral hemispheres are a major olfactory center, the optic lobes a sight center, and the cerebellum a center for equilibrium. These major areas to which the special senses are projected have to some extent "attracted" related sensory impulses, and thus serve as important correlation centers. The various sensory data that are projected to these regions are integrated, and appropriate impulses leave on efferent pathways to the motor nuclei and columns. The brain, however, does not exert as much influence on the motor columns of the cord in the lower vertebrates as it does in the higher vertebrates.

The cerebellum, which is believed to have evolved as an elaboration of the acousticolateral area, is a center for muscular coordination. Its major afferent inflow is from the inner ear, the lateral line, and the proprioceptive organs. It may initiate motor impulses related to the maintenance of body orientation and equilibrium, but much of its activity is to insure that motor directives beginning in other areas are carried out with reference to the existing orientation of the body.

The optic lobes receive projections not only from the eyes but from many other sense organs and sensory centers in the brain: olfactory, gustatory, acousticolateral, and general cutaneous. Motor impulses initiated here pass to the floor of the mesencephalon and back to the motor nuclei and columns in the medulla and spinal cord. In short, the optic lobes of fishes are the master integration center, and are analogous to the cerebral cortex of higher vertebrates.

The cerebral hemispheres of lower vertebrates, such as fishes, are not highly developed, and function primarily as olfactory centers. There may be some projection of other sensory fibers to the hemispheres through the thalamus, but this is not great in lower forms, and the cerebral hemispheres do not have the dominant integrating role that they have in the higher. Most motor activity initiating in this region would be in response to olfactory stimuli. Efferent pathways lead back to the habenula, hypothalamus, and optic lobes. The thalamus, which relays impulses to and from the cerebral hemispheres, is not large in fishes.

The Spinal Cord and Spinal Nerves

The spinal cord lies in the vertebral canal of the vertebral column. To see it and the spinal nerves, remove the muscles overlying an inch or two of the vertebral column, and carefully shave away the dorsal part of the neural arch. You will soon see the **spinal cord** and the **dorsal roots** of the spinal nerves through the cartilage. A **dorsal root ganglion** is present on each, but is rather small. Each dorsal root passes through a foramen in the dorsal intercalary plate. A dorsal root lies slightly posterior to its corresponding ventral root, so to find the **ventral root** it is necessary to cut away the lateral wall of the neural arch anterior to the dorsal root. The ventral root arises from the cord by a fan-shaped group of rootlets, and passes out of the neural canal through a foramen in the neural plate. The two roots unite in the dogfish in the musculature lateral to the vertebral column, but this union, and the subsequent splitting of the spinal nerve into dorsal, ventral, and communicating rami, is difficult to find.

The components in the roots of the dogfish spinal nerves appear to be similar to those in the nerves of higher vertebrates — sensory neurons in the dorsal roots, motor in the ventral. Autonomic fibers present in the dorsal root of many fishes and amphibians have not been positively identified in selachians, but some may be present. Certainly most of the preganglionic visceral motor fibers of the autonomic system extend through the ventral roots to synapse with postganglionic fibers in a series of ganglia lying dorsal to the posterior cardinal sinus and kidneys. The autonomic system of the dogfish thus has cranial and spinal contributions, but the visceral organs apparently do not have the double innervation that they do in higher vertebrates. Also, autonomic fibers do not go to the skin in the dogfish.

PRIMITIVE TETRAPODS

The nervous system of such primitive tetrapods as the amphibians has progressed little beyond the condition seen in fishes. As a matter of fact, the cerebellum is not as well developed in living amphibians as in active fishes like *Squalus*. *Necturus* may be studied as an example of the amphibian condition, but it need not be studied in detail as the changes are not great.

Dorsal Surface of the Brain

Expose the dorsal surface of the cranium by removing the skin and muscles overlying it, and then carefully chip away the bone that forms its roof. The meninx primitiva of fishes is represented by two layers in primitive tetrapods — a tough outer **dura mater**, and an inner, vascular **pia-arachnoid mater** applied to the surface of the brain. If the dura was not removed with the skull bones, it will have to be cut off in order to see the brain.

The anterior region of the brain is formed by the paired **cerebral hemispheres** and **olfactory bulbs** (Fig. 8-7). The olfactory bulbs lie anterior to the hemispheres, but are not clearly separated from them. At most, only a slight lateral indentation may be seen between bulbs and hemispheres. Since the olfactory bulbs are adjacent to the rest of the telencephalon rather than to the olfactory sacs, there is no long, narrow olfactory tract as there was in the dogfish. The bands of nervous tissue that extend between the olfactory sacs and the olfactory bulbs are not the tracts, but the **olfactory nerves**. The small, dark body that lies between the posterior ends of the cerebral hemispheres is the **paraphysis**. This is an evagination from the telencephalon.

The **epithalamic** region of the diencephalon appears as a small, trian-

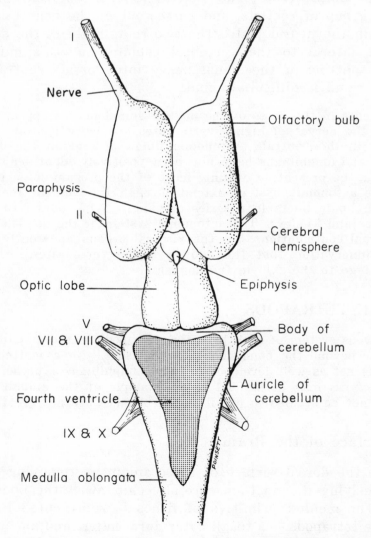

Figure 8-7. Dorsal view of the brain and cranial nerves of *Necturus*.

gular, light area posterior to the paraphysis. As usual, the **epiphysis** and **habenula** are present in this region, but they are hard to distinguish. From the tela chorioidea in this region the **anterior chorioid plexus** has invaginated into the third ventricle, and has even extended caudad into the cavity of the midbrain. It can be seen by cutting open the epithalamus.

The optic lobes lie posterior to the epithalamus. They are relatively small in *Necturus* and do not bulge dorsally to the extent that they do in *Squalus*. The line of separation between optic lobes and epithalamus is not sharp.

The **medulla oblongata** lies posterior to the optic lobes and can be recognized by the very dark tela chorioidea that forms most of its roof. Remove the tela with its **posterior chorioid plexus** and note the triangular, or heart-shaped, **fourth ventricle**.

As mentioned at the outset, the **cerebellum** is very small in amphibians. Its **body** consists of only a narrow, transverse band lying between the optic lobes and the medulla. This fold may be difficult to distinguish from the optic lobes. The **auricles** of the cerebellum are represented by the tissue that surrounds the anterolateral corners of the fourth ventricle.

Cranial and Occipital Nerves

The **olfactory nerve** (I) was observed coming into the brain from the olfactory sac. The **optic nerve** (II) from the eye crosses the floor of the cranial cavity and attaches to the anterior end of the floor of the diencephalon.

The nerves to the eye muscles — **oculomotor** (III), **trochlear** (IV), and **abducens** (VI) — are present, but can be seen only under magnification. They have the same relationships as in *Squalus,* and supply the same muscles. The abducens, however, also goes to the newly evolved **retractor bulbi** (Table 2).

The large trunk arising from the lateral surface of the anterior end of the medulla is the **trigeminal nerve** (V); the trunk posterior to it the common attachment of the **facial** (VII) and **statoacoustic** (VIII) **nerves**. The roots posterior to this trunk represent the origin of the **glossopharyngeal** (IX) and **vagus** (X). Glossopharyngeal and vagus run together until they leave the cranium, then they separate. The distribution, and composition, of these nerves is substantially the same as in *Squalus,* but, of course, there would be some modification in the gill region. Lateral line fibers are present in *Necturus,* but are lost in metamorphosed Amphibia.

Necturus, like other salamanders, has a hypobranchial nerve to the hypobranchial musculature. This nerve is formed primarily of fibers from the first spinal nerve, but it also receives contributions from the second spinal nerve and, in some salamanders, a small twig from the glossopharyngeal-vagus trunk. Certain of these contributions would obviously be homologous to the occipital nerves of *Squalus,* but just which ones is uncertain. It is of interest in this connection that the first, and sometimes the second, spinal nerve consists of only the ventral root.

Ventral Surface of the Brain

Cut across the posterior end of the medulla, across the cranial nerves, and remove the brain. Try to lift the hypophysis from the sella turcica without breaking it off. Note the major regions of the brain, the stumps of the cranial nerves, and study the floor of the diencephalon (hypothalamus) in more detail. The optic nerves cross to form an **optic chiasma** at the anterior end of the hypothalamus, but the chiasma is small and inconspicuous in *Necturus.* As in *Squalus,* the greater part of the hypothalamus consists of a large, posteriorly projecting **infundibulum** to which the hypophysis is attached. A small **saccus vasculosus** is present in the roof of the infundibulum, but it is difficult to see grossly.

MAMMALS

In the evolution through reptiles to mammals, numerous changes occur in the nervous system which are related in large measure to the increased activity and flexibility of response of mammals. The major change in the brain is the evolution of a **neopallium** in the cerebral hemispheres to which sensory impulses are projected. The neopallium first appears in reptiles (Fig. 8–8), and enlarges considerably in the evolution to mammals. It pushes the original olfactory portions of the cerebral hemispheres (**paleopallium** and **archipallium**) apart, and comes to form the greater part of the cerebrum. As it enlarges, it gradually assumes the dominant integrating role, and the tectum is left as a relatively minor optic and auditory reflex center.

Another important change is the great enlargement of the cerebellum. This is correlated with the increased complexity of muscular movement, an increased projection of sensory data to this region, and an interconnection of cerebral hemispheres and cerebellum. To the original auricles (vestibular in nature) and body (proprioceptive) are added a pair of **hemispheres** in which much of the final integration in respect to muscular coordination occurs.

Other changes in the brain tend to be correlated with the increased importance of the neopallium and cerebellum. The thalamus enlarges as it becomes an important pathway and relay station between the cerebrum and other parts of the central nervous system. Other interconnections between the thalamus and cerebrum suggest that these two parts of the brain may form a complex interacting mechanism, so that the thalamus appears to be more than just a relay station. Important centers develop in the ventrolateral portions of the midbrain (**red nucleus**) and metencephalon (**pons**) for the interconnection of the cerebellum and cerebrum. More and larger fiber tracts, both afferent and efferent, evolve in the mesencephalon and hindbrain, for the cerebrum, the cerebellum, and the brain in general exert more influence over the body than they did in lower vertebrates.

The basic pattern of the cranial nerves of mammals remains much the same as in fishes (Table 3), but a few changes have been superimposed upon this plan during the evolution to mammals. The loss of the lateral line system in lower tetrapods entails the loss of lateral line neurons from nerves that previously carried them (VII,IX,X). Except for the trigeminal outflow, mammals retain the autonomic fibers present in the cranial nerves of fishes and, in addition, acquire autonomic fibers in the facial and glossopharyngeal nerves. Those in the facial go to the tear and anterior salivary glands; those in the glossopharyngeal to the parotid gland. With the elaboration of the cucullaris to form the trapezius and sternocleidomastoid complex, we find that the visceral motor fibers that supply these muscles separate from the vagus to form a new cranial nerve, the **accessory** (XI). This change occurs first in reptiles. Finally the posterior limit of the cranium becomes fixed, and the occipital nerves (plus, possibly, an anterior spinal nerve or two) form a definite cranial nerve, the **hypoglossal** (XII). Among living vertebrates, this change is first seen in the reptiles, but it probably occurred earlier, for there is a foramen for such a nerve in the skulls of crossopterygians and labyrinthodonts. In modern amphibians, the hypobranchial musculature is supplied by fibers that travel in the first spinal nerve.

Meninges

The mammalian brain and the stumps of the cranial nerves should be studied from isolated sheep brains. As explained, the peripheral distribution and composition of the nerves is, with a few exceptions, essentially the same as in fishes. The foramina through which they leave the cranium are described on page 80, and certain of the nerves will be seen during later dissections. If it is not possible to study the brain from isolated specimens, it may be removed from your own specimen. But if this is to be done, postpone this study until the end of the course. To remove the brain, first make a sagittal section of the head, and then carefully loosen the halves of the brain and pull them out. Leave the tough membrane covering the brain (dura mater) on as you take it out. The cranial nerves will of course have to be cut, but leave as long stumps as possible.

The tough outer membrane that covers the brain is known as the **dura mater**. Actually it represents the dura of lower vertebrates united with the periosteum lining the cranial cavity. Carefully remove the dura in order to see the other membranes. As you do so note that it sends one extension down between the cerebrum and cerebellum, and another between the two cerebral hemispheres. The former extension is called the **tentorium**; the latter the **falx cerebri** (Fig. 9-14, p. 249). These membranes help to stabilize the brain in the cranial cavity and protect it from distortion during sudden rotational movements of the head. The pia-arachnoid of lower tetrapods has separated into two layers in mammals, but they are hard to distinguish grossly. The **pia mater** is the vascular layer that closely invests the surface of the brain. The **arach-**

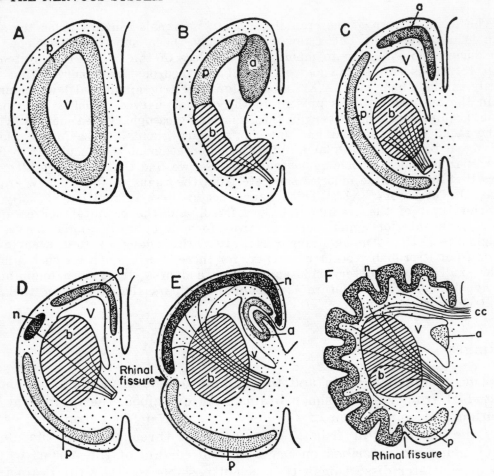

Figure 8–8. Diagrammatic cross sections of the left cerebral hemisphere showing important stages in its evolution. *A*, a primitive fish stage in which the hemisphere is entirely olfactory, and the gray matter is internal and little differentiated; *B*, a modern amphibian in which the gray matter is still internal, but is beginning to differentiate; *C*, a primitive reptile in which the pallial portions of the gray matter are moving toward the surface, but the corpus striatum remains internal and acquires many fiber connections (the lines to and from b); *D*, an advanced reptile in which a neopallium is beginning to appear; *E*, a primitive mammal in which the neopallium is becoming more important and is pushing the original pallial areas apart; *F*, a placental mammal in which the neopallium is highly developed, and connected with the neopallium of the opposite side by the corpus callosum. Abbreviations: *a*, archipallium; *b*, basal nuclei, or corpus striatum, *cc*, corpus callosum; *n*, neopallium; *p*, paleopallium; *V*, lateral ventricle. Major fiber tracts are shown by groups of lines. The lines passing through the corpus striatum represent the internal capsule. (From Romer, The Vertebrate Body. *F*, redrawn and slightly modified.)

noid membrane lies between the pia and dura, and is most easily distinguished from the pia in the region overlying the grooves on the brain surface, for the arachnoid does not dip into them, whereas the pia does. In life the dura and arachnoid adhere to each other, but there is a definitive **subarachnoid space** between the arachnoid and pia which is crisscrossed by weblike strands of connective tissue — a feature that gives the name to the arachnoid membrane. Cerebrospinal fluid circulates in the subarachnoid space, where it forms a liquid cushion around the central nervous system.

The liquid cushion around the central nervous system is very important, for it helps to protect and support the exceedingly soft and delicate nervous tissue. Cerebrospinal fluid gives the brain a great deal of buoyancy. It has been calculated, for example, that a human brain weighing 1500 g. outside the body has an effective weight of only 50 g. in situ. The cerebrospinal fluid also helps to provide the brain with carefully selected nutrients and other substances, for the capillaries within the central nervous system are less pervious to many substances that those elsewhere.

External Features of the Brain and the Stumps of the Cranial Nerves

(A) *Telencephalon*

The paired cerebral hemispheres and the cerebellum are so large that little else is at first apparent in a dorsal view. Notice that the **cerebral hemispheres** are separated from each other by a deep longitudinal furrow which is known as the **longitudinal cerebral fissure**. Spread the dorsal parts of the cerebral hemispheres apart and note the thick transverse band of fibers that connect them. This is the **corpus callo-**

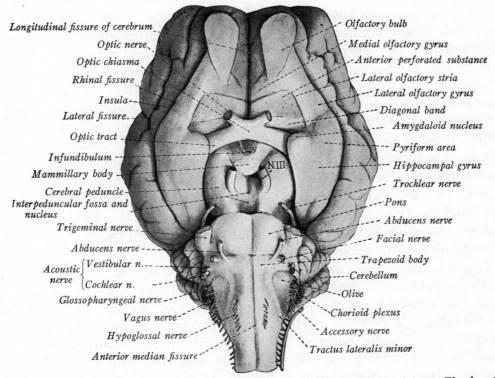

Figure 8–9. Ventral view of the sheep brain. (From Ranson and Clark, Anatomy of the Nervous System.)

sum, a neopallial commissure found only in placental mammals and some marsupials. The surface of each hemisphere is thrown into many irregular folds (the **gyri**) separated from each other by grooves (the **sulci**). A pair of **olfactory bulbs** project from the anteroventral portion of the cerebrum (Fig. 8-9). They can be seen best in a ventral view. The olfactory bulbs lie over the cribriform plate of the mesethmoid and receive the groups of olfactory neurons from the nose. These neurons, which cannot be seen grossly, constitute the **olfactory nerve** (I). A whitish band, the **olfactory tract**, extends at an angle caudally and laterally from each bulb. The ventral portion of the cerebrum, to which each olfactory tract leads, is known as the **piriform area** (**parahippocampal gyrus**). It is separated laterally from the rest of the cerebrum by the **collateral sulcus** or **rhinal fissure**.

The piriform area represents the **paleopallium**, and it is homologous to the olfactory lobes of lower vertebrates. The remaining olfactory portion of the cerebrum, **archipallium**, has been pushed internally and does not show on the surface. Thus all of the superficial part of the cerebrum that lies lateral and dorsal to the collateral sulcus is **neopallium** — the major integrating region of the brain.

(B) Diencephalon

The telencephalon has enlarged to such an extent that it has grown back over and covers the diencephalon and much of the mesencephalon. To see the dorsal portion of the diencephalon (**epithalamus**), it is necessary to spread the cerebral hemispheres apart, cutting the corpus callosum. Pick away the tela chorioidea and its chorioid plexus forming part of the roof of the diencephalon. The longitudinal slit that is then exposed is the **third ventricle**. The knoblike **pineal body** lies posterior to the ventricle. Its function is uncertain, but it is believed to be an endocrine gland. Extracts of the organ (melatonin) do cause retraction of pigment in the chromatophores of lower vertebrates. Secretion of the organ may also suppress the onset of puberty until an age appropriate to the differentiation of the rest of the body. The narrow transverse band of tissue between the pineal body and the ventricle is the **habenular commissure**, and the tissue forming the posterolateral rim of the ventricle is the **habenula** (Fig. 8-11).

Turn the brain over and examine the ventral surface of the diencephalon (**hypothalamus**). The **optic nerves** (II) undergo a partial decussation at the anterior border of the hypothalamus, forming the prominent **optic chiasma**. The rest of the hypothalamus is the oval area lying posterior to the optic chiasma. The **hypophysis** or **pituitary gland** may still be suspended by a narrow stalk, the **infundibulum**, from the hypothalamus. If so, remove it in order to get a clearer view of the region. The cavity in the infundibulum represents an extension of the third

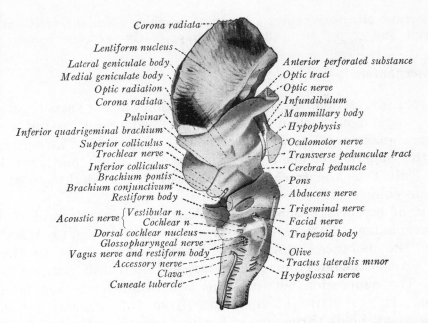

Corona radiata
Lentiform nucleus
Lateral geniculate body
Medial geniculate body
Optic radiation
Corona radiata
Pulvinar
Inferior quadrigeminal brachium
Superior colliculus
Trochlear nerve
Inferior colliculus
Brachium pontis
Brachium conjunctivum
Restiform body
Acoustic nerve { Vestibular n.
Cochlear n.
Dorsal cochlear nucleus
Glossopharyngeal nerve
Vagus nerve and restiform body
Accessory nerve
Clava
Cuneate tubercle

Anterior perforated substance
Optic tract
Optic nerve
Infundibulum
Mammillary body
Hypophysis
Oculomotor nerve
Transverse peduncular tract
Cerebral peduncle
Pons
Abducens nerve
Trigeminal nerve
Facial nerve
Trapezoid body
Olive
Tractus lateralis minor
Hypoglossal nerve

Figure 8–10. Lateral view of the sheep brainstem. (From Ranson and Clark, Anatomy of the Nervous System.)

ventricle. That portion of the hypothalamus adjacent to the infundibulum is known as the **tuber cinereum**. A pair of rounded **mamillary bodies** forms the caudal end of the hypothalamus.

In order to see the **thalamus**, or lateral wall of the diencephalon, you must carefully pull one of the cerebral hemispheres forward and look beneath it. A better view will be had after the cerebrum has been dissected (p. 217), so you may care to postpone a study of the thalamus until then. An **optic tract** leads from the optic chiasma to terminate in an enlargement of the thalamus known as the **lateral geniculate body** (Fig. 8–10). The meninges will have to be removed to see the area clearly. The smaller enlargement posterior to the lateral geniculate body is the **medial geniculate body**, and that portion of the thalamus lying dorsal to the geniculate bodies is known as the **pulvinar**.

The thalamus is an important relay station between the cerebral hemispheres and the rest of the brain. All sensory impulses, except for olfaction, pass through the dorsal portion of the thalamus on the way to the cerebrum. The lateral geniculate body, for example, relays optic impulses, and the medial geniculate body, auditory impulses. The function of the pulvinar is uncertain, but it may be involved in visual and auditory integrations. The ventral portion of the thalamus, which is not exposed on the surface, relays certain efferent impulses on the way back from the cerebral hemispheres. In addition, the thalamus may act as a subcortical center of integration, and its numerous interconnections with the cortex suggest that it plays a role in many cortical functions. The hypothalamus of mammals, as in lower vertebrates, is an important integrating center for many autonomic and visceral functions including sleep, body temperature, blood sugar level, and water balance. The habenula continues to be an important olfactory center. Impulses reach it from the piriform lobe of the cerebrum and go out to nuclei in the floor of the mesencephalon.

(C) Mesencephalon

The roof, or **tectum**, of the mesencephalon can be seen by spreading the cerebrum and cerebellum apart. Four prominent, round swellings (the **corpora quadrigemina**) characterize this region. The larger, anterior pair are the **superior colliculi**; the smaller, posterior pair, the **inferior colliculi**. Note that the **trochlear nerves** (IV) arise slightly posterior to the inferior colliculi (Fig. 8-10).

A pair of **cerebral peduncles** lies along the ventrolateral surface of the mesencephalon. Each emerges from beneath the optic tract and is as wide as the distance from the medial geniculate body to the hypothalamus. An **oculomotor nerve** (III) arises from the surface of each peduncle. The depression between the two peduncles is the **interpeduncular fossa**. If you strip the meninges from this region, you may be able to see small holes through which blood vessels enter the brain. This region constitutes the **posterior perforated substance**. A comparable **anterior perforated substance** lies anterior to the optic chiasma (Fig. 8-9).

The evolution of the neopallium has robbed the tectum of its original importance as the major integrating area. But some optic and auditory fibers are still projected to the superior and inferior colliculi, respectively, and certain reflexes occur in these regions. The peduncles are large bundles of fibers that extend caudally from the cerebral hemispheres. Most efferent impulses from the cerebrum pass back through them.

(D) Metencephalon

The dorsal portion of the metencephalon forms the **cerebellum**. It will be noted that the surface area of the cerebellum is increased by numerous platelike folds (**folia**), separated from each other by **sulci**. The median part of the cerebellum, which has the appearance of a segmented worm bent nearly in a circle, is called the **vermis**; the lateral parts are the **hemispheres**. The lobe of each hemisphere that lies ventral to the main part of the hemisphere, and lateral to the region where the cerebellum attaches to the main part of the hemisphere, is known as the **flocculus (nodulus)**. The flocculi are homologous to the auricles of lower forms and receive vestibular impulses, most of the vermis is homologous to the body and receives proprioceptive impulses, and most of the hemispheres are new additions with cerebral connections.

The cerebellum is connected with other parts of the brain by three prominent fiber tracts, or peduncles (Figs. 8-10 and 8-11). The **middle peduncle**, or **brachium pontis**, lies medial to the anterior half of the **flocculus**. You will have to dissect off the flocculus on one side to see it

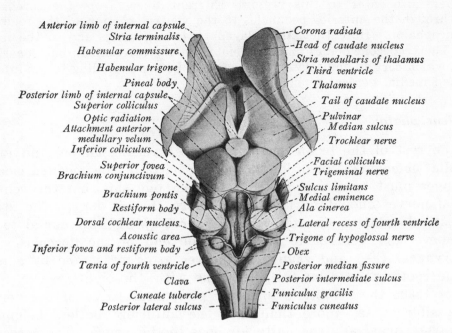

Anterior limb of internal capsule
Stria terminalis
Habenular commissure
Habenular trigone
Pineal body
Posterior limb of internal capsule
Superior colliculus
Optic radiation
Attachment anterior
medullary velum
Inferior colliculus
Superior fovea
Brachium conjunctivum
Brachium pontis
Restiform body
Dorsal cochlear nucleus
Acoustic area
Inferior fovea and restiform body
Tænia of fourth ventricle
Clava
Cuneate tubercle
Posterior lateral sulcus

Corona radiata
Head of caudate nucleus
Stria medullaris of thalamus
Third ventricle
Thalamus
Tail of caudate nucleus
Pulvinar
Median sulcus
Trochlear nerve
Facial colliculus
Trigeminal nerve
Sulcus limitans
Medial eminence
Ala cinerea
Lateral recess of fourth ventricle
Trigone of hypoglossal nerve
Obex
Posterior median fissure
Posterior intermediate sulcus
Funiculus gracilis
Funiculus cuneatus

Figure 8–11. Dorsal view of the sheep brainstem. (From Ranson and Clark, Anatomy of the Nervous System.)

clearly. Note that the middle peduncle connects ventrally with a transverse band of fibers known as the **transverse fibers of the pons**. The tissue posterior and slightly medial to the middle peduncle constitutes the **posterior peduncle**, or **restiform body**. It also continues along the dorsolateral margin of the medulla. The **anterior peduncle**, or **brachium conjunctivum**, lies medial to the middle peduncle, and can be seen by looking in the area between the cerebellum and inferior colliculi.

The ventral portion of the metencephalon has differentiated sufficiently from the medulla oblongata in mammals to be considered a distinct region — the **pons**. Grossly, the pons includes the transverse fibers of the pons already seen, and a narrower transverse band of fibers just posterior to these fibers known as the **trapezoid body**. Several nerves attach on the pons (Fig. 8-9). The very large one that extends anteriorly across the base of the middle peduncle is the **trigeminal** (V). The **facial** (VII) arises just posterior to the trigeminal and extends dorsally, and the **abducens** (VI) arises from the trapezoid body slightly lateral to the midventral line.

The cerebellum is the center for equilibrium and motor coordination. It monitors the motor activity of the body and initiates corrective impulses. In this connection, it receives vestibular fibers from the inner ear, proprioceptive fibers from the muscles of the body, and a variety of impulses from the cerebrum. Fibers from the cerebrum are relayed in the pons, decussate in the transverse fibers of the pons, and ascend to the cerebellum through the middle peduncles. Proprioceptive fibers come in through the posterior peduncles. Vestibular fibers also enter in this region. Efferent fibers from the cerebellum pass out through the anterior peduncles to the ventral portion of the midbrain and metencephalon. Here they are relayed to the thalamus and to the motor columns. The trapezoid bodies are not related to the cerebellum, but constitute a tract in which certain auditory fibers decussate before ascending to the inferior colliculi and medial geniculate bodies.

(E) Myelencephalon

All the rest of the brain belongs to the myelencephalon, and forms the **medulla oblongata**. In order to see the parts of the medulla clearly, the meninges must be stripped off on at least one side, but the remaining cranial nerves should be identified before this is done. The stump of the **vestibulocochlear** or **statoacoustic nerve** (VIII) lies dorsal to the seventh nerve and ventral to the posterior end of the flocculus. The **glossopharyngeal** (IX) and **vagus** (X) **nerves** are represented by a number of fine rootlets posterior to, but in line with, the eighth nerve. Since one cannot trace these rootlets into the peripheral parts of the nerves, it is impossible to say more than that the anterior rootlets belong to the glossopharyngeal and the posterior ones to the vagus. The **accessory nerve** (XI) is the large longitudinal nerve posterior to the vagus. It arises by a number of fine rootlets from the posterior end of the medulla and the anterior end of the spinal cord. The anterior end of the nerve is the end that leads out of the skull to the muscles that the nerve supplies. If the nerve also has a cut posterior end on your specimen, all its origin is not intact. The **hypoglossal nerve** (XII) is represented by the rootlets on the posteroventral portion of the medulla. If any of the cranial nerves cannot be found, refer to Figure 8-9.

Now strip off the meninges on one side of the medulla, and the **tela chorioidea** with its **chorioid plexus** that forms much of the roof of the medulla. Pull the cerebellum forward and note the large **fourth ventricle** extending forward into the metencephalon. The posterior part of the roof of the ventricle was formed by the tela chorioidea, but the anterior part of the roof is formed by a thin layer of fibers termed the **medullary velum**. It can be seen by pushing the cerebellum posteriorly. The trochlear nerve decussates in the anterior part of the velum.

Note the enlargement on the dorsal rim of the medulla just posterior to the point where the restiform body turns into the cerebellum, and dorsal to the vestibulochochlear nerve. Observe that it extends medially to an oval-shaped enlargement in the ventrolateral part of the floor of the fourth ventricle. This enlargement constitutes the **area vestibularis** or **acoustica**.

Examine the dorsal surface of the posterior end of the medulla. (The medulla extends as far caudad as the **first spinal nerve**.) The prominent **dorsal median sulcus** (posterior median fissure, Fig. 8-11) of the cord continues onto the medulla nearly to the fourth ventricle. A less distinct **dorsal intermediate sulcus** lies about 1/8 inch lateral to the preceding, and a **dorsal lateral sulcus** slightly lateral to the dorsal intermediate. These grooves outline two longitudinal fiber tracts, a dorsal **funiculus gracilis** and a more lateral **funiculus cuneatus**. The anterior end of the former tract expands slightly to form a structure called the **tubercle of the nucleus gracilis,** or **clava**. The latter has a comparable enlargement known as the **tubercle of the cuneate nucleus**. As already stated, the restiform body forms the dorsolateral rim of the medulla anterior to these tubercles. It then turns dorsally to enter the cerebellum.

Turn the brain over, and examine the ventral surface of the medulla. The midventral groove is the **ventral median fissure**. The longitudinal bands of tissue on either side of it that are approximately 1/4 inch wide are known as the **pyramids**. Note that some of the pyramidal fibers lie superficial to the trapezoid body. The area lateral to the pyramids, posterior to the trapezoid body, and ventral to the glossopharyngeal and vagus nerves, is known as the **olive**.

The medulla is a transitional region between the anterior parts of the brain and the spinal cord. Most of its gray matter represents the forward continuation of the columns of the cord and the break-up of these into nuclei. Numerous reflex activities occur between these nuclei, and many important visceral activities are controlled here: respiratory movements, salivation, swallowing, rate of heart beat, and blood pressure. Most of the nuclei cannot be seen grossly, but some form bulges on the surface. The area vestibularis is a region in which vestibular and auditory fibers from the ear undergo various relays. Some of the relays extend vestibular impulses to the cerebellum and auditory impulses through the trapezoid body. The olive represents another nucleus, but its function is uncertain. It has connections with the cerebellum via the restiform bodies, and may be related to muscular coordination. Much of the white matter of the medulla represents fiber tracts passing through the region. The pyramids, for example, are a posterior continuation of the fibers that formed the cerebral peduncles. The funiculus gracilis and cuneatus are largely composed of ascending proprioceptive fibers. These relay in the nucleus gracilis and cuneate nucleus before going to the thalamus (for relay to the cerebrum), or to the cerebellum via the restiform bodies.

Sagittal Section of the Brain

Cut the brain in half as close to the sagittal plane as possible. If you deviate from the plane, take the larger half and dissect away enough tissue to be able to see the median cavities of the brain clearly (Fig. 8–12). Many of the features just described can also be seen in this view. Note in particular the way in which the cerebral hemispheres extend back over the diencephalon and mesencephalon. The **corpus callosum** also shows particularly well. Its expanded anterior end is known as the **genu**; its expanded posterior end, as the **splenium**; and the thinner region between, as the **trunk** or **body**. A thin, vertical septum of tissue, **septum**

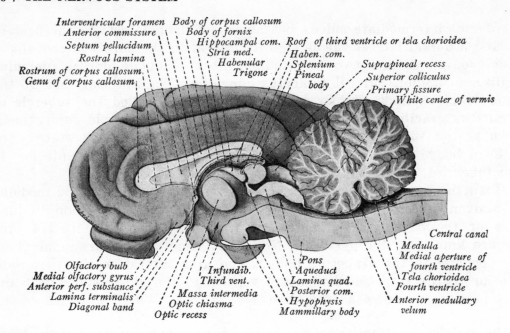

Interventricular foramen Body of corpus callosum
Anterior commissure   Body of fornix
Septum pellucidum   Hippocampal com. Roof of third ventricle or tela chorioidea
Rostral lamina   Stria med.   Haben. com.
Rostrum of corpus callosum. Habenular   Splenium   Suprapineal recess
Genu of corpus callosum   Trigone   Pineal   Superior colliculus
body   Primary fissure
White center of vermis

Olfactory bulb   Pons   Central canal
Medial olfactory gyrus   Infundib.   Aqueduct   Medulla
Anterior perf. substance   Third vent.   Lamina quad.   Medial aperture of
Lamina terminalis   Massa intermedia Posterior com.   fourth ventricle
Diagonal band   Optic chiasma   Hypophysis   Tela chorioidea
Optic recess   Mammillary body   Fourth ventricle
Anterior medullary
velum

Figure 8–12. Sagittal section of the sheep brain. (From Ranson and Clark, Anatomy of the Nervous System.)

pellucidum lies ventral to the anterior part of the corpus callosum. It consists of two thin plates of gray matter with a narrow space (the **fifth ventricle**) between them. The lateral ventricle lies lateral to the septum, and may be seen by breaking it. A band of fibers called the fornix lies posterior to the septum pellucidum. The **body of the fornix** begins near the splenium, and then the band curves forward and ventrally as the column of the fornix. It passes out of the plane of the section posterior to a small round bundle of fibers which is a cross section of the **anterior commissure**, an olfactory decussation. After other features in the section have been seen, carefully dissect away tissue in this region and trace the column of the mamillary body. The fornix is a pathway that connects the cerebrum with the hypothalamus, and will be understood better after the dissection of the cerebrum. The thin ridge of tissue extending ventrally from the anterior commissure to the optic chiasma is the **lamina terminalis** — a landmark representing the anterior end at the embryonic neural tube. The cerebral hemispheres are lateral evaginations that extend anteriorly from the lamina.

The third ventricle and diencephalon lie posterior to the column of the fornix, anterior commissure, and lamina terminalis. Note that the **third ventricle** is very narrow, but that it has a considerable dorsal-ventral and anterior-posterior extent. It is lined by a shiny epithelial membrane (the **ependymal epithelium**), as are all the cavities within the neural tube. The thalamus lies lateral to the third ventricle, but a portion of it, the **massa intermedia**, (**interthalamic adhesion**) extends across the third ventricle and will appear as a dull, circular area not covered by the shiny ependyma. The **interventricular foramen of Monro**, through which each lateral ventricle communicates with the third ventricle, lies in the depression anterior to the massa intermedia. The hypothalamus lies ventral to the third ventricle and the epithalamus dorsal to it. Note again the **pineal body**, the **habenular commissure** (which will show in cross section), and the **habenula**. These features show unusually well in the section. In addition, the epithalamus includes a **posterior commissure**. This is the tissue ventral to the attachment of the pineal body and anterior to the corpora quadrigemina.

A narrow **cerebral aqueduct of Sylvius** leads through the mesencephalon to the fourth ventricle of the hindbrain. The cerebellum lies above the fourth ventricle. Note that most of its **gray matter** (**substantia grisea**) is in the form of a gray **cortex** over the surface of the folia, while the **white matter** (**substantia alba**) is centrally located. The white matter, which represents fiber tracts extending between the cerebellum and other parts of the brain, has the appearance of a tree, and is called the tree of life (**arbor vitae**).

Dissection of the Cerebrum

Take one half of the brain and gradually cut away the dorsal part of the cerebral hemisphere by making thin frontal sections. It will soon be noted that the **gray matter** forms a **cortex** over the surface, while the **white matter** is centrally located. The fibers that make up the white matter interconnect the different parts of the hemisphere, the two hemispheres with each other, and the hemispheres with the rest of the brain. Continue to cut off slices until you nick the lateral ventricle. This will be near the level of the corpus callosum. Now cut off the roof of the lateral ventricle with a pair of fine scissors, but do not remove the part of the corpus callosum near the midline or the fornix (Fig. 8–13). The posterior end of the ventricle extends ventrally into the posteroventral region of the hemisphere (temporal lobe). It may be necessary to cut off some of the lateral surface of the cerebrum in order to trace it. The **chorioid plexus** of the lateral ventricle has invaginated from the thin ventral wall of the cerebral hemisphere, and appears as an irregular, often dark frill in the floor of the ventricle.

The large mass posterior to the chorioid plexus, and curving down into the temporal lobe, is the **hippocampus**. Its thin anterior border, to which the chorioid plexus is attached, is called the **fimbria**. Remove the plexus to see the region more clearly. Cut into the hippocampus and note that it consists of a mass of gray matter covered by white fibers. These fibers lead into the fimbria and thence to the body of the fornix. The part of this fiber tract preceding the body of the fornix is differentiated as the crus of the fornix in some mammals, but this is not the case in the sheep. As was seen in the study of the sagittal section, the

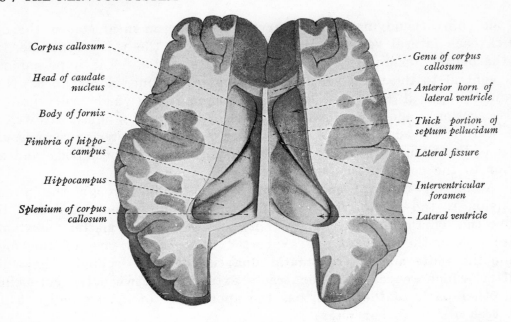

Corpus callosum

Head of caudate
nucleus

Body of fornix

Fimbria of hippo-
campus

Hippocampus

Splenium of corpus
callosum

Genu of corpus
callosum

Anterior horn of
lateral ventricle

Thick portion of
septum pellucidum

Lateral fissure

Interventricular
foramen

Lateral ventricle

Figure 8–13. Dorsal view of a dissection of sheep cerebrum. (From Ranson and Clark, Anatomy of the Nervous System.)

fiber tract that composes the body of the fornix extends ventrally as the column of the fornix. Ultimately it reaches one of the mamillary bodies. A decussation occurs in a band of fibers, the **commissure of the fornix**, located just posterior to the body of the fornix.

The hippocampus is homologous to the archipallium, which has been pushed internally through the enlargement of the neopallium (Fig. 8–8). This was originally an olfactory part of the brain, and it has long been considered to be an olfactory region in mammals. It receives fibers from the piriform lobe, but from a part of the lobe that is believed to be concerned with olfactory association, and it sends fibers, via the fornix, to the hypothalamus. Some other fibers in the fornix go to the habenula. Recent evidence suggests that the hippocampus is concerned with a variety of emotional responses, such as attention, searching, anxiety, and defense.

Cut across the corpus callosum and fornix, and also across the attachment between the lower end of the hippocampus and the piriform area. Then pull off the posterior half of the hemisphere and the hippocampus. (If the features on the lateral surface of the thalamus could not be seen clearly before, they may be reviewed at this time, for the region is now well exposed.) Examine the anterior part of the cerebrum. The tear-shaped enlargement forming the lateroventral part of the floor of the lateral ventricle is one of a group of nuclei and fibers in the base of each cerebral hemisphere known as the **corpus striatum**. This one, the **caudate nucleus**, is the most medial part of the corpus striatum. To see the rest of the group, carefully scrape away the lateroventral portion of the hemisphere between the optic tract and the olfactory bulb

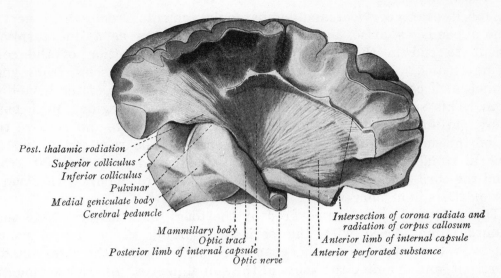

Post. thalamic rodiation
Superior colliculus
Inferior colliculus
Pulvinar
Medial geniculate body
Cerebral peduncle

Mammillary body
Optic tract
Posterior limb of internal capsule
Optic nerve

Intersection of corona radiata and
radiation of corpus callosum
Anterior limb of internal capsule
Anterior perforated substance

Figure 8–14. Lateral view of a dissection of the sheep brain to show the internal capsule. The lateral surface of the cerebrum and the lentiform nucleus have been removed. (From Ranson and Clark, Anatomy of the Nervous System.)

with some blunt instrument. The handle of a scalpel or a pair of forceps works well. After scraping away the superficial gray matter, you will come upon a thin layer of white fibers. Remove these, and you expose a large, oval, gray mass which is the lateral **lentiform nucleus** of the corpus striatum. Remove this, and you expose a thick mass of fibers, the **internal capsule**, passing between the two nuclei of the corpus striatum. Compare your dissection with Figure 8–14.

The internal capsule is the major pathway between the neopallium and the rest of the brain. Most of the afferent impulses, which are relayed in the dorsal thalamus, pass up through the internal capsule, and most of the efferent impulses from the neopallial cortex pass back through the internal capsule, cerebral peduncles, and pyramids. The function of the nuclei of the corpus striatum is not completely understood. An area comparable to the corpus striatum is found in fishes, in which it receives olfactory impulses. In amphibians, and especially in reptiles, it becomes an important integrating region serving to supplement the optic tectum. But in mammals, the neopallium becomes the major integrating region. The nuclei of the corpus striatum receive fibers from the thalamus and cerebral cortex and send fibers to the hypothalamus, floor of the midbrain, and medulla oblongata. There is some evidence that the region influences motor activity, for lesions of these nuclei produce tremor during voluntary activity.

Spinal Cord and Spinal Nerves

(A) The Cord and Roots of the Spinal Nerves

The **spinal cord** (**medulla spinalis**) is a subcylindrical cord lying within the vertebral canal of the vertebral column. It is not uniform in diameter, for it bears cervical and lumbar enlargements from which nerves to the appendages arise, and the caudal end tapers as a fine terminal filament to end in the base of the tail. Only a section of the cord need be studied. To get at it, remove the epaxial muscles from your specimen of a cat or rabbit so as to expose 3 or 4 inches of the vertebral column. This should be done in the posterior thoracic region. With bone scissors, carefully cut across the pedicles of the vertebrae and remove the tops of the neural arches. The spinal cord will be seen lying in the vertebral canal. Continue chipping away bone, and removing fat from around the cord, until you have satisfactorily exposed it along with several roots of the spinal nerves.

The cord and roots are covered by the tough **dura mater**. Note that the dura is not fused with the periosteum lining the vertebral canal. What was the situation in the cranial cavity? Leave the dura on for the present, and examine the roots of the spinal nerves. At each segmental interval, there is a pair of dorsal and ventral roots (Fig. 8-1). Trace a dorsal and ventral root laterally on one side. They pass into the intervertebral foramen before uniting to form a **spinal nerve**. Just before uniting, the dorsal root bears a small round enlargement — the **dorsal root ganglion** (**spinal ganglion**). If you trace the spinal nerve laterally you may see it divide into a **dorsal ramus** to the epaxial regions of the body and a **ventral ramus** to the hypaxial regions. The small **communicating rami** to the sympathetic cord probably will not be seen.

Slit open the dura at one end of the exposed area. This opens the **subdural space**. Observe that the roots of the spinal nerves do not have a simple attachment on the cord, but unite by a spray of fine **rootlets**. Cut out a segment of the spinal cord and strip off the remaining meninges — **arachnoid** and **pia mater**. There is a deep ventral furrow on the cord known as the **ventral median fissure**; a less distinct dorsal furrow, the **dorsal median sulcus**; and a more prominent furrow slightly lateral to the middorsal line, the **dorsal lateral sulcus**. Note that the dorsal rootlets enter along the dorsal lateral sulcus. Recall that these same grooves extend onto the medulla oblongata.

Make a fresh cross section of the cord with some sharp instrument such as a razor blade. Examine the cut surface, and compare it with Figure 8-1. Also look at demonstration slides if possible. The tiny **central canal** can generally be seen grossly, and if you are fortunate you may be able to distinguish the butterfly-shaped central **gray matter** from the peripheral **white matter**. As explained in the introduction to this chapter, the gray matter consists of unmyelinated fibers and the cell bodies of motor and internuncial neurons; the white, of ascending and descending, myelinated fibers. That segment of the white matter that lies between the dorsal median fissure and the dorsal lateral sulcus is called the **dorsal funiculus**; that segment between the dorsal lateral sulcus and the line of attachment of the ventral roots, the **lateral funiculus**;

and that portion between the ventral roots and the ventral median fissure, the **ventral funiculus**.

(B) *The Spinal Nerves and Brachial Plexus*

There are 38 spinal nerves in the cat (8 cervical, 13 thoracic, 7 lumbar, 3 sacral, 7 caudal), and 37 in the rabbit (8 cervical, 12 thoracic, 7 lumbar, 4 sacral, 6 caudal). The dorsal rami of these nerves extend straight out into the epaxial region, but many of the ventral rami unite in a complex manner to form networks, or **plexuses**, before being distributed to the musculature and skin. This is especially true in the region of the appendages. In a typical mammal, the anterior cervical nerves form a **cervical plexus** supplying the neck region; the posterior cervical and anterior thoracic nerves form a **brachial plexus** supplying the pectoral appendage; and the lumbar, sacral, and anterior caudal nerves form a **lumbosacral plexus** supplying the pelvic appendage.

The brachial plexus of the cat or rabbit may be dissected as an example of one of these networks. It lies medial to the shoulder and anterior to the first rib, and should be approached from the ventral surface. If it is still intact on the side on which the muscles were dissected, study it there. Otherwise cut through the pectoralis complex of muscles on the other side. The dissection of the plexus involves the meticulous picking away of fat and connective tissue from around the nerves and the accompanying blood vessels. If you find it necessary to cut any of the larger vessels, do so in such a way that you will be able to appose the cut surfaces when you study the circulatory system. Clean off the nerves from a point as near to the vertebral column as you can reach to the point at which they disappear into the shoulder muscles and brachium.

The brachial plexus is formed by the union of the ventral rami of the sixth to eighth cervical and first thoracic nerves in the cat. The fifth cervical also contributes to it in the rabbit. Considerable variation occurs in the details of the union of these nerves to form the plexus, and in the origin of peripheral branches; but a common pattern for the two mammals under consideration is shown in Figure 8–15. The ventral rami of the nerves that enter the plexus are referred to as the **roots** of the plexus. It will be noted that each root tends to split into two **divisions**, and that the divisions of different nerves unite to form **cords** from which peripheral nerves arise. It will also be noted that the splitting of roots into divisions, and the union of divisions to form cords, occurs in such a way that there tends to be an early segregation of the nerves supplying the dorsal appendicular muscles from those supplying the ventral. The parts of the plexus going to the dorsal musculature have been stippled in the diagram. Many of the nerves are also cutaneous, being distributed to the skin, but only the major cutaneous branches will be described.

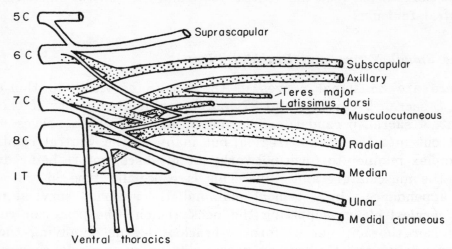

Figure 8–15. Diagram of the major parts of the mammalian brachial plexus based on the cat and rabbit, ventral view of the left side. 5 C, 6 C, 7 C, 8 C, 1 T are ventral rami of the fifth to eighth cervical and first thoracic nerves forming the roots of the plexus. The fifth cervical nerve does not enter the plexus in the cat. Most of the roots split into two divisions, and most of the divisions unite with divisions of adjacent nerves to form cords that lead into the peripheral nerves, or give rise to the peripheral nerves. The divisions, cords, and peripheral nerves to the dorsal appendicular muscles have been stippled.

The most ventral nerves of the plexus are several **ventral thoracics**. They are small nerves which arise from the ventral divisions of the plexus, may or may not unite with each other, and pass to the pectoralis complex.

A large **suprascapular nerve** leaves the front of the plexus, where it arises for the most part from the sixth cervical, and passes between the subscapular and supraspinatus muscles to supply the supraspinatus, infraspinatus, and some of the skin over the shoulder and brachium.

One or more **subscapular nerves** leave posterior to the suprascapular nerve and pass to the large subscapular muscle. They arise, for the most part, from the sixth and seventh cervicals.

A large **axillary nerve**, which lies posterior to the subscapular nerve, arises from the seventh cervical nerve. It passes through the proximal part of the brachium ventral to the long head of the triceps to supply the teres minor and deltoid complex.

Smaller nerves to the **latissimus dorsi** and **teres major** arise from the plexus near the origin of the axillary. Often the teres major nerve springs from the axillary. These are sometimes called subscapular nerves.

The large, deep nerve posterior to the axillary that is formed by the union of the dorsal divisions of the seventh and eighth cervicals and first thoracic is the **radial**. This is the largest nerve of the plexus. It passes between the triceps and the humerus to the lateral surface of the arm and thence down to the extensor portion of the forearm. It supplies the epitrochlearis, triceps, and forearm extensors. A large branch of the nerve is cutaneous.

All the above nerves go to dorsal appendicular muscles, except the ventral thoracics and the suprascapular which supply parts of the ventral musculature. The remaining nerves innervate the rest of the ventral appendicular muscles. It will be noted that all the nerves to the ventral muscles arise from the ventral divisions of the plexus. A small **musculo-cutaneous** springs from the sixth and seventh cervicals, passes superficial to the radial nerve, and enters the biceps. It generally branches before reaching the biceps. It supplies the biceps, coracobrachialis, brachialis, and some of the skin over the forearm.

The ventral divisions of the seventh and eighth cervical and first thoracic combine to form two prominent nerves that run down the medial side of the brachium. The more anterior of these is the **median nerve**, the more posterior the **ulnar nerve**. They are distributed to the forearm flexors and skin of the hand. The median nerve passes through the entepicondylar foramen of the humerus.

A small **medial cutaneous nerve**, which arises from the first thoracic, runs parallel with, and posterior to, the ulnar nerve. It supplies some of the skin over the forearm.

9 / The Coelom and the Digestive and Respiratory Systems

WE NOW TURN from the organ systems concerned with support, movement, and integration to a group concerned with metabolism. The digestive system brings in the raw materials needed by the body; the respiratory system takes care of the essential gas exchanges; the circulatory system transports materials to and from the cells; and the excretory system eliminates all, or much, of the nitrogenous waste products of cellular metabolism, and helps to control the water balance of the body. Much nitrogenous excretion, however, occurs through the gills of fishes. These systems tend to be functionally distinct, but it is convenient to study the digestive and respiratory systems together to some extent, for they are closely associated morphologically. The respiratory system develops embryonically as outgrowths from the digestive system.

The Coelom and Its Subdivisions

The body cavity, or **coelom**, is an epithelium-lined space containing some liquid and surrounding the viscera. Its presence permits a certain amount of movement of the organs, and their change in size and shape. Embryologically, the coelom develops as a pair of mesothelial-lined spaces within the lateral plate mesoderm on either side of the digestive tract (Fig. 9–1). At first there is a right and left coelom that converge above and below the digestive tract to form a **dorsal** and a **ventral mesentery**. Much of the ventral mesentery is ephemeral. It soon disappears in the posterior part of the body, and the originally paired coeloms become continuous. It follows from this development that the viscera are technically outside the coelom, being separated from it by its thin mesothelial wall. However, the visceral organs are generally approached by cutting open the coelom, and it is therefore convenient to speak of them as being within the body cavity.

The coelom of fishes and amphibians becomes divided into two parts. The anterior portion around the heart forms a **pericardial cavity**, aud the rest a **pleuroperitoneal cavity**. In most amniotes, the pleuroperitoneal cavity becomes further subdivided into a pair of **pleural cavities** around the lungs, and a **peritoneal cavity** about the abdominal viscera.

The separation between the pericardial and pleuroperitoneal cavities is by means of a transverse partition known as the **transverse septum**. The ventral

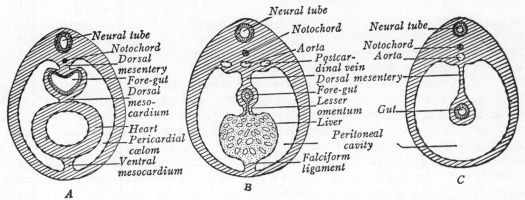

Figure 9–1. Diagrammatic cross section of a vertebrate embryo to show the relationship of the coelom and mesenteries to the visceral organs: *A*, through the level of the heart; *B*, through the liver; *C*, through the intestine. All mesenteries ventral to the digestive tract are parts of the ventral mesentery. (From Arey, Developmental Anatomy. After Prentiss.)

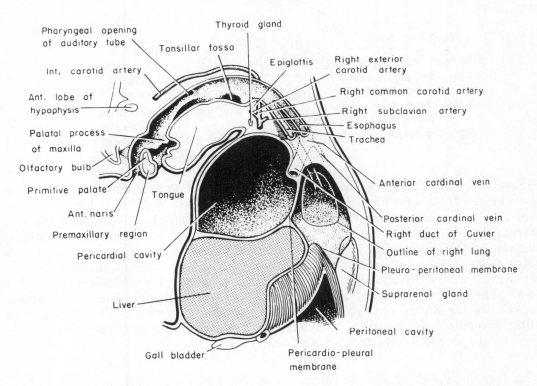

Figure 9–2. Sagittal section of a human embryo to show the subdivisions of the coelom. The pericardial cavity is already separated from the rest of the coelom, for the liver is expanding in the ventral portion of the transverse septum, and the dorsal part of the septum (pericardiopleural membrane) has developed to carry the ducts of Cuvier. The transverse septum has an oblique position, so the future pleural cavities (only the right one is shown) lie dorsal to the pericardial cavity. A pleuroperitoneal membrane, which is shown developing, is largely responsible for the separation of each pleural cavity from the peritoneal cavity. (Redrawn from Hamilton, Boyd, and Mossman, Human Embryology, W. Heffer & Sons, Ltd.)

portion of this septum develops embryologically through the expansion of the liver in the ventral mesentery (Fig. 9–1, *B*) ; the dorsal portion, by a pair of folds (called **pericardiopleural membranes** in mammalian embryology) that carry the ducts of Cuvier down to the heart (Fig. 9–2).

In fishes, the transverse septum is a vertical partition, but the posterior migration of the heart and pericardial cavity in most tetrapods causes the septum to assume a somewhat oblique position. The paired anterior parts of the pleuroperitoneal cavity thus come to lie dorsal to the pericardial cavity (Fig. 9–2). It is in these anterior recesses that the lungs lie. The separation of this pair of recesses as pleural cavities (when it eventually occurs in certain reptiles) is through the development of another pair of folds, the **pleuroperitoneal membranes** (Fig. 9–2), aided by subsidiary folds from the body wall and dorsal mesentery. The **diaphragm** of mammals represents these coelomic folds, plus the ventral part of the transverse septum, plus somatic musculature of cervical origin that has invaded them (Fig. 9–3). In mammals, the pleural cavities subsequently extend ventrally lateral to the pericardial cavity and thus more or less surround the pericardial cavity.

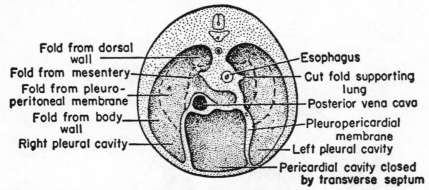

Figure 9–3. Anterior view of the diaphragm of a mammalian embryo to show the various folds of which it is composed. The heart and lungs have been removed to show the posterior portions of the two pleural cavities and the pericardial cavity (darker stippling). At this stage of development, the posterior surface of the pericardial cavity lies against the diaphragm, but it later separates from it in many mammals. Also, the caval fold has not formed, but the separation of the pericardial cavity from the diaphragm would result in such a fold. The pleuropericardial membrane (also called pericardiopleural membrane) represents the dorsal part of the primitive transverse septum. The lighter stippling extending from dorsal to ventral is the mediastinum. (From Romer, The Vertebrate Body. After Broman and Goodrich.)

The Development of the Digestive and Respiratory Systems

The major part of the digestive system develops from the embryonic **archenteron,** and hence is lined with endoderm. However, variable amounts of the front and hind ends of the digestive tract are formed by ectodermal invaginations — the **stomodeum** and **proctodeum,** respectively (Fig. 9–4). The former forms the mouth or **oral cavity**; the latter contributes to the cloacal region. At first the ectodermal invaginations are separated from the archenteron by plates of tissue, but these eventually break down. It is then difficult to determine precisely where ectoderm ends and endoderm begins, and the precise limits of the oral cavity (Fig. 9–4).

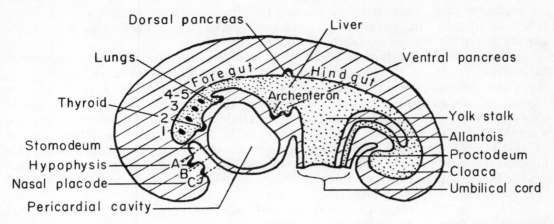

Figure 9–4. Diagrammatic sagittal section of a mammalian embryo to show the development of the digestive and respiratory systems. The points of entrance of the pharyngeal pouches are numbered. Lines *A*, *B*, and *C* indicate the comparative position of the mouth openings of an agnathous fish, *A;* a jawed fish not having internal nostrils, *B;* and fishes and tetrapods with internal nostrils, *C.*

For purposes of description, the archenteron may be divided into a **foregut** and **hindgut**. The former differentiates into the pharynx, esophagus, and, generally, a stomach; the latter differentiates into the intestinal region and much of the cloaca. The main divisions of the gut can be understood from the dissection of the adult, but most of the numerous outgrowths of the archenteron require further elucidation.

In all vertebrates, a series of **pharyngeal pouches** grow out from the side of the pharynx (Fig. 9–4). There are six of these in most fishes, fewer in tetrapods. The tissues between the pouches constitute the **branchial bars**, the first bar being anterior to the first pouch. The skeletal visceral arches, branchial muscles, certain nerves, and the aortic arches grow into these bars. In fishes, the endodermal pharyngeal pouches meet comparable ectodermal furrows, break through to the surface, and form the **gill slits** and **pouches** on whose walls the gills develop. The first pharyngeal pouch and furrow form the **spiracle** of such a fish as *Squalus;* the others form the five definitive gill pouches. The pharyngeal pouches do not normally break through in tetrapods (except for certain ones in larval amphibians), but are present embryonically nonetheless. The first gives rise to the **tympanic cavity** and **auditory tube** in all tetrapods except urodeles; the mammalian **tonsillar fossa**, containing the palatine tonsil, develops at the site of the second pouch; the other pouches give rise to certain glandular structures and then disappear.

The development of endocrine glands, and of glandlike structures, from the pharyngeal pouches varies considerably among vertebrates. A **thymus** develops as epithelial thickenings of the dorsal part of most (or all) of the pharyngeal pouches in fishes and urodeles, but from the ventral part of just the third and fourth pouches in mammals. These primordia may remain as distinct glandlike structures or coalesce into a single one. **Parathyroid glands** are absent in fishes, but present in all tetrapods. They develop as epithelial thickenings from the ventral part of the third and fourth pouches in urodeles, from the dorsal part of the third and fourth in mammals. Finally, **ultimobranchial bodies** develop in all vertebrates from the posterior face of the last pharyngeal pouch. There is evidence that they have a parathyroid-like function in fishes, but their function, if any, in tetrapods is unknown. All these structures lose their connection with the pharynx in the adult.

In addition to the lateral pharyngeal pouches, certain median evaginations arise from the floor of the pharynx. An endocrine **thyroid gland** grows out from the floor between the level of the first and second pouches. In fishes, the thyroid, although losing its connection with the pharynx, tends to remain in this anterior position, but it migrates a variable distance caudad in tetrapods. The thyroid may remain a median organ, or it may bifurcate and form a pair of glands. As discussed earlier (p. 24), the development of the thyroid of the lamprey from a part of the endostyle-like subpharyngeal gland of the Ammocoetes larva points to a possible homology between the vertebrate thyroid and the endostyle of lower chordates.

The **lungs** of sarcopterygian fishes and tetrapods arise as a median bilobed evagination from the floor of the pharynx just posterior to the pharyngeal pouches. The early primordia of the lungs resemble a pair of pharyngeal pouches in the embryos of some amphibians, and it is possible that they evolved from a ventrally displaced pair of posterior pouches.

It is believed that lungs are very primitive. The fossils of some placoderms show evidence of the presence of a lunglike structure, and vestiges of lungs are found in the living sarcopterygian and in primitive members of the actinopterygian fishes, where they serve to supplement gill respiration. In higher actinopterygians, however, the lungs have become transformed into the dorsally placed, hydrostatic **swim bladder**. In most cases, the swim bladder develops as an evagination from the roof of the posterior pharyngeal region, but in at least one fish it arises from the lateral wall. This may be an interesting transitional stage from the ventrally derived lungs.

No other outgrowths arise from the digestive tract of most vertebrates until the level of the front of the hindgut. At this point one finds the liver and pancreas. The **liver** arises embryologically as a prominent ventral diverticulum (Fig. 9-4), which, as mentioned earlier, grows into the ventral mesentery posterior to the heart, and by its expansion forms the ventral part of the transverse septum. During subsequent development, the liver grows caudally in the ventral mesentery, and in the adult remains connected to the septum (or diaphragm) only by a mesentery known as the **coronary ligament**. Functionally the liver is a very diverse organ. It secretes bile (a mixture of excretory products and the fat-emulsifying bile salts), it processes blood that is brought to it from the stomach and intestinal region by the hepatic portal system, and it serves as a storage center for glycogen, and as a site for numerous syntheses and chemical transformations of food and excretory products.

The pancreas is a gross organ in all gnathostomes, but in the agnatha it is represented only by scattered cells in the walls of the intestine and liver (p. 21). The organ arises embryologically from one or more intestinal outgrowths near the liver primordium. Frequently there is both a dorsal evagination (**dorsal pancreas**) and a ventral one (**ventral pancreas**). The latter is often paired, and is associated with the base of the liver anlage and hence with the future bile duct (Fig. 9-4). The ventral pancreas tends to work its way around the intestine, and to fuse and grow up into the dorsal mesentery with the dorsal pancreas. All the stalks of the primordia may persist as ducts, or certain ones may be lost in the adult. A ventral pancreatic duct can be recognized by the fact that it enters the intestine in common with the bile duct; a dorsal pancreatic duct, by its independent entrance on the opposite side of the intestine.

Most of the pancreas secretes digestive enzymes which are discharged into the intestine and act on proteins, carbohydrates, fats, and nucleic acids. But little islands of endocrine tissue (the **islets of Langerhans**) are scattered among the exocrine cells. These islets produce insulin, which is vital in carbohydrate metabolism.

More caudally along the hindgut there is, in the embryos of amniotes and certain fishes, a **yolk stalk** connecting with the **yolk sac** (Fig. 9–4). The yolk stalk and sac become relatively smaller as the embryo grows, and are lost by the adult stage.

A final major outgrowth is the **urinary bladder**, present in most tetrapods. It develops from the embryonic cloaca near the posterior end of the hindgut. In the embryos of amniotes, this structure expands considerably and extends beyond the limits of the embryo as the **allantois**—an extraembryonic membrane which serves for excretion and respiration in the embryos of reptiles and birds, and for vascularizing the fetal portion of the placenta in eutherian mammals.

FISHES

The coelom and the digestive and respiratory systems of *Squalus* are reasonably good examples of the condition of these structures in primmitive jawed fishes. Several qualifications are necessary, however. In respect to the pharyngeal region, the first gill pouch would not have been reduced to a spiracle in such a fish as a placoderm. Moreover, primitive bony fishes ancestral to tetrapods, and possibly their placoderm ancestors, had lunglike outgrowths from the back of the pharynx. In respect to the more caudal parts of the digestive tract, a stomach was probably absent in very primitive fishes. Lower chordates are, and early vertebrates may have been, filter feeders, feeding more or less continuously on minute food particles. Such types need no stomach, but, with the shift from this mode of feeding to one in which large chunks of food are taken irregularly, the stomach is advantageous for temporary storage and preliminary physical and chemical treatment of food.

Pleuroperitoneal Cavity and Its Contents

(A) Body Wall and Pleuroperitoneal Cavity

It is desirable in the fish to study the posterior parts of the digestive system before the mouth and pharyngeal area. The **pleuroperitoneal**, or **perivisceral**, **cavity** should be opened by a longitudinal incision slightly to one side of the midventral line, preferably the right side, if the muscles were dissected on the left. Extend the incision as far forward as the pectoral girdle, and as far caudad as the base of the tail. In doing the latter, cut through the pelvic girdle and continue posteriorly on one side of the cloacal aperture. Make a transverse incision on each side that extends from about the middle of the longitudinal cut to the lateral line. You now have four flaps of body wall that can be turned out to expose the body cavity and its contents, but do not break tissue extending between the anterior part of the liver and the ventral body wall.

Note the layers of the body wall through which you have cut. The outermost is the **skin**. This is followed by a thin and inconspicuous layer of **connective tissue** comparable to the fascia of higher vertebrates, the **hypaxial musculature**, and finally the shiny **epithelium** lining the pleuroperitoneal cavity. That portion of the coelomic epithelium adjacent to the body wall musculature is the **parietal peritoneum**; that portion covering the viscera, the **visceral peritoneum**; and, finally, that portion that extends from the body wall to the viscera contributes to the **mesenteries**. As a result of the way they develop, mesenteries consist of a double layer of epithelium between which lie the vessels and nerves to the viscera imbedded in a certain amount of connective tissue (Fig. 9–1).

In most vertebrates the coelom is a closed cavity having no direct communication with the outside. But in primitive vertebrates, including the dogfish, it may communicate with the exterior by a pair of **abdominal pores** (Fig. 11–3, p. 318). These can be found by probing the most posterior recess of the pleuroperitoneal cavity beside the posterodorsal portion of the **cloaca** (the posterior chamber receiving the intestine and genital ducts), on the side of the body that is still intact. Each opens through the lateral wall of the cloacal aperture, but sometimes the lips of the pores have grown together. Their significance is obscure, but they may serve to eliminate excess coelomic fluid. Some have tried to homologize them with the genital pores of cyclostomes, but the relationships are somewhat different.

(B) Visceral Organs

A large **liver** with a pair of long, pointed lobes occupies most of the anteroventral portion of the pleuroperitoneal cavity. You may cut off the ends of these lobes, but do not injure blood vessels going to the liver, or the bile duct going from the liver to the intestine. The **bile duct** accompanies these vessels for most of its length, but separates from the vessels near the beginning of the intestine (Fig. 9–5). Spread the lobes of the liver apart, and you will see the **esophagus** and **stomach** in a more dorsal position. Both have about the same diameter, so there is no external line of demarcation separating them. If a constriction is seen, it represents a peristaltic contraction fixed at death and during preservation of the animal, not a line of demarcation The posterior end of the stomach curves anteriorly and gives the organ a J shape. The digestive tract then turns posteriorly and forms the straight, **valvular intestine** that continues to the **cloaca**. If part of the intestine has been everted through the cloaca, it must be pulled back into the body cavity.

The large, triangular-shaped organ applied to the posterior end of the stomach (at the region where the stomach turns forward) is the **spleen** — an organ related to the circulatory system. The elongate **pan-**

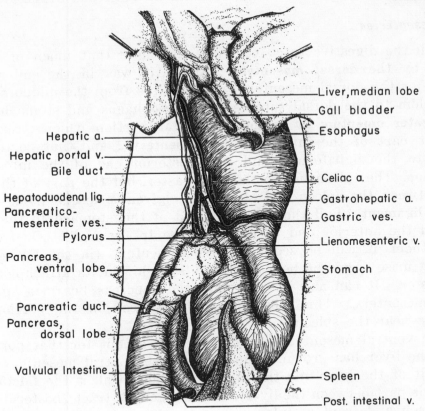

Liver, median lobe
Gall bladder
Esophagus

Hepatic a.
Hepatic portal v.
Bile duct

Celiac a.
Gastrohepatic a.

Hepatoduodenal lig.
Pancreatico-
mesenteric ves.
Pylorus

Gastric ves.

Lienomesenteric v.

Pancreas,
ventral lobe

Stomach

Pancreatic duct
Pancreas,
dorsal lobe

Valvular Intestine

Spleen

Post. intestinal v.

Figure 9-5. Ventral view of the liver, stomach, pancreas, and associated organs of *Squalus*. The left and right liver lobes have been pulled laterally.

creas extends from the right side of the spleen anteriorly toward the beginning of the intestine. An oval-shaped lobe of the pancreas is applied to the surface of the intestine near its junction with the stomach. The finger-like organ extending dorsally from the posterior end of the intestine is the **digitiform**, or **rectal, gland**. It has been shown by Burger and Hess (1959) to be a salt excreting gland, and Oguri (1964) reports that it undergoes regressive changes in sharks that enter fresh water.

Other organs within the pleuroperitoneal cavity are parts of the urogenital system. They will be considered in more detail later, but should be identified at this time. A pair of large **gonads** (**testes** or **ovaries**) lie one on either side of the anterior end of the esophagus and stomach. If your specimen is a mature female, a pair of prominent oviducts will be seen dorsal to the gonads and continuing to the cloaca. Their posterior ends are enlarged, greatly so in pregnant females. The paired **kidneys** are represented by the long bands of dark material located dorsal to the parietal peritoneum on either side of the middorsal line of the pleuroperitoneal cavity. They are very long, and are not uniform in diameter, for the posterior end of each is much wider than the anterior. If the specimen is a mature male, a large, twisted excretory duct (the **archinephric duct**) will be seen on the ventral surface of each one.

(C) Mesenteries

Pull the digestive tract ventrally and note that much of it is supported by the **dorsal mesentery**. All of it was in the embryo. That portion of the dorsal mesentery which passes from the middorsal line of the coelom to the dorsal surface of the esophagus and stomach is called the **greater omentum** or **mesogaster**; that portion which passes to the anterior part of the intestine, the **mesentery**[13]; and that portion which passes to the digitiform gland and posterior end of the intestine, the **mesocolon**. The spleen lies in the mesogaster, but the part of this mesentery between the spleen and stomach is given a special name — **gastro-splenic ligament**. Pull the posterior end of the stomach to the animal's left and the anterior end of the intestine to the right. On looking between them, it will be seen that the mesentery (in the limited sense) does not arise from the body wall, but has shifted its attachment to the mesogaster. It can also be seen that the pancreas lies in a special fold of the mesentery. This relationship shows best on the portion of the pancreas near the spleen.

The ventral mesentery has disappeared except for that portion into which the liver has grown. The part of the ventral mesentery between the front of the liver and the midventral body wall is the **falciform ligament**; the part between the liver and the digestive tract (posterior portion of stomach and front of intestine) is the **lesser omentum** or **gastrohepatoduodenal ligament**. The latter mesentery, which contains the bile duct and the blood vessels going to the liver, is unusually complex in the dogfish. Near the liver, the mesentery is a unit, but it divides near the digestive tract. Part of it (the **gastrohepatic ligament**) passes into the angle formed by the bending of the stomach, and part of it (the **hepatoduodenal ligament**) carries the bile duct to the beginning of the intestine. A part of the dorsal mesentery supporting the pancreas extends between these two limbs of the lesser omentum and brings the larger vessels to the omentum.

Subsidiary mesenteries support certain of the genital organs. Each testis is supported by a **mesorchium**, each ovary by a **mesovarium**, and each oviduct in a mature female by a **mesotubarium**. What is the relation of these mesenteries to the dorsal mesentery?

(D) Further Structure of the Digestive Organs

Cut open the esophagus and stomach by a longitudinal incision that extends all the way to the intestine. Remove the contents, if any, and

---

[13] Unfortunately, the term mesentery is used in two ways — in a broad sense for all membranes passing to the viscera, and in a limited sense for the mesentery of most of the intestine.

wash out these organs. (One can often determine the feeding habits of animals by analysis of the stomach contents.) The esophagus can be recognized internally by the **papillae** in its lining; the stomach by longitudinal folds called **rugae**. If the stomach was greatly distended, the rugae will have stretched out and most of the lining will be smooth. The anterior end of the stomach, which would be adjacent to the esophagus and often near the heart, is called the **cardiac region**; the main part, the **corpus**, or **body**; and the portion that turns forward, the **pyloric region**. Note that the last has a thicker muscular wall, especially just before the intestine, where it forms a muscular valve, the **pylorus** or **pyloric sphincter**. It should be pointed out that these topographic regions in different vertebrates may or may not correspond with glandular regions given the same names. They do not in most lower vertebrates, for cardiac glands are absent. The surface of the stomach along which the mesogaster and gastrosplenic ligament attach is called its **greater curvature**; the opposite surface, the **lesser curvature**. The greater curvature represents the original dorsal side of the stomach.

The bile and pancreatic duct, which will be seen presently, enter the anterior end of the valvular intestine; the digitiform gland, the posterior end. Aside from this slight modification at either end, the intestine is undifferentiated. All of it contains a complex spiral fold, the **spiral valve**, which is similar to that of other primitive fishes (Fig. 9–6). The line of attachment of the valve shows as the spiral line on the surface of the intestine. If a special preparation of the valve is not available, part of it can be seen by cutting a tangential slice from the intestine wall. The spiral valve slows down the passage of food through the intestine, thus increasing digestive efficiency, and it also increases the absorptive area.

It is difficult to compare the intestine of primitive fishes with the small and large intestine of tetrapods. The anterior end of the fish intestine, which receives the bile and pancreatic ducts, would be roughly comparable to the duodenum; most of the remainder, to the rest of the small intestine; and the posterior end receiving the digitiform gland, to part, at least, of the large intestine. But in the absence of clear lines of separation between these potential regions, it is best to think of the primitive fish intestine as a unit — the valvular intestine.

Examine the liver in more detail. It consists of a long **right** and **left lobe**, and a smaller **median lobe** containing the elongate, thin-walled **gall bladder**. A part of the bladder shows on the surface, but most is imbedded within the lobe. Expose it by scraping away liver tissue, and find out where the bile duct unites with it. In the dogfish, the bile leaves the liver through a number of inconspicuous **hepatic ducts** that enter the gall bladder; from here it goes to the intestine through the **bile duct**. The point of entrance of the bile duct into the lumen of the intestine is

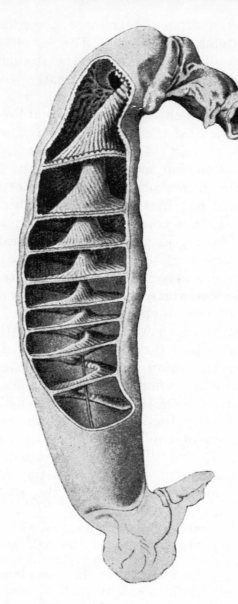

Figure 9–6. Valvular intestine of the skate, *Raja*, cut open to show the spiral valve. A bristle is shown passing through the central lumen of the intestine. (From Paul Mayer.)

some distance caudad to its point of attachment on the intestine. The duct can be traced through the wall by removing the visceral peritoneum and longitudinal muscle layer.

As in other vertebrates, the dogfish's liver is an important site for the metabolism and storage of food products. Glycogen is present, but an unusual feature is that much of the food is stored as oil. This lowers the specific gravity of the fish and makes the animal more buoyant.

The pancreas is divided into two lobes, an elongate **dorsal lobe** that extends from the spleen anteriorly, passing dorsal to the pylorus, and an oval-shaped **ventral lobe** adhering to the beginning of the intestine. The two are connected by an isthmus of pancreatic tissue. Both lobes drain by a common **pancreatic duct** that leaves the ventral lobe and travels obliquely caudad in the wall of the intestine for a short distance before entering the lumen (Fig. 9-5). To see the duct, it is necessary to cut the attachment of the mesentery, and carefully remove the visceral peritoneum and longitudinal muscle from the wall of the intestine adjacent to the posterior part of the ventral lobe.

Relationships at the front of the intestine are misleading, for this portion of the digestive tract has undergone a rotation of nearly 180 degrees during development. This has brought about an apparent reversal of normal relationships. In the adult, the bile duct enters on what would at first seem to be the dorsal surface of the intestine, and the dorsal mesentery appears to attach along the ventral surface. The pancreas of the dogfish is entirely a dorsal pancreas; however, its duct enters on what would seem to be the original lateral surface of the intestine.

Pericardial Cavity

The second division of the coelom, the **pericardial cavity**, is located far forward in fishes, for it lies just anterior and dorsal to the pectoral girdle and deep to the posterior hypobranchial musculature (Fig. 10-7, p. 279). To get at it, continue your original ventral incision forward through the pectoral girdle and the posterior hypobranchial musculature. Veer toward the midventral line as you go, but do not cut the falciform ligament. Additional transverse cuts will have to be made just anterior to the pectoral girdle.

Spread open the flaps thus formed and study the cavity. The only organ within it is the **heart**. That portion of the coelomic epithelium adjacent to the tissues surrounding the cavity is the **parietal pericardium**; that covering the heart, the **visceral pericardium**. The heart developed embryonically in the ventral mesentery beneath the pharynx (Fig. 9-1, *A*) and at one time was supported by parts of the ventral mesentery, termed mesocardia. The mesocardia disappear in the adult, and the heart remains attached only at its anterior and posterior ends.

The vertical septum separating the pericardial and pleuroperitoneal cavities is the **transverse septum**. You can now see that the liver is attached to the posterior face of the septum by the **coronary ligament**. This ligament is continuous ventrally with the anterior part of the falciform ligament. As in many other primitive fishes, the separation of the pericardial and pleuroperitoneal cavities is not complete in the dogfish, for a **pericardioperitoneal canal** connects the two ventral to the esophagus. It will be seen later after the heart is studied (p. 279).

Oral Cavity, Pharynx, and Respiratory Organs

The pharynx should be opened on the same side of the body that the pleuroperitoneal cavity was opened. With a strong pair of scissors, cut through the angle of the mouth and continue posteriorly through the external gill slits and visceral arches. Then extend your cut medially and caudally through the pectoral girdle and body wall to intersect your previous longitudinal incisions in the body wall and esophagus. Go ventral to the pectoral fin and dorsal to the anterior end of the liver. It is not necessary to cut the liver. Swing open the floor of the oral cavity and pharynx. If the esophagus has everted into the pharynx, it must be pulled back.

The demarcation between the **oral**, or **buccal**, **cavity** and the **pharynx** is not clearly definable in the adult (p. 226), but the pharynx is approximately that portion of the digestive tract into which the gill slits enter. The pharynx, of course, leads to the narrower esophagus. A tongue-like structure supported by the hyoid arch lies in the floor of the mouth and front of the pharynx. It is not a true tongue, as no muscle extends up beneath its epithelium, but sometimes it is referred to as a **primary tongue**, for it is destined to contribute to the true tongue of tetrapods (Fig. 9–7).

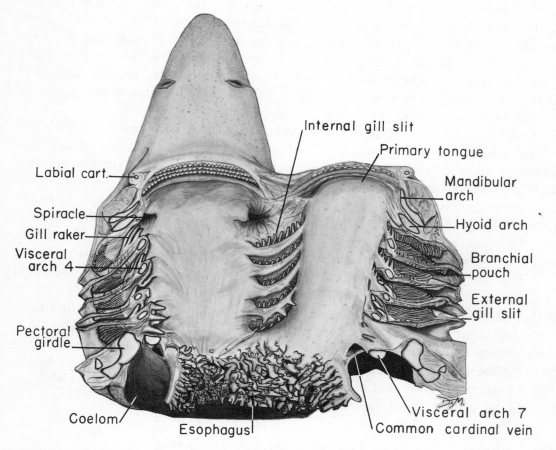

Figure 9–7. Oral cavity and pharynx of *Squalus*. The floor of the mouth and pharynx has been swung open to the right.

The entrance of the **spiracle** will be seen at the front of the pharynx roof, and this is followed by the five definitive **internal gill slits**. The homologue of the spiracle was a complete gill slit in placoderms. A number of papilla-like **gill rakers** project across the internal gill slits and act as strainers. Note that each internal gill slit leads into a large **gill** or **branchial pouch**. The portion of the pouch lateral to the gills is the **parabranchial chamber** (Fig. 9–8), and it opens to the surface by the **external gill slit**. The tissue between the gill slits and pouches constitutes the **interbranchial septa**. The outermost portion of each septum is thin and flaplike, and constitutes a **valve** that can close and open the external gill slits. The actual gills are composed of a number of plate-like **primary lamellae**, which are attached to the surface of the septa. Examine the primary lamellae with a hand lens, and note that each bears many small, closely packed, **secondary lamellae** extending perpendicularly from the surface of the primary lamella. All of the lamellae on one surface of an interbranchial system constitute a half gill, or **hemibranch**; those on both the anterior anb posterior surfaces of the same septum, a complete gill, or **holobranch**.

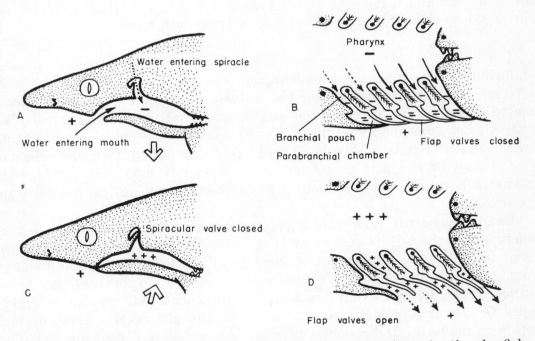

Figure 9–8. Diagrams to show the mechanics of respiration in the dogfish as seen in lateral views (*A* and *C*) and in frontal sections (*B* and *D*) of the pharynx. *A* and *B*, inspiration; *C* and *D*, expiration. Relative pressures are indicated by + and −. Open arrows indicate the direction of movement of the floor of the mouth and pharynx, solid arrows the course of the current of water entering the mouth, and dotted arrows the course of the water entering the spiracle. (Slightly modified after Hughes, Comparative Physiology of Vertebrate Respiration, Harvard University Press.)

Dr. G. M. Hughes of Cambridge University has recently determined the biomechanics of respiration in dogfish (Fig. 9–8). The most active phase is expiration. Contraction of the branchial muscles compresses the buccopharyngeal cavity and branchial pouches and expels water through the gill slits. During inspiration, these chambers are expanded, largely by the elastic recoil of the branchial skeleton, although the lowering of the floor of the mouth and pharynx by the contraction of hypobranchial muscles does play a part. Internal pressure is reduced relative to the external pressure, the lowest pressure being found within the parabranchial chambers. The flap valves over the external gill slits close. Water enters the mouth and spiracle and follows the pressure gradient across the gills and into the parabranchial chambers. During expiration, a pressure gradient is developed that closes the spiracular valves and forces water out the external gill slits. Water that entered the spiracle leaves through the more anterior gill slits, whereas water that entered the mouth leaves through the more posterior slits.

The details of the flow of water across the gills is unknown, but it probably passes between the secondary lamellae in a direction counter to the flow of blood, even though such a flow would be hindered to some extent by the well developed interbranchial septum. Teleosts have lost this septum, and are able to extract a greater percentage of the oxygen from the water than cartilaginous fishes. Satchell (1960) has further shown that the heartbeat in the dogfish is coordinated with the respiratory movements in such a way that there is a rapid flow of blood through the gills at the time that water is expelled across them. This would increase the efficiency of gas exchange still more.

Cut open the spiracle on the side you have been dissecting. A minute hemibranch, known as the **pseudobranch**, can be found on the valve-like flap (**spiracular valve**, p. 32) on the anterior wall of the spiracle. Since aerated blood passes through the pseudobranch it is often regarded as vestigial, but some believe that it has an accessory respiratory function. In some marine teleosts its homologue contains secretory cells believed to be a part of the salt excretory mechanism. It may also produce or activate a hormone, for its removal in certain bony fishes causes the skin to darken.

Examine the cut surface of a representative holobranch and note the structures of which it is composed (Fig. 9–9). A supporting **visceral arch** lies at its base. (Be sure you can identify all these arches — mandibular, hyoid, and five branchial.) You may also be able to see one or more cartilaginous **gill rays** extending into the interbranchial septum at right angles to the arch. Cartilage also supports the **gill rakers**. Much of the septum is made up of branchial muscles. An **adductor** lies medial to each arch, and an **interbranchial** and a **superficial constrictor** extend out into the septum. Close examination will also reveal several blood vessels. An **afferent branchial artery** lies near the middle of the septum just lateral to the visceral arch. It brings unoxygenated blood from the heart and ventral aorta to the capillaries in the gills, and probably is not injected. An **efferent branchial artery**, which will be injected if the arteries have been injected, lies at the base of the primary lamellae on each surface of the septum.   These vessels, of which there are two in a typical septum, drain the gill capillaries and carry oxygenated blood to the dorsal aorta.   Also recall (p. 192) that each septum would contain a **pretrematic** and **posttrematic** branch of a cranial nerve, but these are hard to find in a section. Which nerve, or nerves, are associated with the septum you have been studying?

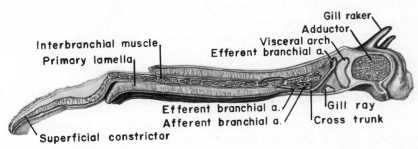

Interbranchial muscle
Primary lamella

Gill raker
Adductor
Visceral arch
Efferent branchial a.

Efferent branchial a.
Afferent branchial a.

Gill ray
Cross trunk

Superficial constrictor

Figure 9-9.   A frontal section through a holobranch of *Squalus*.   The anterior surface is toward the top of the page. Twice natural size.

A thyroid gland, a series of thymus bodies, and an ultimobranchial body are derived from the pharynx, but are difficult to see. The thyroid may have been seen during the dissection of the hypobranchial muscles (p. 115), and can be seen again during the dissection of the branchial blood vessels (p. 271).

PRIMITIVE TETRAPODS

The coelom of primitive tetrapods is very similar to that of fishes except that, with the beginning of neck formation, the pericardial cavity has moved caudad a short distance. This causes the transverse septum to assume a somewhat oblique position.

More conspicuous changes are seen in the respiratory and digestive systems. In respect to respiration, tetrapods retain the internal nostrils and elaborate upon the lungs of their piscine ancestors, but, excepting larval amphibians, lose the gills. Larval amphibians have gill slits and gills of two types. **External gills**, which are also present in some fish larvae, develop as outgrowths from the neck surface near the gill slits in urodele and frog larvae. These are later replaced in frog larvae by **internal gills** that develop closer to the gill arch, hence may be homologous to the internal gills of adult fishes. At metamorphosis, they too are lost. The lungs and methods of air exchange between the lungs and outside (a pumping action of the oral cavity and pharynx) are very primitive in amphibians, and most members of this class supplement pulmonary respiration with some other form such as cutaneous or buccopharyngeal.

Major changes in the digestive tract of primitive tetrapods are the evolution of small **oral glands** and a muscular **tongue**, both being correlated with the problem of food manipulation in a terrestrial environment. The amphibian tongue consists of little more than the primary tongue of fishes overlying the hyoid apparatus, plus a swelling (**gland field**) that develops between the hyoid and mandibular arches. The whole is invaded by certain prehyoid hypobranchial muscles (p. 122). The intestinal region has changed in that the primitive spiral valve has been lost. The resulting reduction in internal surface area is compensated for by other folds and an increase in the length of the intestine. A differentiation into **small** and **large intestines** also occurs. Finally a **urinary bladder** has evolved as an outgrowth from the ventral surface of the cloaca, but this is related to the urinary system.

Necturus illustrates the early tetrapod stage well save for the retention of many larval features, the major ones being a vertical transverse septum, the presence of gill slits and external gills, and a poorly formed tongue. Moreover, all urodeles differ from most other tetrapods in lacking the auditory tube and tympanic cavity.

Pleuroperitoneal Cavity and Its Contents

(A) *Body Wall and Pleuroperitoneal Cavity*

If your specimen of *Necturus* has been injected, a partial incision will have been made through the body wall on one side of the midventral line (probably the right side). Continue this incision anteriorly to the pectoral girdle, and posteriorly through the pelvic girdle and along one side of the cloacal aperture. The layers of the body wall through which you have cut are similar to those of *Squalus*, i.e., **skin**, **connective tissue**, **hypaxial muscles**, and **parietal peritoneum**. How many muscle layers were cut through? The part of the coelom opened is, as in fishes, a **pleuroperitoneal cavity**. It is lined with the parietal peritoneum, and the viscera are covered with **visceral peritoneum**.

(B) *Visceral Organs*

The **liver** is the largest of the visceral organs, and lies in an anteroventral position (Fig. 9–10). Notice that it is displaced toward the right side of the cavity. It is not as obviously subdivided into lobes as it is in *Squalus*. A portion of the ventral mesentery, the **falciform ligament**, extends from the midventral line of the pleuroperitoneal cavity to attach along the entire length of the ventral surface of the liver. It will have to be cut to approach the left side of the cavity, but cut it in such a way that you can later reconstruct the major veins that pass through it. Pull the left side of the body wall and the liver apart, and you will see the elongate **stomach**. Notice that it is straight as it is in larvae rather than J-shaped as it is in fully metamorphosed amphibia. The **esophagus** is a short connecting piece between the pharynx and stomach, and will be seen more clearly later. The **spleen** is the elongate, oval organ attached to the left side of the stomach. The long, finger-like left **lung** lies dorsal to the spleen; a similarly shaped right lung lies in a comparable position on the other side of the body.

The **small intestine** begins at the posterior end of the stomach, and

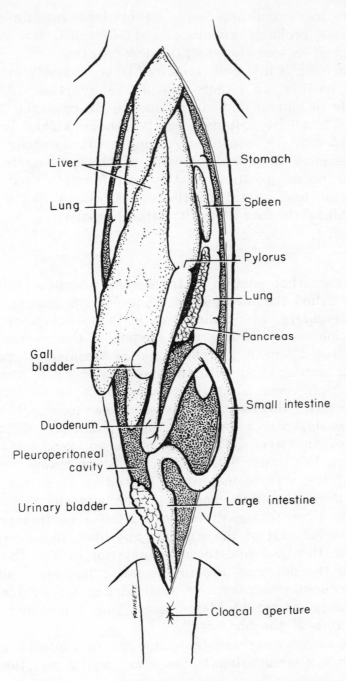

Figure 9–10. Ventral view of the digestive tract and associated organs of *Necturus*. Part of the liver has been reflected, and is seen in a dorsal view.

makes one anterior loop (the **duodenum**) before passing into a number of convolutions. An irregular-shaped **pancreas** lies in the vicinity of this anterior loop. Part of it is in contact with the intestine, part with the liver, and one part (the **tail**) passes dorsal to the stomach and nearly reaches the posterior tip of the spleen. Shortly before entering the

cloaca, the intestine widens and forms a short **large intestine**. The **urinary bladder**, which is probably collapsed and shriveled, lies ventral to the large intestine. It enters the cloaca independently.

If your specimen is a female, a pair of large, coarsely granular **ovaries** lie in a dorsal position on either side of the intestine. A pair of convoluted **oviducts** lie dorsal and lateral to the ovaries, and extend nearly the length of the pleuroperitoneal cavity. Each **kidney** is an elongate, dark organ that may be seen by spreading apart an ovary and oviduct.

If your specimen is a male, a pair of large, elongate, oval-shaped **testes** lie in a dorsal position in the anterior intestinal region. The **kidneys** lie dorsal and lateral to the testes. Each has a conspicuous, convoluted **archinephric duct** along its lateral border.

(C) Mesenteries

As in the fish, that portion of the dorsal mesentery which passes to the stomach is called the **mesogaster**; that portion passing to the small intestine the **mesentery**, in a limited sense; and that portion passing to the large intestine the **mesocolon**. The portion of the mesogaster between the spleen and the stomach is called the **gastrosplenic ligament**. Most of the left lung is connected with the mesogaster by a short **pulmonary ligament**. The right lung is also supported by a comparable but wider pulmonary ligament. However, an accessory mesentery (the **ligamentum hepatocavopulmonale**) passes from the right pulmonary ligament to the dorsal surface of the liver. The large posterior vena cava approaches the liver through the posterior margin of this mesentery.

The part of the ventral mesentery extending from the liver to the body wall (**falciform ligament**) has been seen and cut. Two other parts extend from the liver to the digestive tract — a **gastrohepatic ligament** between the anterior part of the stomach and liver, and a **hepatoduodenal ligament** between the liver and the most anterior loop of the small intestine. A part of the pancreas lies in the latter ligament, but most of it extends into the dorsal mesentery. A final part of the ventral mesentery, the **median ligament of the bladder**, passes from the urinary bladder to the midventral line of the body wall.

The genital mesenteries are the same as in *Squalus,* i.e., a **mesorchium** to the testis, a **mesovarium** to the ovary, and a **mesotubarium** to the oviduct.

(D) Further Structure of the Digestive Organs

Cut open the stomach and wash it out if necessary. Its lining is thrown into a number of irregular, longitudinal folds (**rugae**). As in other vertebrates it can be divided into several gross regions (a **cardiac region** next to the esophagus, a central **corpus**, and a posterior **pyloric region**), but these are not sharply demarcated. The pyloric region ends in a **pyloric sphincter**.

Cut open the intestine at several points and notice that a spiral valve is lacking, but that its internal surface area is increased through many small, wavy, longitudinal folds (**plicae**).

Most of the bile is drained from the liver by several **hepatic ducts** that unite to form a **common bile duct** that enters the duodenum (Fig. 9-11). Some of the bile backs up into the **gall bladder**, where it is temporarily stored, through a **cystic duct**. The gall bladder can be found on the dorsal surface of the liver near the apex of the anterior loop of the intestine, but most of the ducts are imbedded within pancreatic tissue. Some can be found by careful dissection in this region.

The pancreas of urodeles is formed from a dorsal and a pair of ventral primordia. All fuse in the adult, but each retains its duct (Fig. 9-11). The pancreatic ducts cannot be found by gross dissection.

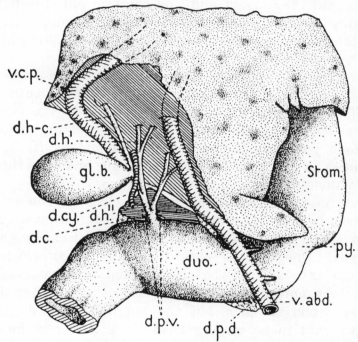

Figure 9–11. Ventral dissection of the bile and pancreatic ducts of *Salamandra*. Part of the liver and pancreas have been cut away. Abbreviations: *d.c.*, common bile duct; *d.cy.*, cystic duct; *d.h.'*, *d.h."*, hepatic ducts; *d.h-c.*, hepato-cystic duct; *d.p.d.*, dorsal pancreatic duct; *d.p.v*, ventral pancreatic ducts; *duo.*, duodenum; *gl. b.*, gall bladder; *stom.*, stomach; *py.*, pylorus; *v.abd.*, ventral abdominal vein; *v.c.p.*, posterior vena cava. (From Francis, The Anatomy of the Salamander, Oxford University Press.)

Pericardial Cavity

The second division of the coelom, the **pericardial cavity**, is located further forward in *Necturus* than in most amphibians, for it still occupies the fish position anterior to the pectoral girdle and deep to the posterior hypobranchial musculature. Expose it by removing the musculature in this region. It is lined with **parietal pericardium** and contains only the **heart** which is covered with **visceral pericardium**.

Cut through that portion of the ventral body wall lying between the pericardial cavity and the anterior end of your incision into the pleuroperitoneal cavity. The two divisions of the coelom are completely separated by a small, vertical **transverse septum**. A **coronary ligament** runs from the posterior surface of the transverse septum to the anterior end of the liver. A large vein (the posterior vena cava) lies within it.

Oral Cavity, Pharynx, and Respiratory Organs

Since *Necturus* is neotenic, its major respiratory organs are not the lungs, but the three pairs of larval **external gills** arising from the back of the head. Spread them apart and find the two **gill slits** between their bases. If a living specimen is available in an aquarium, notice the great vascularity of the numerous **gill filaments**, and the way in which the gills are waved back and forth.

Open the oral cavity and pharynx on the side on which you have been dissecting by cutting through the angle of the mouth, posteriorly through the gill slits, the side of the neck lateral to the pericardial cavity, the ventral portion of the pectoral girdle, and on to intersect the incision by which you opened the pleuroperitoneal cavity. The deeper part of the incision should pass dorsal to the lung and into the esophagus. Swing open the floor of the mouth and pharynx.

As explained in the introduction to this chapter, the breakdown of the plate of tissue between the stomadeum and archenteron makes it difficult to draw a sharp line between the anterior **oral cavity** and the more posterior **pharynx**. One merges with the other. Posteriorly, however, the pharynx leads into a somewhat constricted, short passage (the **esophagus**), which soon enters the stomach. There is no sharp demarcation between esophagus and stomach, but the longitudinal folds in the lining of the esophagus tend to be smaller than those in the stomach.

A **tongue-like fold** supported by the hyoid arch is located in the floor of the mouth (Fig. 9-12). It has developed little beyond the primary tongue of fishes, but a few hypobranchial muscle fibers enter its base (p. 122). The gland field of the tongue of adult amphibians (p. 239) is barely developed. The third to fifth visceral arches lie posterior to the hyoid arch, and can be palpated. Between which do the gill slits lie? With which slits of *Squalus* are these slits homologous? Vestiges of **gill rakers** can be seen on the pharyngeal surface of the gill slits.

Necturus seldom breathes air, but the structures needed for pulmonary respiration are present. Air can enter the oral cavity and pharynx by way of the external nares, nasal passages, and internal nares. If a nasal

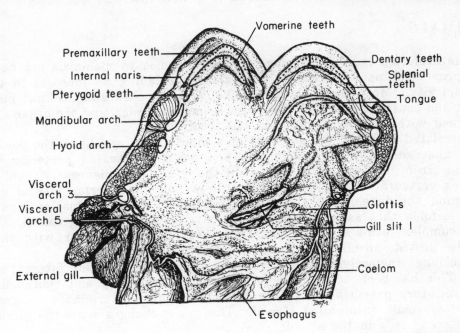

Figure 9–12. Oral cavity and pharynx of *Necturus*. The floor of the mouth and pharynx has been swung open to the right.

passage was not dissected in connection with the sense organs (p. 172), expose it now. An **internal naris** can be seen in the roof of the mouth lateral to the most posterior, and shortest of the tooth rows (pterygoid teeth). The air leaves the pharynx through the **glottis** — a short longitudinal slit in the center of the floor of the pharynx about the level of the posterior gill slits. The glottis leads into a median **laryngotracheal chamber** from whose posterior end a pair of openings (the **bronchi**) pass to the lungs. These structures can best be seen by cutting one lung near its anterior end, passing a blunt probe forward to the glottis, and then cutting open the floor of the esophagus and pharynx along the course of the probe. A pair of small cartilages (**lateral cartilages**), which are difficult to see but may be felt, support the wall of the laryngotracheal chamber (Fig. 4–9). They are probably derived from the sixth and seventh visceral arches (p. 61). Cut open more of the lung, and observe that it is an empty sac with a smooth internal surface. In adult amphibians, the internal surface tends to be increased by pocket-like folds. If possible, examine a demonstration of a lung of a frog.

A thyroid and a series of parathyroids and thymus glands develop from the pharynx, but are difficult to find. The thyroid may be seen during the dissection of the arteries (p. 286).

MAMMALS

The evolution from amphibians through reptiles to mammals has been one of improvement upon terrestrial adaptations, including an increase in general body activity. Most mammals are active, warm-blooded animals that maintain a high and constant rate of metabolism (**homoiothermic**), and the organ systems concerned with metabolism are adapted accordingly.

Correlated with the development of a mobile neck and head, the pericardial cavity undergoes a caudad migration and becomes situated posterior to the pectoral girdle in the thoracic region of the body. The pleuroperitoneal cavity of lower vertebrates becomes divided into a pair of anterior **pleural cavities** containing the lungs, and a posterior **peritoneal cavity**. Muscles invade the coelomic folds that separate the pleural cavities from the peritoneal, and the whole complex forms a diaphragm whose movements, together with those of the ribs, provide an efficient means of ventilating the lungs.

The lungs themselves have increased tremendously in their internal surface area. This has been accomplished through the branching and rebranching of the respiratory passages within the lungs, so that all terminate in grapelike clusters of small, thin-walled **alveoli**. The gas exchanges with the circulatory system occur only in the alveoli.

The more anterior respiratory passages have also become modified. With the evolution of a distinct neck, the laryngotracheal chamber of amphibians becomes divided into an anterior **larynx** from which a **trachea** descends to the lungs. And, as explained in connection with the skeleton, the evolution of a bony secondary palate and fleshy soft palate separates a respiratory passage from the original oral cavity and anterior part of the pharynx. This permits simultaneous respiration and manipulation of food within the mouth. Food and air passages cross only in the posterior part of the pharynx, and, even here, food is normally prevented from entering the larynx through the evolution of a flaplike **epiglottis**.

Many changes also occur in the digestive system. Conspicuous **salivary glands** evolve from certain of the small oral glands of primitive tetrapods. In addition to providing secretions for lubricating the food, these glands secrete a digestive enzyme, **ptyalin**, that acts on carbohydrates.

The primitive tongue becomes a prominent muscular organ, for a pair of large, **lateral lingual swellings** are added anterior to the primary tongue and gland field (**tuberculum impar** of mammalian embryology), and the invasion of the organ by hypobranchial muscles has continued. The sensory innervation of the tongue correlates with the nerves of those visceral arches above which it develops. Thus the anterior part of the tongue, which develops from the lateral lingual swellings overlying the mandibular arch, is innervated by the trigeminal nerve; the posterior part (primary tongue) overlying the hyoid apparatus, by the facial and glossopharyngeal nerves; and the intermediate portion of the tongue (tuberculum impar), by both the trigeminal and facial nerves. But the muscles of the tongue, since they belong to the hypobranchial rather than to the branchial group, are innervated by the hypoglossal nerve.

The most notable change in the intestinal region is a great increase in internal surface area. This is accomplished in part through an increase in the length of the intestine, and in part through the evolution of numerous, minute, finger-like **villi**. Another major change is the division of the cloaca into dorsal and ventral portions. The ventral portion contributes to the urogenital passages; the dorsal portion forms the **rectum**. The mammalian rectum is therefore not homologous to the large intestine of lower vertebrates in which a cloaca is still present.

Superimposed upon these general features are numerous modifications of the digestive tract correlated with the diverse diets and modes of life of the various mammalian groups. The digestive tract of the cat is a good example of the more primitive carnivore pattern, while that of the rabbit is illustrative of one of several herbivore groups.

Digestive and Respiratory Organs of the Head and Neck

(A) Salivary Glands

The salivary glands of your cat or rabbit may be studied on the same side of the head that was used for the dissection of the muscles, provided the glands were not injured. If they were destroyed, carefully remove the skin on the opposite side of the head overlying the cheek, throat, and side of the neck ventral to the pinna. The superficial integumentary muscles (platysma and facial muscles, p. 129) must also be taken off. Be especially careful in the cheek region, for one of the salivary ducts is very superficial.

Pick away the connective tissue ventral to the auricle and you will expose the large **parotid gland**. It can be recognized by its lobulated texture. The gland is somewhat oval in the cat (Fig. 9–13), but shaped like a dumbbell in the rabbit. In both animals, the **parotid duct** emerges from the front of the gland, crosses the large cheek muscle (masseter), and perforates the mucous membrane of the upper lip opposite the last premolar (cat) or molar (rabbit) tooth. Frequently accessory bits of glandular tissues are found along the duct. Two branches of the facial nerve going to facial muscles emerge from beneath the parotid gland and cross the masseter, one dorsal and one ventral to the parotid duct. Do not confuse the duct with these.

A **submandibular**, or **submaxillary**, **gland** lies posterior to the angular process of the jaw and deep to the ventral border of the parotid. The submandibular is a large oval gland having the same lobulated texture as the parotid. Do not confuse it with smaller, smoother textured **lymph nodes** in this region. The **submandibular duct** emerges from the front of the gland and passes forward, first going lateral to the digastric muscle (cat) or tendon (rabbit). The digastric is the large muscle arising from the base of the skull and inserting along the ventral border of the lower jaw (p. 157). Cut and reflect the digastric, and you can see that the duct then passes medial to the posterior border of the mylohyoid — the thin transverse sheet of muscle lying between the paired digastric muscles. Cut and reflect the mylohyoid, and follow the duct forward as far as you can. It is crossed anteriorly by the **lingual branch** of the trigeminal nerve returning sensory fibers from the tongue. The **hypoglossal nerve** carrying motor fibers to the tongue musculature lies posterior and dorsal

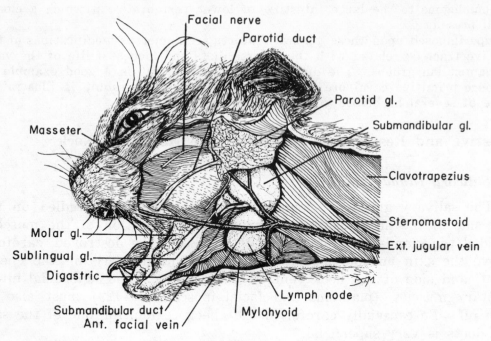

Figure 9–13. Lateral view of the salivary glands of the cat.

to the submandibular duct. The submandibular ducts of opposite sides converge and enter the floor of the mouth by a pair of inconspicuous openings situated just anterior to the midventral septum of the tongue, the **lingual fenulum**. The openings are borne on flattened papillae in the cat.

A small, elongated **sublingual gland** is located beside the submandibular duct. In the cat, the sublingual lies along the posterior one third of the duct and generally abuts against the submandibular gland. But it lies along the anterior one third of the duct in the rabbit. The sublingual is drained by a minute duct which parallels the submandibular duct, but is hard to distinguish grossly. The ducts enter the floor of the mouth. In the cat the sublingual duct accompanies the submandibular duct.

The parotid, submandibular, and sublingual glands are the most common of the salivary glands of mammals, but others are present in certain species. Both the cat and rabbit have an **infraorbital salivary gland** that has been seen if the eye was dissected (p. 170). In addition, the cat has a small, elongate **molar gland** situated between the skin and mucous membrane of the posterior half of the lower lip. Several small ducts, which cannot be seen grossly, lead from it to the inside of the lip. The rabbit has elongated **buccal glands** beneath the skin of the upper and lower lips.

(B) Oral Cavity

Open the **oral cavity** by cutting through the floor of the mouth with a scalpel. Do this from the external surface, cut on each side, and keep as close to the mandible and chin as possible. Then cut through the symphysis of the mandible with bone scissors, spread the two halves of the lower jaw apart, and pull the tongue ventrally. The anterior part of the roof of the oral cavity is formed by the **bony**, or **secondary, palate**; the posterior part by the fleshy **soft palate** (Fig. 9–14). A pair of small openings will be seen at the very front of the hard palate just posterior to the incisor teeth. These are the openings of the **naso-palatine ducts**. These ducts pass through the incisive foramina of the skull to Jacobson's organs in the nasal cavities (p. 81). The lateral walls of the oral cavity are bounded by the teeth, lips, and cheeks. That portion of the cavity lying between the teeth and cheeks is called the **vestibule**. A well developed, muscular **tongue** (**lingua**) lies in the floor of the cavity, and is connected to the floor by the vertical lingual frenulum previously seen.

Pull the tongue ventrally sufficiently far to tighten, and bring into prominence, a pair of lateral folds that extend from the sides of the posterior portion of the tongue to the soft palate. These folds constitute the **glossopalatine arches**, and they represent the boundary between the adult oral cavity and **pharynx**. However, part of the back of the oral cavity as thus defined probably develops from the embryonic pharynx. The passage between the glossopalatine arches is called the **fauces**. Notice that the very back of the tongue lies within the pharynx.

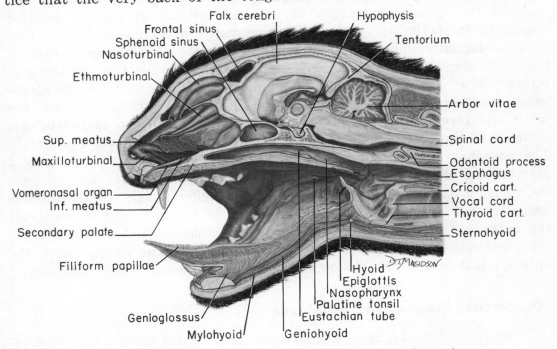

Figure 9–14. Sagittal section of the head of a cat.

Cut through the glossopalatine arch on one side, pull the tongue down more, and examine its dorsal surface. The dorsum of the tongue is covered with papillae, the most numerous of which are the pointed **filiform papillae**. The anterior ones of the cat bear spiny projections with which the animal grooms its fur or rasps flesh from bones, but the posterior ones are soft. Small, rounded **fungiform papillae** are interspersed among the filiform, especially along the margins of the tongue. You may have to use a hand lens to see them. Several **circumvallate (vallate) papillae** are located near the back of the tongue. Each papilla is a relatively large, round patch set off from the rest of the tongue by a circular groove. There are two in the rabbit, but four to six in the cat, distributed in a V-shaped line. The apex of the V is directed caudad. Leaf-shaped **foliate papillae** can be found along the side of the tongue lateral to the circumvallate papillae. Those of the rabbit are collected into a pair of patches. Microscopic taste buds are found on the sides and base of all the papillae, except for most of the filiform papillae.

(C) Pharynx

With a pair of scissors, cut posteriorly through the lateral wall of the pharynx on the side on which the glossopalatine arch was cut. Follow the contour of the tongue to the laryngeal region, then extend the cut dorsal to the larynx and back into the esophagus. Do not cut into the soft palate or larynx. Swing open the floor of the mouth and pharynx. The pharynx may be divided somewhat arbitrarily into oral, nasal, and laryngeal portions. The **oropharynx** lies between the glossopalatine arches and the free posterior margin of the soft palate. A pair of **palatine tonsils** lie in its lateral (cat) or laterodorsal (rabbit) walls. Note that each is partially imbedded in a **tonsillar fossa**. The **laryngopharynx** is the space dorsal to the enlargement, the **larynx**, in the floor of the posterior part of the pharynx. It communicates posteriorly with the **esophagus**, and ventrally with the larynx. The slitlike opening within the larynx is termed the **glottis**. A trough-shaped fold, the **epiglottis**, lies anterior to the glottis and acts to deflect food around or over the glottis. The **nasopharynx** lies dorsal to the soft palate. Open it by making a longitudinal incision through the middle of the soft palate. Spread open the incision as wide as possible and try to shine a light down into the nasopharynx. The pair of slitlike openings in the laterodorsal walls are the entrances of the **auditory**, or **eustachian, tubes**. What do these tubes represent phylogenetically? The choanae, or internal nares, enter the anterior end of the nasopharynx, but cannot be seen in this view.

(D) Larynx, Trachea, and Esophagus

Approach the laryngeal region from the ventral surface of the neck.

Several muscles will have to be removed, but do not injure any of the larger blood vessels. The **larynx** is the chamber whose walls are supported by relatively large cartilages. The **hyoid bone**, which forms a sort of sling for the support of the base of the tongue, is imbedded in the muscles anterior to the larynx. Its parts have been described elsewhere (p. 83). The larynx is continued posteriorly as the windpipe, or **trachea**, whose walls are supported by a series of cartilaginous rings. Actually these rings are incomplete, for their sides do not quite meet dorsally. The **esophagus** is a collapsed, muscular tube lying dorsal to the trachea. Why is it advantageous for the lumen of the trachea to be held open, while this is not the case for the esophagus?

The dark **thyroid gland** lies against the anterior end of the trachea. In the cat it consists of **lobes**, one on either side of the trachea, that are connected across the ventral surface of the trachea by a very narrow band of thyroid tissue called the **isthmus**. The isthmus is frequently destroyed. The thyroid of the rabbit differs in having a wide, prominent isthmus. Two pairs of **parathyroid glands** are imbedded in the dorsomedial surface of the thyroid, but they cannot be seen grossly.

Return to the larynx and study it more thoroughly. Expose the laryngeal cartilages by stripping off all the surrounding tissues on its dorsal, ventral, and one of its lateral surfaces. The large anterior cartilage that forms much of the ventral and lateral walls of the larynx is called the **thyroid cartilage**. It is this cartilage that forms the projection in the neck of man known as Adam's apple. The ring posterior to the thyroid cartilage is the **cricoid cartilage**. The cricoid is shaped like a signet ring, for its dorsal portion is greatly expanded and forms most of the dorsal wall of the larynx. Careful dissection will reveal a pair of small, triangular cartilages anterior to the dorsal part of the cricoid. These are the **arytenoid cartilages**. Additional, minute cartilages are frequently associated with the arytenoids, but are seldom seen. An **epiglottic cartilage** supports the epiglottis.

Cut open the larynx along its middorsal line. The pair of whitish, lateral folds that extend from the arytenoids to the thyroid cartilage are the true **vocal cords** (**plica vocalis**). They are set in vibration by the movement of air across them and are controlled by the movement of the arytenoids, the action of muscles within them, and slight changes in the shape of the larynx. The **glottis** is the space between them. In the cat, an accessory pair of folds, the **false vocal cords**, extend from the arytenoids to the base of the epiglottis. In the rabbit, a small pair of bumps, the **epiglottic hamuli**, lie at the base of the epiglottis.

Most of the laryngeal cartilages develop from certain of the visceral arches, but there is some doubt as to the precise homologies. The arytenoids and cricoid are the first to appear phylogenetically, being represented by the lateral cartilages of the laryngotracheal chamber of amphibians (p. 245). They appear to develop from the sixth, or from the sixth and seventh, visceral arches. The fourth and fifth visceral arches are incorporated in the hyoid apparatus in amphibians. In mammals, the hyoid apparatus involves only the second and third visceral arches, and in most mammals the fourth and fifth arches form the newly evolved thyroid cartilage. Some believe that the tracheal rings may evolve from a splitting and multiplication of the seventh arch, but this is doubtful. The epiglottic cartilage is apparently a new structure.

Thorax and Its Contents

(A) *Pleural Cavities*

Open the thorax by making a longitudinal incision about 3/4 of an inch to the right of the midventral line, and extending the length of the sternum. Use a strong pair of scissors. Spread open the incision and look into the **right pleural cavity**. A dome-shaped, transverse, muscular partition (the **diaphragm**) will be seen at the posterior end of the cavity Make another cut just anterior to the diaphragm that extends laterally and dorsally to the back. Follow the line of attachment of the diaphragm, but keep on the pleural side. Spread the right thoracic wall laterally, breaking the dorsal parts of the ribs as you do so.

The right pleural cavity and its **lung (pulmo)** are now well exposed. The coelomic epithelium lining the walls of the pleural cavity is called the **parietal pleura**; that covering the surface of the lung, the **visceral (pulmonary) pleura**. The right lung is divided into three **lobes — anterior, middle**, and **posterior**. The posterior is the largest, and is subdivided into a **lateral** and **medial lobule**. The medial lobule is the smaller. It extends dorsal to a large vein (**posterior vena cava**) and then ventrally into a pocket on the medial side of the mesentery (**caval fold**) attaching to the ventral surface of the vein (Fig. 9-15). Tear the caval fold near the vein and you can see this portion of the lobule. The lobes of the lung are attached to the medial wall of the pleural cavity by a pleural fold known as the **pulmonary ligament**. The blood vessels connected with the heart, and the **bronchi** from the trachea to the lung, pass through part of this ligament, but they should not be dissected at this time. These structures constitute the **root of the lung**. Cut into a part of the lung and notice that it is not an empty organ, but a very spongy one. The numerous respiratory **alveoli** are not visible grossly.

The medial wall of each pleural cavity (right and left) is formed only of a layer of parietal pleura. The space, or potential space, between the medial walls of the two pleural cavities constitutes the **mediastinum**. This space, however, is largely filled with structures that lie between the two cavities. For example, the pericardial cavity and heart, which form the large bulge medial and ventral to the lung, lie in the mediastinum. In places the medial walls of the pleural cavities meet and form a mes-

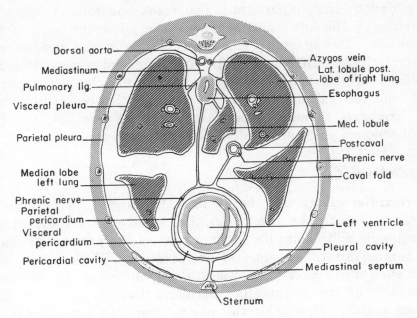

Figure 9–15. A diagrammatic transverse section through the thorax of the cat at the level of the ventricles to show the coelomic epithelium and its relation to the thoracic viscera. The section is viewed from behind so that left and right sides of the animal and drawing coincide.

entery-like structure termed the **mediastinal septum**. The mediastinal septum can be seen posterior to the heart, and medial to the medial lobule of the posterior lobe of the right lung. The caval fold is an evagination from this portion of the septum. In the cat, the mediastinal septum continues anteriorly ventral to the heart, but the medial walls of the pleural cavities do not meet in this region in the rabbit.

Break the line of attachment of the parietal pleura to the ventral thoracic wall. You will have to cut some blood vessels passing to the ventral thoracic wall, but do so in such a way that their ends can later be apposed. Cut laterally and dorsally through the left thoracic wall close to the diaphragm in the same way that you did on the right. Turn back the body wall and examine the **left pleural cavity**. (A good view of the third muscle layer of the thorax, transversus thoracis, referred to on page 153, can now be had.) The left lung, like the right, is divided into three lobes, but the anterior lobe in the rabbit is little more than a bump on the middle lobe. The left posterior lobe is not divided into lobules.

Pull the left lung ventrally, and examine the region dorsal to it. A large artery, the **dorsal aorta**, and the **esophagus** can be seen passing through the dorsal portion of the mediastinum. The aorta lies to the left of the vertebral column; the esophagus, more ventrally. Let the lungs fall back into place. A pair of white strands, the **phrenic nerves**, can be seen in the central portion of the mediastinum on each side of the pericardial cavity and heart. They lie ventral to the roots of the lung and pass caudad to the diaphragm. The right one follows the posterior vena cava closely; the left passes through the posterior part of the mediastinal septum. The origin of these nerves from the ventral rami of the fourth (rabbit), or fifth and sixth (cat) cervical nerves is indicative of the cervical derivation of the diaphragmatic muscles. The portion of the mediastinum ventral and anterior to the heart is occupied by the dark, irregularly lobulated **thymus**. The thymus varies considerably in size, being best developed in young individuals.

(B) Pericardial Cavity

The **pericardial cavity** and **heart** are the largest structures within the mediastinum. Cut open the pericardial cavity by a midventral incision. The coelomic epithelium lining the cavity is the **parietal pericardium**; that covering the heart, the **visceral pericardium**. These layers are continuous with each other over the vessels at the anterior end of the heart. The relationships of the mammalian pericardial cavity are very different from those of a fish. How does the cavity come to assume these relationships, and what is the derivation of its wall?

Peritoneal Cavity and Its Contents

(A) Body Wall and Peritoneal Cavity

Make a longitudinal incision through the abdominal wall slightly to the right of the midventral line. Extend the cut from the diaphragm to the pelvic girdle. Then cut laterally and dorsally along the attachment of the diaphragm to the body wall. Do this on both sides, thereby freeing the diaphragm as far as the back. Reflect the flaps of the abdominal wall. What layers constitute the wall, and how do they compare with those of the thoracic wall? The portion of the coelom exposed is the **peritoneal cavity**. Its walls are lined with **parietal peritoneum**, and its viscera covered with **visceral peritoneum**. Wash out the cavity if necessary.

(B) Abdominal Viscera and Mesenteries

The concave surface of the dome-shaped diaphragm forms the anterior wall of the peritoneal cavity, and the large **liver** (**hepar**) lies just posterior to it, and is shaped to fit into the dome. Pull the liver and diaphragm apart. It can now be seen that the central portion of the diaphragm is formed by a tendon (the **central tendon**), into which its muscle fibers insert. A vertical **falciform ligament** extends between the diaphragm, liver, and ventral wall. A thickening, which represents a vestige of the embryonic umbilical vein, may be seen in its free edge. It is known as the **round ligament** (**ligamentum teres hepatis**) of the liver. Diaphragm and liver are closely apposed dorsal to the falciform ligament, and the reflections of the peritoneum from the one to the other in this region constitute the **coronary ligament**.

The liver can be divided into right and left halves at the cleft into which the falciform ligament passes. Each half is divided into a lateral and a medial lobe, thus making a **left lateral**, **left medial**, **right medial**, and **right lateral lobe**. The right lateral lobe of the cat is split into two **lobules** by a deep cleft. In both animals, the right **kidney** lies dorsal and posterior to the right lateral lobe. A **hepatorenal ligament** extends from this lobe to the parietal peritoneum near the right kidney. The left kidney lies in a slightly more posterior position on the opposite side of the body. The **gall bladder** (**vesica fellea**) lies in a depression on the dorsal surface of the right medial lobe. Lift up the left lateral lobe, and you will see dorsal to it a small **caudate lobe** lying deep to a mesentery, the **lesser omentum**, or **gastrohepatoduodenal ligament**, extending from the liver to the stomach and beginning of the intestine.

As in the majority of vertebrates, most of the **stomach** (**ventriculus**) lies on the left side of the peritoneal cavity, and the organ is more or less J-shaped. Cut through the left side of the diaphragm to find the point at which the **esophagus** enters. There is an abrupt change in the diameter of the digestive tract at this point. The portion of the stomach adjacent to the esophagus is the **cardiac region**; the dome-shaped portion extending anteriorly to the left of the cardiac region, the **fundus**; the main part of the stomach, the **corpus**; and the narrow posterior portion, the **pyloric region**. But, as pointed out elsewhere, these gross regions do not necessarily correspond with glandular regions bearing the same names. The stomach ends in a thick muscular **pyloric sphincter** which can be seen if you cut open the stomach. Do not injure any mesenteries. In the cat, you will also see longitudinal ridges in the lining that are called **rugae**. The long left and posterior margin of the stomach, which represents its original dorsal surface, constitutes its **greater curvature**; the shorter right and anterior margin, which represents its original ventral surface, the **lesser curvature**.

Notice that the lesser omentum, which represents a part of the ventral mesentery, attaches along the lesser curvature of the stomach. The mesentery that attaches along the greater curvature is the **greater omentum** or **mesogaster** — a part of the dorsal mesentery. The greater omentum of mammals does not extend directly to the middorsal line of the peritoneal cavity, as it does in lower vertebrates, but is modified to form a saclike structure (the **omental bursa**), which drapes down over the intestines (Fig. 9–16).

The omental bursa of the cat is very large, contains considerable fat in its wall, and extends over the intestine nearly to the pelvic region. It is often entwined with the intestines, and must be untangled carefully.

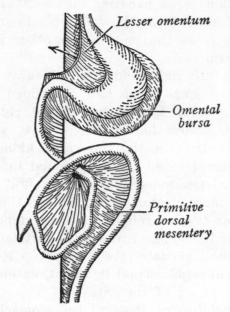

Lesser omentum

Omental bursa

Primitive dorsal mesentery

Figure 9–16. Diagrammatic ventral view of the mesenteries of a mammalian embryo to show the formation of the omental bursa from the posterior extension of the mesogaster. The liver would attach to the lesser omentum along its cut edge (double line). The arrow indicates the approximate position of the epiploic foramen. (From Arey, Developmental Anatomy.)

The rabbit's bursa is much smaller. In both animals, the greater omentum extends posteriorly from its line of attachment on the stomach. It then turns upon itself and extends anteriorly and dorsally to attach onto the dorsal wall of the peritoneal cavity. The **spleen** (**lien**), which is enormous in the cat, lies in the wall of the omentum on the left side of the stomach. That portion of the greater omentum between the spleen and stomach is known as the **gastrolienic ligament**. A small, triangular mesentery, the **gastrocolic ligament**, passes from the part of the greater omentum lying dorsal to the spleen over to the mesentery of the large intestine.

A part of the peritoneal cavity that is called the **lesser peritoneal cavity** lies between the descending and ascending walls of the omental bursa. Tear the bursa and verify that it contains a space. At one stage of development, the lesser peritoneal cavity would have a wide communication with the main part of the peritoneal cavity. But subsequent adhesions of the liver to the dorsal part of the diaphragm and adjacent body wall reduce this to a relatively small **epiploic foramen of Winslow** (Fig. 9–17). The epiploic foramen lies dorsal to the lesser omentum, and between the right lateral lobe of the liver and the mesentery to the duodenum. If your specimen is large enough, you can pass a finger through the epiploic foramen, and extend it dorsal to the stomach and into the omental bursa.

Carefully dissect that portion of the lesser omentum lying between the caudate lobe of the liver and the epiploic foramen in order to expose the system of bile ducts extending from the liver and gall bladder to the beginning of the duodenum. Some lymphatic vessels, which look like chains of small nodules, may have to be removed. A **cystic duct** comes down from the gall bladder, and unites with several **hepatic ducts** from various parts of the liver to form a **common bile duct** (**ductus choledochus**) which passes to the duodenum. One particularly prominent hepatic duct comes in from the left lobes of the liver, and another from the right lateral lobe.

The **small intestine** extends caudad from the stomach, passes through numerous convolutions, and eventually enters a **large intestine**. Certain features of the intestinal region and the pancreas differ sufficiently in the cat and rabbit to warrant separate treatment, but a few common features may be observed at this time. The small intestine of mammals has differentiated into an anterior **duodenum**, and a more posterior **jejunum** and **ileum**. The duodenum, as will be seen presently, is the first, approximately U-shaped loop of the intestine, but there is no sharp transition between the jejunum and ileum. About all that can be said in respect to their gross anatomy is that the jejunum is the anterior half of the post-duodenal small intestine, and the ileum the posterior half. The small intestine is supported by a part of the dorsal mesentery — that portion passing to the duodenum being the **mesoduodenum**, and that portion to the jejunum and ileum the **mesentery**.

Cut open a part of the small intestine, and examine its lumen. The lining has a velvety appearance which results from the presence of numerous, minute, finger-like projections called **villi**. These greatly increase the internal surface area. Parasitic roundworms and tapeworms are often found in the intestine of the cat.

The large intestine of mammals is much longer than that of lower vertebrates, and generally has a greater diameter than the small intestine. Most of it constitutes the **colon**, and it is supported by a portion of the dorsal mesentery termed the **mesocolon**. Skip the pattern that the anterior colon assumes for a moment, and examine the posterior part. This portion extends caudad against the dorsal wall of the peritoneal cavity and enters the pelvic canal. This portion of the colon also lies dorsal to the pear-shaped **urinary bladder** and, if your specimen is a female, to the Y-shaped **uterus**. Notice that the bladder is supported by a vertical **median ligament**, which, being a part of the ventral mesentery, extends to the midventral body wall, and by a pair of **lateral ligaments**. The latter often contain wads of fat. Cut open the posterior part of the colon, clean it out, and notice that it lacks villi. Also notice the extension of the coelom into the pelvic canal. That portion of the coelom in the male that extends posteriorly between the large intestine and the urinary bladder is called the **rectovesical pouch**. The comparable coelomic extension in the female is divided by the uterus into a shallow **vesico-uterine pouch** between the bladder and uterus, and a deep **rectouterine pouch** between the uterus and large intestine.

Deep within the pelvic canal, the colon passes into the terminal segment of the large intestine, the **rectum**, which, in turn, opens on the

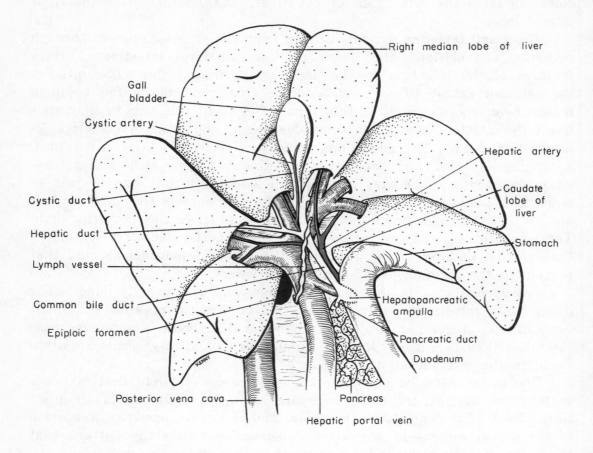

Figure 9–17. Ventral view of the vessels and ducts in the lesser omentum in the cat. The right median and part of the right lateral lobe of the liver have been turned forward.

body surface through the **anus**. The rectal region will be seen later when the pelvic canal is opened.

(C) Further Structure of the Digestive Organs

Keeping these general features in mind, examine the intestine and pancreas in more detail.

Cat

The intestine and pancreas of the cat are relatively simple. The **duodenum** curves caudad from the pylorus on the right side of the body. Then it bends toward the left and ascends nearly to the stomach. It is arbitrarily considered to end at its next major bend. The duodenum is about 7 inches long in an adult cat. A small, triangular-shaped **duodenorenal ligament** passes from the mesoduodenum to the dorsal wall medial to the right kidney.

The **pancreas** can be recognized by its lobulated texture. A part of it lies against the descending portion of the duodenum, and a part of it extends as a tail-like process transversely across the body to the spleen. This portion lies in the dorsal wall of the omental bursa. Two pancreatic ducts are present, for the ducts of both the dorsal and the single ventral primordium are retained. The **pancreatic duct** unites with the common bile duct as the latter enters the duodenum. It can be found by carefully picking away pancreatic tissue in this region. The enlargement on the duodenum where the two unite is known as the **hepatopancreatic ampulla of Vater**. An **accessory pancreatic duct**, which would be the duct of the dorsal pancreas, enters the duodenum about 1/4 of an inch posterior to the main duct, but it is small and very hard to find.

Follow the coils of the rest of the small intestine (jejunum and ileum) until it enters the much wider colon. A short, blind diverticulum, the **cecum**, extends caudally from the beginning of the colon. The vermiform appendix of man is located at the end of the cecum, but an appendix is absent in the cat. Cut open the wall of the cecum and colon opposite the entrance of the ileum. Notice how the ileum projects slightly into the lumen of the colon. The musculature of the intestine in this region forms a sphincter-like **ileocecal** valve that prevents the backing up of colic material into the small intestine. The colon itself extends forward on the right side of the body for a short distance (**ascending colon**), crosses to the left side (**transverse colon**), and extends back into the pelvic canal (**descending colon**) to the rectum and anus.

Rabbit

The **duodenum** of the rabbit extends posteriorly in small convolutions nearly to the pelvic region. It then twists to the left and, continuing to convolute, ascends nearly to the stomach. It is arbitrarily considered to end where it next turns caudad. The duodenum is approximately 20 inches long in an adult rabbit.

Stretch out the mesentery (mesoduodenum) that lies between the descending and ascending limbs of the duodenum. The dark, lobulated tissue that you see is the **pancreas**. It is very diffuse in the rabbit (Fig. 9-18). The single **pancreatic duct**, which is the duct of the embryonic dorsal pancreas, enters the ascending limb about 2 inches anterior to the most posterior loop of the duodenum.

Follow the coils of the rest of the small intestine (jejunum and ileum) until it enters the large intestine. The posterior end of the ileum is modified to form a round, muscular enlargement known as the **sacculus**

rotundus. Cut open the large intestine opposite the entrance of the ileum, clean out its lumen, and find the orifice of the. ileum (**ileocecal valve**). The ileocecal valve, together with the sacculus rotundus, prevents material in the large intestine from backing up into the small.

A large cecum extends in one direction from the sacculus rotundus; the colon, which can be distinguished by its more wrinkled appearance, in the other. First follow the **cecum**. It is a wide, thin-walled, blind sac that extends for about 15 inches in a circular course and ends in a thicker-walled **vermiform appendix**. The appendix is about 5 inches long. Cut open the cecum and notice that the spiral line that can be seen on its surface marks the point of attachment of a **spiral valve**. The cecum contains a symbiotic bacterial colony that breaks cellulose down into simpler compounds that can be utilized by the rabbit. Presumably these compounds are absorbed from the cecum, as has been shown to be the case in certain rodents (Voge and Bern, 1955). Cut open the appendix and note the relative thickness of its wall. Large amounts of lymphoid tissue accumulate in the wall, and may be concerned with neutralizing bacterial toxins.

Return to the **colon** and follow it. Parts of the mesentery may have to be torn. The longitudinal muscle layer of the first part of the colon is limited to two or three bands, called **taeniae coli**, one of which lies

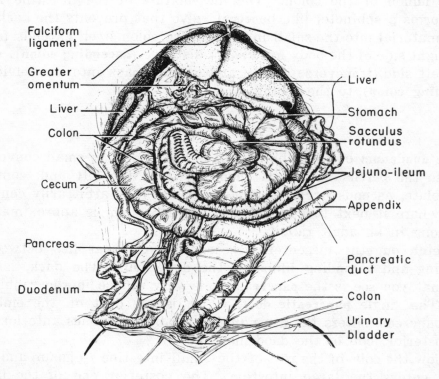

Figure 9–18. Ventral view of the abdominal viscera of the rabbit, showing in particular the large cecum and appendix characteristic of many herbivores.

along the line of attachment of the mesocolon. The wall of the colon between the taeniae protrudes to form little sacculations (**haustra coli**). More caudally, the wall of the colon is smooth, bulging only where there are pellet shaped-feces within it. After a very circuitous course, the colon descends into the pelvic canal to the rectum and anus.

10 / The Circulatory System

Functions of the Circulatory System

Continuing with the organ systems concerned with metabolism, we will next consider the circulatory system. This system is primarily the great transport system of the body. Oxygen and food are carried from the respiratory and digestive organs to all the tissues and cells; carbon dioxide and other excretory products are carried from the tissues to sites of removal; and hormones are transported from the endocrine glands to the responding tissues. But the system has other functions. It aids in combating disease and in repairing tissues, and helps to maintain the constancy of the internal environment in many ways.

Parts of the Circulatory System and the Course of the Circulation

The blood and lymph and their cells are functionally the most important part of the circulatory system, but only the vessels that propel and carry these can be studied in a course of this scope. These vessels may be subdivided in most vertebrates into (1) a **cardiovascular system**, consisting of the **heart, arteries, blood capillaries**, and **veins**; and (2) a **lymphatic system**, consisting of closed **lymphatic capillaries** and **lymphatic vessels**. In addition, the lymphatic system of higher vertebrates includes many **lymph nodes** located at strategic junctions along the course of the vessels. These nodes act as filters and as sites for the production of certain white blood cells (**lymphocytes**). The **spleen** is similar to a lymph node, but is interposed in the cardiovascular system. It is at different times and in different vertebrates a site for the production, storage, and elimination of blood cells.

Briefly, the course of the circulation through these two systems of vessels is as follows. **Blood** leaves the heart and travels through the arteries to the capillaries in the tissues. At this point, some of the blood plasma leaves the capillaries to circulate among the cells as **tissue fluid**. Much of the tissue fluid re-enters the capillary bed and returns to the heart through veins. But, in all but the most primitive vertebrates, some enters the lymphatic capillaries and is carried as **lymph** by the lymphatics to the larger veins. Primitive jawless fishes and cartilaginous fishes lack a true lymphatic system. Blood pressure in the veins is very low in these fishes, and the veins with their large sinuses adequately drain the tissues. Venous blood pressure tends to be higher in other vertebrates, and the lymphatics serve as a low pressure drainage system for the tissues. Particulate matter and colloids that may be in the tissue fluid can also enter the lymphatics.

The heart is the major pump in causing the fluids to circulate, but by the time the blood reaches the veins the pressure is relatively low. Moreover, pressures created by the heart do not directly affect the lymphatic system. Thus the return of lymph, and to some extent the return of blood, are implemented by other forces. Among these, the contraction and tonus of the surrounding body muscles play a major role. The veins and lymphatics also contain **valves** that prevent a back flow of the fluids.

The lymphatic system is rather obscure and not commonly seen in gross dissections. Parts of it will be studied in the mammal, but the emphasis throughout will be on the more conspicuous cardiovascular system.

Development of the Cardiovascular System

Early in the development of a fish embryo, a system of blood vessels is established that provides for the metabolic needs of the embryo. Higher vertebrates have inherited this pattern of development, but it has been modified somewhat to fit their particular requirements. Since the early basic pattern is repeated to a large extent in all vertebrates during the ontogeny of their diverse adult patterns, it forms a necessary basis for understanding the adult cardiovascular system.

The first vessels to take definite form in the embryo are a pair of **vitelline veins** (Fig. 10–1) that lie beneath the embryonic gut and carry blood and food forward from the yolk-laden archenteron, or yolk sac if such a sac is present. The anterior parts of these vessels fuse to form the **heart and ventral aorta**. Posterior to the heart, they are engulfed and broken up into a capillary network by the enlarging liver. The portion of the vitelline veins lying posterior to the liver become the **hepatic portal system**; the portion from the liver to the heart, the **hepatic veins**. (**Portal veins** are simply veins that, after draining one capillary bed, pass to another capillary bed in a different organ. Veins going directly to the heart are referred to as **systemic veins**.)

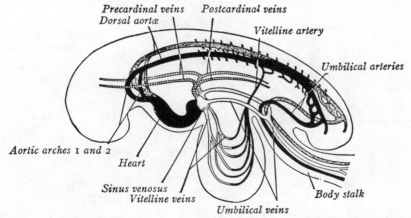

Figure 10–1. A diagrammatic lateral view of the major blood vessels of an early amniote embryo. Four more posterior aortic arches, and other vessels, would be present in later stages. The arteries have been drawn darker than the veins. Arteries are defined as vessels flowing away from the heart; veins as vessels flowing toward the heart. (From Arey, Developmental Anatomy. After Felix.)

Meanwhile, a series of six aortic arches develop in the first six branchial bars, which lie between the gill pouches, to carry blood from the heart and ventral aorta up to the **dorsal aortae**. The dorsal aortae, which are first paired but later fuse, carry the blood caudally and, by **vitelline arteries**, back to the archenteron and yolk sac. This completes one circuit.

It will be noted that the early circulation is largely a visceral circulation to and from the "inner tube" of the body, but a somatic circulation to the "outer tube" soon appears. Other branches from the dorsal aorta carry blood into the body wall and out to an allantois (by **umbilical** or **allantoic arteries**) if one is present. Blood from the dorsal portions of the body returns by way of **anterior** and **posterior cardinals**. The cardinals of each side unite anterior to the liver, and turn ventrally to the heart as **common cardinals** or **ducts of Cuvier**. It will be recalled that the common cardinals pass through the transverse septum which they helped to form. Blood from the more lateral and ventral portions of the body wall, and from the allantois, returns by a pair of more ventrally situated vessels. These are called **lateral abdominal veins** in lower vertebrates and **umbilical** or **allantoic veins** in amniotes. These vessels enter the base of the common cardinals in early embryos, but in the later embryos of the higher vertebrates they acquire a connection with the hepatic portal system, and are drained through the liver.

As development proceeds and sites of nutrition, gas exchange, and excretion change, the pattern of vessels changes; new channels appear and some old ones atrophy. Thus in the development of an embryo, especially embryos of the higher vertebrates which have had a long and complex phylogenetic history, we see a succession of vessels. Much of the variation seen in the vessels of the adult can be attributed to the persistence of embryonic channels that normally atrophy and to the failure of certain later channels to develop. Other variation results from the enlargement of one channel over another in a primordial capillary plexus that exists in many parts of the embryo (Fig. 10–2). The relative

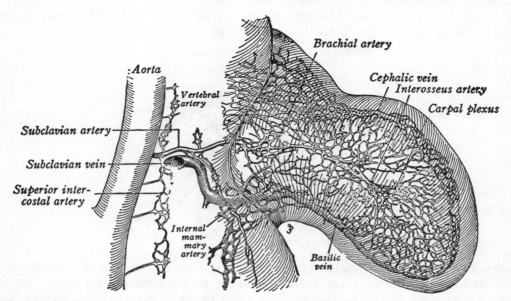

Figure 10–2. An early stage in the development of the pectoral limb bud of a pig embryo, to show the early plexus of vessels. The definitive vessels arise by the enlargement of certain channels. (From Woolard.)

rate of blood flow, as well as hereditary and other factors, is an important factor in determining which channels will enlarge.

The Study of Blood Vessels

While dissecting the vessels, the student should keep in mind the fact that they are subject to considerable variation, and may not be just as described. Since the veins of the higher vertebrates have had a more complex ontogeny than the arteries and the venous blood has a more sluggish flow, it is to be expected that more variations will be found in the venous system of these animals than in the arterial system.

Odd as it may at first seem, the peripheral parts of the vessels are subject to less variation than are certain of the more central and larger channels. For example, the left ovarian vein of a mammal always drains the ovary, but it may enter either the left renal vein or the posterior vena cava. Hence if a vessel cannot be identified from its point of connection with a major vessel, it can be identified if its peripheral distribution can be established.

Another fact to bear in mind is that the arteries and veins tend to follow each other. Thus if the ovarian vein cannot be found, look for it again when the ovarian artery is studied. A trace, at least, of the vein will probably be found beside the artery.

The blood vessels of vertebrates are exceedingly numerous, and not all can be studied in a course of this scope. Emphasis has been placed on the major channels in the axis of the body, and the vessels connecting with them. In general, the blood vessels of the head, appendages, and the more distal vessels in the intestinal region are treated superficially.

FISHES

The cardiovascular system of adult fishes is in a very primitive stage, for its development does not proceed far from the early embryonic pattern described above. Fishes in general have a very low pressure and sluggish circulation, for there are several capillary beds interposed along the course of the vessels, and there is a pressure loss through friction in each. Thus the pressure built up by the heart is reduced almost immediately when the blood flows through the gill capillaries interposed in the aortic arches; again in the capillaries in the tissues; and, for certain circuits, still again in the liver (hepatic portal system) or kidney (renal portal system). Recent measurements made by Satchell (1960) show that the mean blood pressure in the ventral aorta of the dogfish is about 20 mm. Hg. This drops to about 15 mm. Hg in the dorsal aorta and to 0 mm. in the veins near the heart. Venous return to the heart is facilitated by many large venous sinuses associated with the veins, for these offer very little resistance to blood flow.

The blood vessels of *Squalus* are a good example of those of most fishes, but it must be remembered that fishes ancestral to tetrapods would have had lungs as well as gills, and hence a pulmonary circulation. The cardiovascular system should be studied on specimens in which at least the arteries and hepatic portal system have been injected. Triply injected specimens are necessary if all the veins are to be studied.

External Structure of the Heart

The pericardial cavity of the dogfish has been opened, and the heart observed. Return to the cavity and examine the external features of the heart. The heart of fishes is essentially an S-shaped tube that receives venous blood at its posterodorsal end, increases the blood pressure, and sends the blood out to the gills and body at its anteroventral end. Four chambers have differentiated in linear sequence along this tube. These are, from posterior to anterior, the **sinus venosus, atrium,**[14] **ventricle**, and **conus arteriosus**. The last two lie along the ventral half of the S, and so are the first to be seen. The ventricle is the thick-walled, muscular, oval-shaped structure lying in the posteroventral portion of the pericardial cavity; the conus arteriosus, the tubelike chamber extending from the front of the ventricle to the anterior end of the cavity. Lift up the pointed posterior end of the ventricle (the **apex** of the heart). The thin-walled, bilobed chamber dorsal to it is the atrium. The sinus venosus is the triangular shaped chamber lying posterior to the atrium and extending between its two lobes. The posterior surface of the sinus venosus adheres to the transverse septum, and receives through it the various veins draining the body.

Venous System

(A) *Hepatic Portal System and Hepatic Veins*

The principal vein of the hepatic portal system is the **hepatic portal vein** going to the liver. It can be found lying beside the bile duct in the gastrohepatoduodenal ligament. The hepatic portal vein receives small **choledochal veins** from the bile duct, but is formed by the confluence, near the anteroventral tip of the dorsal lobe of the pancreas, of three large tributaries. A **gastric vein** comes in from the central part of the stomach; a **lienomesenteric vein**, from along the line of attachment of the mesentery to the dorsal lobe of the pancreas; and a **pancreaticomesenteric vein**, from the deep side of the ventral lobe of the pancreas. The gastric and lienomesenteric veins often come together and form a short common trunk before uniting with the pancreaticomesenteric, which is the conspicuous vein easily seen passing just dorsal to the pylorus (Fig. 9-5). Follow each one of these three veins toward its periphery, noting

---

[14] The terms atrium and auricle are variably used; sometimes they are used synonymously, and sometimes a subtle difference is made between them. Some anatomists use the term atrium for the undivided fish chamber, and auricle for the pair of tetrapod chambers derived from the splitting of the atrium. But in human anatomy, auricle is used for an ear-shaped appendage on the atrium. The author will use the term atrium throughout.

the organs that it drains. Each has numerous subsidiary tributaries, but only the principal ones will be mentioned. The lienomesenteric vein receives many little vessels from the dorsal lobe of the pancreas as it courses along this organ. At the posterodorsal tip of this part of the pancreas, it receives a prominent **posterior lienogastric vein** from the spleen and adjacent parts of the stomach, and a **posterior intestinal vein** from the posterior part of the valvular intestine. The pancreaticomesenteric receives vessels from the pyloric part of the stomach, anterior part of the spleen, ventral lobe of the pancreas, and from the spiral valve inside the intestine. But its main tributary is the **anterior intestinal**, which can be seen running along the surface of the intestine near the line of attachment of the mesentery.

The hepatic portal vein can be traced into the substance of the liver where it breaks up into many branches that lead to the capillary-like sinusoids. The sinusoids in turn are drained by a system of hepatic veins that lead ultimately to a pair of large **hepatic veins**, or **sinuses**. The hepatic veins are systemic veins. Although not injected, one of these large hepatic sinuses can be found by cutting the transverse septum down to the coronary ligament on the side of the body previously cut open, and dissecting away liver tissue just posterior to the coronary ligament. Do not injure the falciform ligament or oviduct in this region. You will soon enter the large hepatic sinus. Pass a probe forward through the vessel, and it will be seen to go through the coronary ligament and transverse septum into the sinus venosus. The same is true for the hepatic sinus on the other side, but it should not be dissected. If desired, the hepatic sinus can also be traced caudally well into the liver.

(B) Renal Portal System

Cut a cross section through the tail just posterior to the cloacal aperture. The caudal **artery** and **vein** will be seen within the hemal arch, the artery lying dorsal to the vein and being injected. The vein will not be injected, unless it happened inadvertently during the injection of the artery. Now make a frontal section through the base of the tail and the posterior end of the trunk that lies in the plane of the caudal vein. This section will also cut into the posterior end of each kidney. Clean out the caudal vein, and it will be seen to bifurcate at the level of the middle of the cloaca. Each branch of the bifurcation is a renal **portal vein**, and can be seen extending anteriorly dorsal to the medial border of the kidney (Fig. 10-3, B). Blood in these veins is carried to the capillaries associated with the kidney tubules by inconspicuous **afferent renal veins**. These capillaries are drained by **efferent renal veins** that enter the posterior cardinals (see below).

A renal portal system is not present in primitive agnathous fishes such as

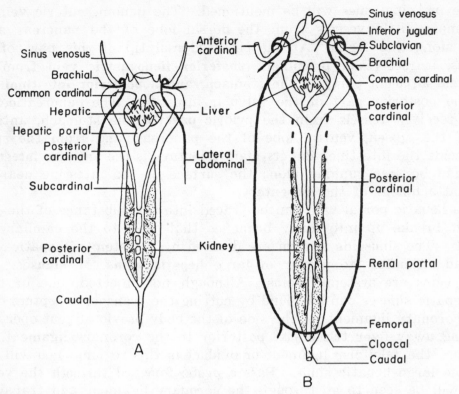

Figure 10–3. Diagrams from a ventral view of the major veins of *A*, a late embryo, and *B*, an adult dogfish. These diagrams show in particular the development of the adult renal portal system from the posterior parts of the embryonic posterior cardinals, and the development of the posterior parts of the adult posterior cardinals from the embryonic subcardinals. The broken lines in *B* indicate the portions of the embryonic posterior cardinals that disappear in the adult. (*A*, modified after Hochstetter; *B*, modified after Daniel.)

the lamprey, and in such forms the caudal vein leads directly into the two posterior cardinals. Such a condition is also present in the embryonic dogfish (Fig. 10–3, *A*). During subsequent development, a pair of new veins (called **subcardinals** in the embryo) appear ventral to the kidneys. These veins tap into the front of the posterior cardinals. Meanwhile, a portion of the posterior cardinals caudal to this union atrophies (dotted line in Fig. 10–3, *B*). The posterior portions of the embryonic posterior cardinals are now renal portal veins, for all their blood necessarily goes to the kidneys. The adult posterior cardinals are composed of the embryonic subcardinals plus the anterior end of the embryonic posterior cardinals. The reasons for the evolution of the renal portal system are not clear: possibly, it insures the passage of a large volume of blood through the kidneys, thus compensating for a small and low pressure arterial supply.

(C) Systemic Veins

For purposes of description, the systemic veins of fishes may be arranged in five groups: (1) the hepatic veins already seen, (2) the cardinal system draining most of the trunk and head, (3) the inferior jugulars draining the ventral portion of the head, (4) the lateral abdominal system draining the appendages and lateroventral portion of the body wall, and (5) the coronary veins draining the heart wall. Reference should be made to Figure 10–3, *B*, during the dissection of these veins, for many are hard to find unless you have a triply injected specimen. The following directions are based on specimens in which these veins have not been injected.

Open the sinus venosus by a transverse incision and wash it out. The small entrances of the pair of hepatic sinuses can now be seen in the central part of the sinus venosus. The large, round openings at the posterolateral angles of the sinus are the entrances of the paired **common cardinal veins**, or **ducts of Cuvier**. Pass a probe into the common cardinal on the side of the body that has been opened, and cut the vessel open along the course of the probe. The common cardinal passes along the lateroventral wall of the esophagus within the transverse septum, and becomes continuous with the large **posterior cardinal sinus** — the large, membrane-covered space that lies against the dorsolateral surface of the esophagus and curves toward the middorsal line of the pleuroperitoneal cavity. Cut open the posterior cardinal sinus and follow it posteriorly. The sinuses of opposite sides are interconnected dorsal to the gonads, and receive in this region genital vessels (**ovarian** or **testicular veins**) from **gonadial sinuses** beside the gonads, and also veins from the esophagus. It is difficult to see these vessels clearly. More caudally, each sinus narrows to form a **posterior cardinal vein** which continues posteriorly along the dorsomedial border of each kidney. The posterior cardinal sinus is actually just the expanded anterior part of this vein. Each posterior cardinal receives numerous **efferent renal veins** from the kidneys and **intersegmental** or **parietal veins** from between the myomeres. Renal and intersegmental veins are hard to see unless they happen to be filled with blood.

One of the paired **anterior cardinal sinuses** was seen on the large head of *Squalus* during the dissection of the nervous system (p. 193). Expose one on your present specimen by making a deep longitudinal cut extending ventrally and medially from that portion of the lateral line overlying the branchial region. Open it on the side on which you have been working. Pass a probe posteriorly through the vessel, and it will be seen to turn ventrally and unite with the posterior cardinal sinus. The union of these two vessels marks the beginning of the common cardinal. The anterior cardinals drain the brain and all the head except the floor of the branchial region. If desired, the anterior cardinal can be traced forward by probing and cutting to a large **orbital sinus** around the eye.

The floor of the branchial region is drained by a pair of **inferior jugular veins**. The entrance of one can be seen in the anterior wall of the common cardinal just before the common cardinal enters the sinus venosus. The vein extends forward dorsal to the pericardial cavity. The anterior cardinal and inferior jugular of each side are interconnected by a **hyoidean sinus** lying posterior to the hyoid arch. The dorsal end of the hyoidean sinus can be seen in the floor of the anterior cardinal just posterior to the hyomandibular. The hyoidean sinus can be traced ventrally to the inferior jugular, and the inferior jugular posteriorly to the common cardinal, by probing the vessels and cutting them open along the course of the probe.

The appendages, and some of the lateroventral portion of the trunk, are drained by a pair of **lateral abdominal veins**. These veins are the pair of dark, longitudinal lines that one sees on the inside of the body wall beneath the parietal peritoneum. Examine the posterior end of one by probing and cutting open the vessel. After passing dorsal to the pelvic girdle, the vessel receives two tributaries. One (the **cloacal vein**) comes in from the lateral wall of the cloaca; the other (the **femoral vein**), from the pelvic fin. There is also an anastomosis between the cloacal veins of the opposite sides of the body. You may not be able to find these vessels in an uninjected specimen. Anteriorly, the lateral abdominal is joined by a **brachial vein** from the pectoral fin. The brachial can be found by cutting off the pectoral fin close to the body on the side on which you have been working. The vein lies ventral to the middle of the basal cartilages. Trace it to its union with the lateral abdominal by probing and cutting along the course of the probe. A **subscapular vein** may be seen entering the proximal end of the brachial. The vessel formed by the union of the brachial and lateral abdominal is known as the **subclavian vein**. Trace it in the same way, and it will be seen to enter the front of the common cardinal beside the entrance of the inferior jugular. If it is necessary to find any of these veins on the opposite side, avoid injuring the falciform ligament.

Coronary veins can be seen on the surface of the heart, especially the ventricle. They enter the sinus venosus by a common aperture which cannot be found at this stage of the dissection.

Arterial System

(A) Branchial Arteries

The branchial arteries should be exposed on the intact side of the pharynx. If the interbranchial septa have not been mobilized, this should be done by cutting through the top and bottom of each of the five external gill slits, and continuing the incision to, but not through, the internal gill slits.

Lay the specimen on its back, and note again the conus arteriosus. After this passes through the front of the pericardial cavity, the vessel is known as the **ventral aorta**. Trace the ventral aorta anteriorly. It is generally not injected, so be careful. It passes between the hypobranchial musculature and basibranchial cartilages (Fig. 10–7, p. 279). Much of the musculature must be removed, but do not injure the major injected vessels in this region. They are derived from the efferent branchial arteries, and will be considered shortly.

As the ventral aorta passes forward, it gives off five pairs of **afferent branchial arteries** that extend up into the interbranchial septa (Figs. 9–9 and 10–4). The posterior two pairs of afferent branchial arteries (the **fourth** and **fifth**) leave the ventral aorta just anterior to the pericardial cavity, and from the dorsolateral side of the aorta. They come off very close together, and sometimes by a short, common trunk. The middle pair of afferent branchial arteries (the **third**) leaves slightly anterior to the posterior two pairs. The ventral aorta then passes anteriorly for some distance without further branches. Slightly posterior to the level of the basihyal, it bifurcates. (The dark **thyroid gland** may be seen anterior to this bifurcation and ventral to the prehyoid muscles.) Trace one of the bifurcations, and it will be seen to subdivide again. These subdivisions are the **first** and **second** afferent branchial arteries. Trace each of the

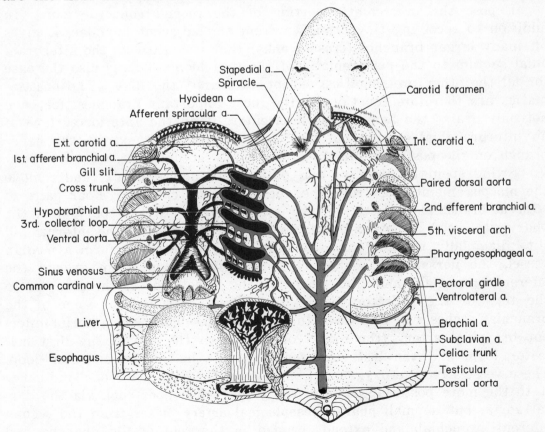

Figure 10–4. A dissection of the branchial arteries of *Squalus*. In this case the afferent branchial arteries, which have been drawn very dark, have been approached through the floor of the pharynx, and are seen in a dorsal view. The efferent branchial arteries, which are shown in red, are seen in a ventral view.

five afferent branchial arteries far enough into the interbranchial septa on the intact side of the specimen to ascertain which septa and gills they are supplying, and to observe the numerous small branches that they send into the gill lamellae. What does this system of vessels do?

Oxygenated blood is collected from the gill lamellae by a system of efferent branchial arteries. The first portion of this system is a series of four and one-half **collector loops**. A representative collector loop can be seen by spreading open the first definitive external gill slit, and looking into the first pouch. A circle of mucous membrane will be seen at the base of the gill lamellae lateral to the internal gill slit. Strip off this membrane, and a vascular circle, or loop (the first collector loop), receiving tiny vessels from the lamellae, will be seen beneath it. The vessel forming the anterior half of the loop is the **pretrematic branch** of an efferent branchial artery, and it is noticeably smaller than the vessel forming the posterior half of the loop, the **posttrematic branch**. A second, third, and fourth loop can be found in the second to fourth definitive gill pouches, but only the pretrematic half of a loop is present in the last pouch.

Expose the pretrematic portion of the second collector loop. In addition to receiving tiny branches from the adjacent lamellae, it gives off many larger branches (**cross trunks**) that pass through the interbranchial septum to the posttrematic of the first loop. This is also the case for all the other pretrematics, except, of course, the first. The posttrematics are therefore functionally the most important branches, for each not only drains the hemibranch adjacent to it on the anterior surface of an interbranchial septum but also drains, to a large extent, the hemibranch on the posterior surface of the same septum.

Swing open the floor of the oral cavity and pharynx, and remove the mucous membrane from the roof of these cavities. Four pairs of **efferent branchial arteries** will be seen on the posterior portion of the pharyngeal roof. They extend from the dorsal angles of the internal gill slits diagonally medially and caudally, and converge to form a median vessel, the **dorsal aorta**. Remove connective tisssue from around the efferent branchials, and trace them on the intact side of the pharynx to the tops of the collector loops. The dorsal cartilages of some of the branchial arches will also have to be removed. The most anterior efferent branchial artery (the **first**) connects with the top of the first collector loop, and the last (the **fourth**) with the fourth collector loop. These vessels receive oxygenated blood from the collector loops, and carry it to the more posterior parts of the body. Most goes back via the dorsal aorta, but a small **pharyngoesophageal artery** arises from the second efferent branchial, and extends caudad in the roof of the pharynx and esophagus.

Arterial blood to the anterior parts of the body travels in other vessels. A **hyoidean efferent artery** arises from the top of the first collector loop anterior to the first efferent branchial, and passes forward in the roof of the pharynx. Opposite the spiracle, it receives, on its medial side, a small vessel that extended forward from the medial end of the first efferent branchial. This artery and its mate of the opposite side represent some of the anterior portions of the embryonically paired dorsal aorta. They are simply called the **paired dorsal aortae**. The vessel anterior to the union of the hyoidean efferent with a paired dorsal aorta also develops embryonically from the front of the dorsal aorta. In the adult, it is called the **internal carotid artery**. The internal carotid continues forward, curves toward the middorsal line, crosses its mate of the opposite side, uniting with its mate in varying degrees as it crosses, and enters the chondrocranium through the carotid foramen. Follow the internal carotids by chipping away cartilage. They soon diverge and then unite, at the level of the hypophysis, with the arteries on the ventral surface of the brain. The internal carotids are the major arteries supplying the brain.

At the point at which the internal carotid curves toward the middorsal line to unite with its fellow, it gives off a **stapedial** artery from its anterolateral surface. Note the proximity of this artery to the point of union of the hyomandibular (the future stapes) with the otic region of the chondrocranium. Follow the stapedial forward. It passes dorsal to a prominent artery extending medially from the spiracle (see below), and is distributed to the orbit and snout.

Return to the pretrematic of the first collector loop. A vessel arises near the middle of this pretrematic that at first resembles a cross trunk. This is the **afferent spiracular artery**. Trace it forward by cutting through the skin ventral and posterior to the spiracle (Fig. 6–3, p. 116). The afferent spiracular crosses the lateral surface of the hyomandibular, and extends to the pseudobranch in the spiracle. It would of course normally carry arterial blood to this structure. An **efferent spiracular artery** continues from the pseudobranch medially to unite with the internal carotid within the cranial cavity. This portion of the vessel can be found by removing the mucous membrane lining the front of the spiracle. Approach the spiracle from its pharyngeal entrance. It was the efferent spiracular that passed ventral to the stapedial.

An **external carotid artery** arises from the ventral end of the first collector loop and passes forward to supply the lower jaw region. It is best approached from the ventral surface of the specimen where it will be seen dorsal to the proximal end of the first afferent branchial.

Another vessel, the **hypobranchial artery**, usually arises from the ventral end of the second collector loop, but it may receive contributions from any of the other loops. The vessel supplies most of the hypobranchial musculature, and then bifurcates at the front of the pericardial cavity. One branch, the **coronary artery**, is distributed to the heart. The other, the **pericardial artery**, extends caudad in the dorsal wall of the pericardial cavity.

Primitive fishes retain parts, at least, of all six embryonic aortic arches, but the arches are obviously modified in the adult for the interposition of the gills and the supply of blood to the head. The developmental history by which this modification comes about is complicated, but the essentials can be summarized. At an early stage in development, six complete aortic arches are present (Fig. 10–1). During subsequent development, the ventral portion of the first disappears, but the ventral portion of the remaining five forms the afferent branchial arteries (Fig. 10–5, *B*). The dorsal part of the first arch, together with a new connection that early develops between the first and second arches (Fig. 10–5, *A*), forms the spiracular artery. The dorsal part of the second arch forms

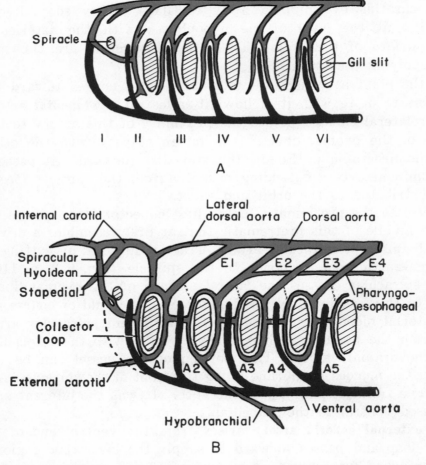

Figure 10–5. Diagrams of *A*, the embryonic, and *B*, the adult aortic arches and their derivatives in a dogfish as seen from a lateral view. The embryonic stage is one in which the modification of the earlier, uninterrupted aortic arches (Fig. 10–1) has begun. All the vessels would be paired except the posterior parts of the dorsal and ventral aorta. The afferent vessels are drawn in black; the efferent in red. Abbreviations: *A1* to *A5*, first to fifth afferent branchial arteries; *E1* to *E4*, first to fourth efferent branchial arteries; *I* to *VI*, first to sixth aortic arches.

Table 5.   Derivation of the Branchial Vessels of the Dogfish

The relation of the adult branchial vessels of a dogfish to the embryonic aortic arches is shown in the following table. The derivation of each collector loop from outgrowths of two aortic arches is explained in the text, and not shown in this table.

Afferent Vessels	Efferent Vessels	Embryonic Origin
	Efferent spiracular	Aortic arch 1
	Afferent spiracular	Cross connection between aortic arches 1 and 2
	Hyoidean efferent	Aortic arch 2
Afferent branchial 1	Efferent branchial 1	Aortic arch 3
Afferent branchial 2	Efferent branchial 2	Aortic arch 4
Afferent branchial 3	Efferent branchial 3	Aortic arch 5
Afferent branchial 4	Efferent branchial 4	Aortic arch 6
Afferent branchial 5	Internal carotid	Anterior dorsal aorta and forward outgrowth
	Stapedial	Anteroventral outgrowth from dorsal aorta
	External carotid	Anterior outgrowth from the ventral extension of the part of aortic arch 2
	Hypobranchial	Ventral outgrowth from ventral part of second collector loop (aortic arches 3 and 4)
	Pharyngoesophageal	Posterior outgrowth from dorsal part of aortic arch 4

the hyoidean efferent; the dorsal parts of the remaining arches, the four efferent branchials. The external carotid, distal part of the internal carotid, stapedial, hypobranchial, and pharyngoesophageal arteries represent new outgrowths from the dorsal aorta, or various arches as shown in Figure 10-5 and Table 5.

The derivation of these vessels is reasonably straightforward, but the formation of the collector loops is more complicated. At any early stage in development, the five posterior arches break near their dorsal end. The dorsal part of the second arch then grows ventrally in front of what is to be the first definitive gill slit, and the posterior four bifurcate and grow ventrally between the gill slits (Fig. 10-5, A). The posterior bifurcation of one arch unites above and below a gill slit with the anterior bifurcation of the next posterior arch to form a collector loop. This process is repeated for the other loops. Thus each collector loop is formed from outgrowths of two successive aortic arches — the pretrematic portion of the loop from the posterior branch of the bifurcation of the anterior arch, and the posttrematic portion from the anterior branch of the bifurcation of the next posterior arch. At one stage of development, each loop would be drained by the dorsal parts of both parent arches. This double drainage persists in the adult for the first collector loop (Fig. 10-5, B). But the second and succeeding loops lose their original connection (xxx in Figure 10-5, B) with their anterior parent, and would be drained exclusively by the posterior parent if it were not for the development of new connections (cross trunks) with the anterior parent.

(B) Dorsal Aorta and Its Branches

The basic pattern of the branches of the dorsal aorta is shown in Figure 10-6. As can be seen, there are three major categories of vessels. (1) Paired **intersegmental arteries**[15] pass between each of the body segments, and each soon bifurcates into a **dorsal ramus** to the epaxial region and a **ventral ramus** to the hypaxial region. Longitudinal anastomoses may occur at various points between successive intersegmentals. Vessels to the appendages are simply enlarged intersegmental arteries. (2) Paired **lateral visceral arteries** pass at intervals to such dorsolateral organs as the kidneys, gonads, and suprarenal glands. (3) Unpaired **ventral visceral arteries**, which develop from the embryonic vitelline arteries, pass through the dorsal mesentery to the viscera, "within" the coelom. This basic pattern of the branches of the aorta shows very well in the dogfish, and in the embryos of higher vertebrates, but it is modified somewhat in the adults of the latter.

Working in so far as possible from the side of the body that has been opened (the right), trace the dorsal aorta caudad. The anterior portion of the esophagus will have to be separated from the body wall, and some mesenteries will have to be torn. A pair of **subclavian arteries** arise from the aorta, generally between the third and fourth efferent branchials (Fig. 10-4). Follow one as it curves ventrally against the lateral wall of the posterior cardinal sinus. It follows the posterior margin of the scapula and gives off several branches to the adjacent musculature.

---

[15] These vessels are often called segmental arteries, but the term intersegmental is preferable, for the vessels occupy an intersegmental position.

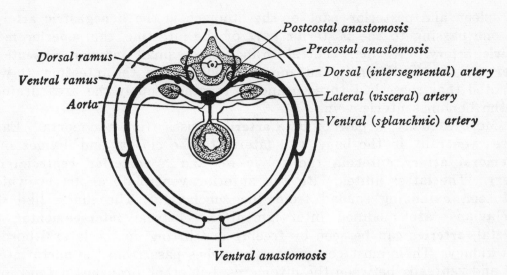

Postcostal anastomosis
Precostal anastomosis
Dorsal ramus
Ventral ramus
Aorta
Dorsal (intersegmental) artery
Lateral (visceral) artery
Ventral (splanchnic) artery
Ventral anastomosis

Figure 10–6. Diagram of the dorsal aorta and its branches as seen in a cross section through the trunk. Longitudinal anastomoses appear as beadlike enlargements on the vessels. Although based on a mammalian embryo, this basic pattern is seen in the adults of the lower vertebrates. (From Arey, Developmental Anatomy.)

Its major branches are a **brachial artery**, which enters the pectoral fin, and an **anterior ventrolateral artery**, which continues anteriorly and ventrally along the posterior margin of the scapula and coracoid. The ventrolateral then curves posteriorly and extends caudad in the ventrolateral portion of the body wall. This part of the vessel can be seen beneath the parietal peritoneum between the lateral abdominal vein and the midventral line.

The subclavians are modified intersegmental vessels. Other intersegmentals will be seen presently, but the ventral visceral arteries should be examined next. The first of these, the lagre **celiac trunk**, enters the front of the pleuroperitoneal cavity, and extends ventrally and posteriorly along the right side of the stomach to the anteroventral tip of the dorsal lobe of the pancreas (Fig. 9–5). Here it bifurcates. One branch, the **pancreaticomesenteric artery**, follows the pancreaticomesenteric vein dorsal to the pylorus and onto the intestine as the **anterior intestinal artery**. In addition, it sends smaller branches to the pyloric region of the stomach and the ventral lobe of the pancreas, and into the spiral valve. In general, it supplies the area drained by the pancreaticomesenteric vein. The other branch of the celiac, the **gastrohepatic artery**, soon divides into a small **hepatic artery** that follows the hepatic portal vein to the liver, and a **gastric artery** that passes to the stomach following closely the gastric vein and its tributaries.

The next ventral visceral branches of the aorta will be found in the free posterior edge of the dorsal mesentery. Two vessels arise close together in this region and pass through the mesentery. The one going to the spleen and posterior part of the stomach is the **lienogastric artery**; the one passing to the posterior part of the intestine, the **superior mesenteric artery**. A final ventral branch is a small **inferior mesenteric artery** to the digitiform gland and posterior end of the intestine. It will be noted that these last three branches together supply the area drained by the lienomesenteric vein.

More caudally, a pair of **iliac arteries** arises from the aorta. Each passes ventrally in the body wall lateral to the cloaca, and divides into a **femoral artery** entering the pelvic fin and a **posterior ventrolateral artery**. The latter unites with the anterior ventrolateral artery which was seen extending caudad from the subclavian. The iliacs, like the subclavians, are modified intersegmentals. Typical **intersegmental**, or **parietal, arteries** can be seen by freeing and lifting up the lateral border of a kidney. Their most conspicuous branches pass from the aorta laterally and ventrally between the myomeres, but other branches extend into the epaxial region of the body (Fig. 10-6). Lateral visceral arteries include a number of small **renal arteries** and a pair of **gonadial arteries** (**ovarian** or **testicular**). The former arise from the aorta close to, or in common with, the intersegmentals and pass to the kidneys, the latter from the very base of the celiac. Posterior to the iliacs, the dorsal aorta enters the tail as the **caudal artery**.

Internal Structure of the Heart

The general structure of the heart and its coronary vessels has been seen in preceding dissections. Remove the heart by cutting the attachments of the **sinus venosus** to the transverse septum, and cutting across the posterior end of the ventral aorta, last afferent branchials, and coronary arteries. Wash out the sinus venosus and look inside it. This thin-walled chamber receives venous blood from the body by way of the paired hepatic veins and common cardinals already observed. Venous blood then enters the **atrium** by way of the slitlike **sinuatrial aperture** at the front of the sinus. This opening is guarded by a pair of lateral folds, the **sinuatrial valve**, which prevents the backflow of blood. One or more openings of the coronary veins may be seen in the wall of the sinus near the sinuatrial aperture.

Cut forward through the dorsal end of the sinuatrial aperture and the dorsal wall of the atrium, and clean out this chamber. It, too, is thin-walled, but muscular strands can be seen on the inside of its walls. Note that, despite its two lobes, its cavity is undivided. Find the **atrioventricular aperture** in the floor of the atrium. The opening is guarded by a pair of folds, the **atrioventricular valve**.

Insert a pair of scissors into the stump of ventral aorta, and cut back through the ventral surface of the **conus arteriosus** and continue to the posterior end of the **ventricle**. The thicker, muscular walls of

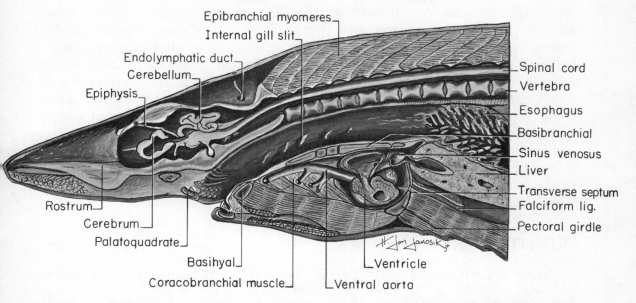

Epibranchial myomeres
Internal gill slit
Endolymphatic duct
Cerebellum
Epiphysis

Spinal cord
Vertebra
Esophagus
Basibranchial
Sinus venosus
Liver
Transverse septum
Falciform lig.
Pectoral girdle

Rostrum
Cerebrum
Palatoquadrate
Basihyal
Coracobranchial muscle
Ventral aorta
Ventricle

Figure 10–7. Sagittal section through the head of *Squalus* showing the structure of the heart and its relation to surrounding structures. A bristle passes through the sinuatrial aperture.

these chambers, especially of the ventricle, will be noted. The ventricular wall is also very spongy, for it is criss-crossed by numerous muscular strands. Notice that the lumen of the ventricle is U-shaped, for the entrance from the atrium and the exit to the conus lie nearly side by side. Backflow of blood from the front of the heart is prevented by the presence of three rows of **semilunar valves** within the conus. Each row consists of three pocket-shaped valves. One row lies at the very anterior end of the conus, and the other two lie close together near the posterior end of this chamber. The entrances to the pockets face anteriorly, so that they can be found and opened by probing posteriorly.

The dogfish heart lies in a pericardial cavity that is solidly encased by the cartilaginous pectoral girdle, the large basibranchial cartilage, and the hypobranchial musculature (Fig. 10–7). Its location within this relatively rigid box permits it to act as an aspiratory pump. When the ventricle contracts and blood is pumped out of the heart, pressure is lowered in the pericardial cavity, for the walls of the cavity do not collapse. This, in turn, lowers the pressure within the heart, and blood is sucked, so to speak, out of the veins and into the sinus venosus. This mechanism is of considerable importance in the return of blood from veins and sinuses in which the pressure is exceedingly low.

Pericardioperitoneal Canal

Now that the heart has been removed, it is possible to find the communication between the pericardial and the pleuroperitoneal cavities that was referred to on page 235. Pass a blunt probe into the recess in the transverse septum dorsal to the line of attachment of the sinus venosus. The probe will be seen to pass into a canal beneath the visceral peritoneum on the ventral surface of the esophagus, and after passing a distance of about 1 inch it will emerge into the pleuroperitoneal cavity through a semilunar opening. This entire passage is the **pericardioperitoneal canal**. Frequently the canal bifurcates at its caudal end, in which event there would be a pair of openings into the pleuroperitoneal cavity. Inconspicuous valves within the canal permit liquid to escape from the pericardial cavity into the pleuroperitoneal cavity. This would seem to be important for the maintenance of the low pressure within the pericardial cavity which is necessary for the aspiratory action of the heart.

PRIMITIVE TETRAPODS

Important changes occur in the cardiovascular system during the transition from water to land. The most conspicuous of these are in the heart and aortic arches, for the branchial circulation is lost, and a **pulmonary circulation** evolves. In this connection, the heart tends to become divided into a right and left side — the former receiving depleted blood from the body and sending it out to the lungs; the latter receiving aerated blood from the lungs and sending it out to the body. The separation is nearly complete in certain living lungfishes, but only the atrium is divided into right and left chambers in most amphibians. Reptiles also have a partial division of the ventricle. In most amphibians some gas exchange occurs in the skin as well as in the lungs. Therefore, aerated blood from the skin and depleted blood from the body both enter the right atrium; whereas only aerated blood from the lungs enters the left atrium. It has been found that in the frog (Rana) there is some mixing of the two circulations in the single ventricle, but the anterior aortic arches going to the head and body contain a greater percentage of left atrial blood, while the most posterior arch going to the lungs and skin contains a greater percentage of right atrial blood (DeLong, 1963). This, seemingly, is a less efficient arrangement than in a lungfish, but does allow an amphibian to remain submerged in the water for long periods. The only aerated blood returning to the heart during this period would be from the skin to the right atrium, but the absence of an interventricular septum permits some of this blood to be distributed to the tissues of the body. It is possible that the ancestors of modern amphibians had a more completely divided heart, and that the interventricular septum was lost in part as an adaptation for cutaneous respiration. Many urodeles, which depend to a greater extent than frogs on cutaneous respiration, have a reduced interatrial septum as well.

As regards the aortic arches, at least the first two are lost in primitive tetrapods, but (excepting the arch to the lung) those that remain are complete and are not interrupted by gill capillaries. An important corollary of the changes in the heart and aortic arches is a relative increase in the blood pressure in the dorsal aorta, and hence in the general efficiency of circulation. A frog, for example, has a systolic pressure in the dorsal aorta of about 30 mm. Hg compared to a mean pressure in the dorsal aorta of a dogfish on the order of 15 mm. Hg. The frog, moreover, is a much smaller animal, so that the relative efficiency of its circulation is even greater than these figures imply.

The venous system remains somewhat fishlike in early tetrapods, but important changes are seen in the primitive lateral abdominal system, and the right hepatic vein and parts of the cardinal veins. Parts of the last two have been transformed into a short **posterior vena cava**.

Necturus is a good example of the early tetrapod condition, excepting certain features of its heart and aortic arches. Being neotenic, this urodele retains larval methods of respiration. The cardiovascular system should be studied on doubly or triply injected specimens.

Heart and Associated Vessels

The pericardial cavity was noted during the study of the coelom. Open it wider if necessary, and identify the chambers of the heart. The most conspicuous chamber in a ventral view is the large, muscular **ventricle** which occupies the posteroventral portion of the cavity. The narrow vessel that emerges from the right side of the front of the ventricle is the **conus arteriosus**. This soon expands into a wider vessel, the **bulbus arteriosus**, lying at the front of the pericardial cavity. The conus arteriosus is a chamber of the heart; the bulbus arteriosus, the modified ventral aorta, which in this case lies within the pericardial cavity.[16] In many vertebrates that have it, the bulbus arteriosus acts to even out the pressure waves of ventricular contraction, and this is presumably its function in *Necturus*. The **atrium** lies dorsal to the conus and bulbus, and appears as the lobes on either side of these structures. It is partially divided internally into **right** and **left atria**. Lift up the posterior end of the ventricle (**apex** of the heart) and you will see the small thin-walled **sinus venosus**.

The pair of large veins that enter the sinus venosus posteriorly are the **hepatic sinuses**. **Common cardinal veins** enter the posterodorsal corners of the sinus venosus just anterior to the hepatic sinuses. A pair of **pulmonary veins** from the lungs passes dorsal to the hepatic sinuses and unites to form a single vessel which enters the left atrium. This vessel may be seen between the hepatic sinuses.

Since *Necturus* is a permanent larva respiring primarily by means of external gills, its heart is not a good example of the adult heart of a primitive tetrapod. The venous drainage of the body enters the sinus venosus, which in turn enters the right atrium. Arterial blood from the lungs enters the left atrium. Considerable mixing must occur in the atria, for the **interatrial septum** is poorly developed, and the two atria have a common opening into the single ventricle. Further mixing must occur in the ventricle and conus. This mixing is unimportant in *Necturus* for all the blood that leaves the heart passes through

---

[16] There is some terminological confusion of structures in this region. The anterior chamber of the heart of lower vertebrates is the conus arteriosus, but embryonically this region is called the bulbus cordis. The ventral aorta is the median vessel anterior to the heart from which the aortic arches arise. It generally lies anterior to the pericardial cavity. Truncus arteriosus is a synonym for the ventral aorta, and bulbus arteriosus is an expanded ventral aorta.

the gills before it is distributed to the body and lungs. Thus the heart of *Necturus* is very similar to that of *Squalus* functionally, and its structure is not sufficiently different to warrant its dissection.

Venous System

(A) *Hepatic Portal System*

Functionally, the hepatic portal system is the same in *Necturus* as in fishes, but the pattern of its tributaries differs somewhat. The major features of the pattern seen in *Necturus,* however, are very representative of those of tetrapods in general. Stretch out the mesentery of the intestine and you will see a longitudinal vessel, the **mesenteric vein**, passing forward and disappearing in the pancreas. Notice that it receives numerous **intestinal veins** from the intestine. Next look on the tail of the pancreas near the spleen. The vessel seen is the **gastrosplenic vein**, and it is formed by the confluence of a **splenic vein** from the spleen and several **gastric veins** from the stomach. Carefully dissect away pancreatic tissue and find the point where the gastrosplenic and mesenteric unite. The common vessel that passes forward from here to the liver is the **hepatic portal vein** (Fig. 10-8).

(B) *Ventral Abdominal Vein*

The median, longitudinal vessel that lies in the falciform ligament posterior to the liver is the **ventral abdominal vein**. This vessel has evolved from the ventral migration and fusion of the paired lateral abdominal veins of fishes, and its posterior relationships are still very similar to those of the lateral abdominal. It receives several small **vesical veins** from the urinary bladder (a ventral outgrowth of the embryonic cloaca), and then bifurcates into **pelvic veins**. Each pelvic vein extends laterally and caudally, and after about 1/4 inch receives, on its lateral side, a **femoral vein** from the hind limb. The vessel that continues from the pelvic and femoral caudally and dorsally to the renal portal vein (see below) is the **common iliac vein**. Blood from the leg may pass forward on one of two routes — the ventral abdominal (the primitive route) or the common iliac and renal portal (a new route).

The anterior relationships of the ventral abdominal differ from those of the lateral abdominal of fishes, for the ventral abdominal has lost its primitive connection with the common cardinals and has developed a new one with the hepatic portal system. In this respect it resembles its homologue in late amniote embryos — the umbilical vein (p. 264).

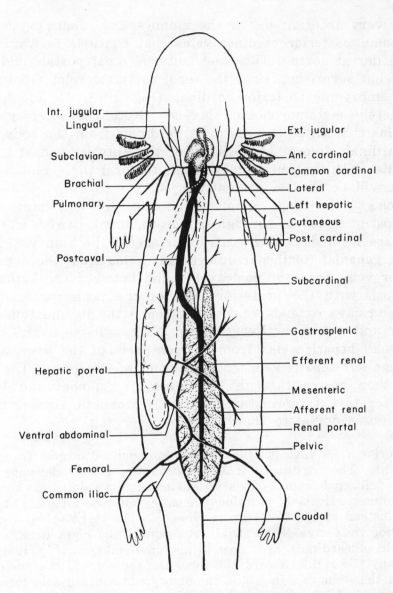

Figure 10–8. Diagrammatic ventral view of the venous system of *Necturus*. The outline of the liver is shown by broken lines.

(C) Renal Portal System

The vessel that runs along the lateral margin of each kidney dorsal to the prominent archinephric duct in the male, or oviduct in the female, is the **renal portal vein**. As already described, a common iliac enters each. Trace the two renal portals caudad and try to find the point where they unite and receive the median **caudal vein** from the tail. The renal portals receive blood from the tail and some from the legs and adjacent body wall, and carry it to the kidneys.

(D) Posterior Systemic Veins

At the very anterior end of the kidneys, the renal portals lead into a pair of small **posterior cardinal veins** that continue forward on either side of the dorsal aorta. This continuity of renal portals and posterior cardinals is not surprising, since the renal portals develop from the caudal end of the embryonic posterior cardinals (Fig. 10-3, *A*), and *Necturus* is an incompletely metamorphosed species. Note the **intersegmental**, or **parietal**, **veins** that enter the posterior cardinals from the body wall. The posterior cardinals diverge at the level of the anterior end of the esophagus and unite with the anterior cardinals to form the common cardinals. This region will be studied presently.

The blood in the renal portals that passes into the kidneys, and this would be most of it, leaves through numerous, small, paired, **efferent renal veins** that are located on the ventral surface of the kidneys. These, together with **gonadial** (**testicular** or **ovarian**) **veins** from the gonads, enter the **posterior vena cava** (or **postcaval**) lying between the kidneys. After an anastomosis with the posterior ends of the posterior cardinals, the posterior vena cava extends ventrally through the ligamentum hepatocavopulmonale and enters the liver. Trace it through the liver. It receives numerous small **hepatic veins** from various parts of the liver, and a particularly large **left hepatic vein** from the front of the liver. The posterior vena cava then passes through the coronary ligament and transverse septum. After this it bifurcates into the two **hepatic sinuses** that enter the sinus venosus dorsal to the ventricle.

The posterior vena cava is a new vessel compounded largely from ones previously present. The portion of it anterior to the kidney develops from the right hepatic vein, and from a caudal extension of this vein. This explains why there is only one particularly prominent hepatic vein (the left one) at the front of the liver instead of the two seen in *Squalus*. The right one is incorporated in the posterior vena cava. The caudal extension of the right hepatic taps into the embryonic subcardinals, as it does in mammal embryos (Fig. 10-15, *B*), and these (especially the right subcardinal) form the segment of the posterior vena cava between the kidneys. In fishes, the embryonic subcardinals form the posterior portions of the adult posterior cardinals (p. 268). Although they now contribute to the posterior vena cava, they still retain a connection with the posterior cardinals. Blood that has passed through the kidneys may take one of two routes forward. It either can go over to the posterior cardinals, its primitive route, or it can stay in the posterior vena cava. Most of it does the latter, as the posterior cardinals are reduced in size.

(E) Anterior Systemic Veins

The anterior systemic veins are hard to dissect unless they are filled with blood. The common cardinal veins were seen entering the sinus venosus. Each receives a number of tributaries, most of which can best be found peripherally and then traced to the common cardinal. They should be studied on the side of the body opposite to the one on which the muscles were dissected. Do not injure arteries while dissecting the veins.

Carefully remove the skin from the lateral surface of the brachium and shoulder, and you will see the **brachial vein**. It soon joins with a **cutaneous vein** from the skin to form the **subclavian vein**. Trace the subclavian forward. It turns into the musculature, and enters the **common cardinal vein** near the latter's dorsal end. Now remove the skin ventral to the gill slits and separate the hypobranchial muscles from the branchial region. The longitudinal vessel lying on the hypobranchial musculature in this region is the **lingual vein**. Trace it caudally, and it will be seen to enter the common cardinal beside, or in common with, the subclavian. Having located these two veins, cut through the muscles ventral to the union of these two veins with the common cardinal, and thereby expose the common cardinal descending to the heart.

Next remove the skin from the side of the trunk posterior to the shoulder. The longitudinal vessel lying between the epaxial and hypaxial muscles is the **lateral vein**. Remove the scapula and its muscles, and trace the vein forward. It enters the top of the common cardinal slightly dorsal to the preceding two vessels. The **posterior cardinal** can now be traced forward, and it will be seen to enter the common cardinal posterior to the entrance of the lateral vein. The vessel entering the top of the common cardinal anterior to the lateral and posterior cardinals is the **common jugular**, or **anterior cardinal**. Try to trace it forward. The largest part of the vessel (**external jugular vein**) passes dorsal to the gills, but a small branch (**internal jugular vein**) goes into the roof of the mouth.

The common and internal jugulars are homologous to the anterior cardinals of fishes. The subclavian and brachial represent the anterior part of the primitive lateral abdominal system, and they still have the same essential relationships as their homologues in *Squalus*. The lingual is homologous to the internal jugular of fishes. Although there has been some change in terminology, it is obvious that the major anterior veins of *Necturus* are very similar to those of fishes.

Arterial System

(A) *Aortic Arches and Their Branches*

Return to the pericardial cavity, and carefully dissect away the muscular tissue (mostly rectus cervicis) that lies between the cavity and the external gills on the side of the pharynx that has not been cut open. Two arteries leave from each side of the front of the bulbus arteriosus. Trace them laterally on the intact side. They are probably not injected, so be careful. They cross the transversi ventrales muscles (p. 128) and then pass deep to the subarcuals (Fig. 10-9). The more anterior one, known as the **first afferent branchial artery**, follows the first branchial arch (third visceral arch), and enters the first external gill. A small, probably well injected, artery lies just anterior to the distal portion of the first afferent branchial. This is the **external carotid artery**. Notice that it has numerous branches supplying the muscles in the floor of the pharynx and mouth. The external carotid is a branch of the efferent branchial system, hence its good injection, but it generally has tiny anastomoses with the first afferent branchial. One of the paired **thyroid glands** lies in the angle formed by the meeting of the first afferent branchial and external carotid. It can be recognized by its texture, for it is composed of many follicles which appear grossly as small vesicles.

The more posterior vessel leaving the bulbus arteriosus soon bifurcates. One branch, the **second afferent branchial artery**, follows the second branchial arch and enters the second gill. The other branch,

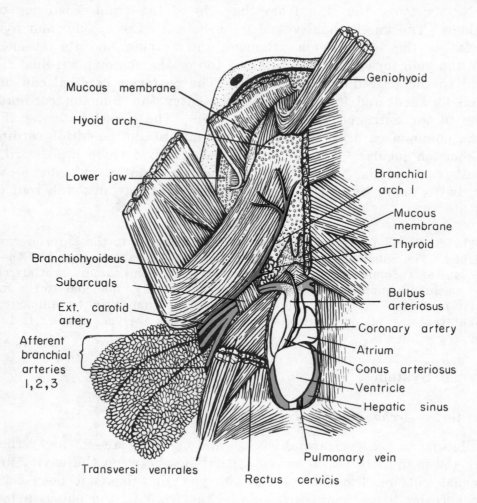

Figure 10–9. Ventral view of the afferent branchial arteries of *Necturus*. Much of the rectus cervicis has been cut away.

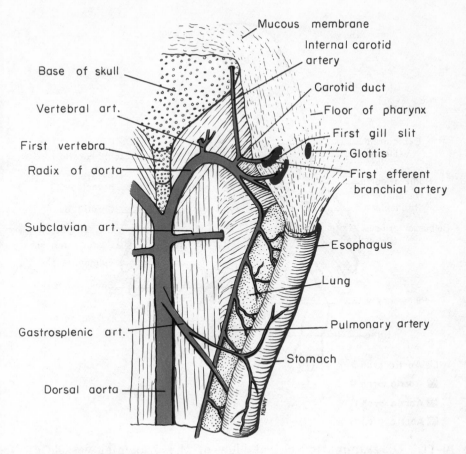

Figure 10–10. Ventral view of the efferent branchial arteries of *Necturus*.

the **third afferent branchial artery**, follows the third branchial arch and enters the third gill.

Swing open the floor of the mouth and pharynx, and carefully remove the mucous membrane from the roof of the pharynx. A large pair of vessels will be seen converging toward the middorsal line where they unite to form the **dorsal aorta**. They are the **radices** of the aorta. Carefully trace the radix on the intact side toward the gill slits. Slightly lateral to the vertebral column, it gives off a small, anterior branch (the **vertebral artery**), which soon disappears in the musculature at the base of the skull. Further laterally, the radix curves posteriorly. Another vessel lies anterior to this portion of the radix, and it is connected with the radix by a short, stout anastomosis called the **carotid duct**. That portion of the anterior vessel which extends forward in the roof of the pharynx from the carotid duct is the **internal carotid artery**; that portion which extends laterally from the carotid duct represents the entrance of the **first efferent branchial artery**. The internal carotid is distributed to the facial region, and enters the skull to supply the brain. Trace the first efferent artery as far laterally as feasible, noting that it comes

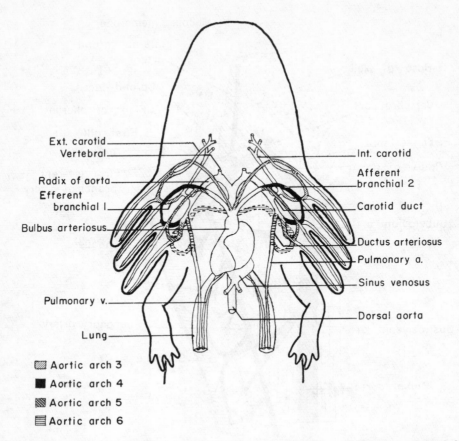

Ext. carotid
Vertebral
Radix of aorta
Efferent branchial I
Bulbus arteriosus
Pulmonary v.
Lung

Int. carotid
Afferent branchial 2
Carotid duct
Ductus arteriosus
Pulmonary a.
Sinus venosus
Dorsal aorta

▨ Aortic arch 3
■ Aortic arch 4
▩ Aortic arch 5
▤ Aortic arch 6

Figure 10–11. Diagrammatic ventral view of the branchial vessels of *Necturus* to show their derivation from the aortic arches. (Modified after Figge.)

from the first external gill. Continue to trace the radix laterally from the carotid duct. You will soon see a posterior branch. This is the **pulmonary artery**, and it should be traced to the lung. If difficulty is encountered, find the pulmonary artery on the lung and trace it forward. (The **pulmonary vein** lies on the opposite side of the lung. Its entrance into the heart has already been noticed.) Lateral to the pulmonary artery, the radix bifurcates and receives the **second** and **third efferent branchial arteries** from the respective gills.

It is difficult to get a clear view of the efferent branchials from the roof of the pharynx. To get a better view, approach the region from the outside by carefully dissecting away the skin from the posterior surface of the base of each external gill. The three efferent branchial arteries lie near the dorsal edge of the gills. Trace them to the point reached in the previous dissection. The origin of the external carotid from the first efferent branchial should also be found during this dissection.

Necturus, like all tetrapods, has lost the first two aortic arches, but the dorsal aorta and the ventral aortae as far caudad as the third arch persist as parts of the internal and external carotids, respectively (Fig. 10–11). The third and fourth aortic arches have given rise to the first two branchial arteries, but there has been some doubt concerning the derivation of the third branchial artery. It is often assumed that the fifth aortic arch is lost, and that the third branchial represents the sixth arch. The argument for this is that most tetrapods do lose the fifth arch while retaining the sixth, or all but the dorsal portion of it (ductus arteriosus), as the point of origin of the pulmonary artery. But Figge (1930) has shown that this is not the case in *Necturus*. The relationships of the third afferent branchial of *Necturus* to the third branchial arch (fifth visceral arch), its location just posterior to the second gill slit, and the fact that it leads into the third external gill (as the fifth arch does in other larval urodeles) all indicate that it represents the ventral half of the fifth aortic arch, and that the ventral half of the sixth has been lost. The dorsal relationships are not as clear, but if the first part of the argument is valid, it would seem that the third efferent branchial is the dorsal part of the fifth aortic arch, and that the pulmonary artery arises from the radix of the aorta by the dorsal part of the sixth aortic arch (ductus arteriosus)!

Figge has also shown that if the ventral part of the sixth aortic arch of a larval *Ambystoma tigrinum* is ligated, the larvae cannot be made to metamorphose unless they are in an unusually oxygen-rich environment. He believes that, since the lungs do not become functional, the oxygen level of the blood flowing through the gills cannot be raised to the level apparently necessary for their reabsorption. Some other investigators do not agree with this interpretation. Figge further postulates that the absence of the ventral part of the sixth aortic arch in *Necturus* explains its failure to metamorphose.

(B) Dorsal Aorta and Its Branches

A pair of **subclavian arteries** arises from the dorsal aorta just posterior to the union of the radices of the aorta. Trace one laterally. At the base of the appendage, it divides into a **brachial artery** that continues out into the arm, and a **cutaneous artery** to the skin and adjacent muscles. In amphibians with cutaneous respiration, the cutaneous artery is very large and carries blood to the skin where some gas exchange occurs.

Continue to follow the aorta posteriorly. It next gives off a median **gastrosplenic artery** which soon branches to go to various parts of the stomach, and to the spleen. The next median branch, the **celiacomesenteric artery**, arises some distance caudad. It passes ventrally to the tail of the pancreas where it divides into a **splenic artery** to the spleen, a **hepatic artery** to the liver, and a **pancreaticoduodenal artery** to the pancreas and duodenum. Much of the pancreas will have to be dissected away to see all of these branches. The remaining median visceral arteries are a number of **mesenteric arteries** to the intestine, and a pair of cloacal arteries to the cloaca (see page 290). You will have to separate the posterior vena cava from the aorta to see the point of origin of the mesenterics. Notice that the posterior vena cava lies toward the right of the aorta and the mesenteric arteries. Why?

The lateral visceral arteries consist of a number of paired **gonadial** (**testicular** or **ovarian**) **arteries** to the gonads, and very small **renal arteries** to the kidneys. The latter can be found by dissecting away the posterior vena cava between the caudal ends of the kidneys.

Paired intersegmental arteries include the subclavians already seen; a number of typical **intersegmental**, or **parietal**, **arteries** that arise from the dorsal surface of the aorta and pass into the body wall; and the **iliac arteries**. The iliacs can be found dorsal to the posterior ends of the kidneys. After traveling a short distance, each gives off anteriorly an **epigastric artery** that ascends in the body wall, posteriorly a **hypogastric artery** that supplies the urinary bladder and cloaca, and then the iliac continues as the **femoral artery** into the hind leg.

Cut through the body wall and muscles lateral and posterior to the cloaca, and trace the aorta caudad. It gives off the paired **cloacal arteries** referred to above, and then enters the hemal canal of the caudal vertebrae as the **caudal artery**.

MAMMALS

Changes that occur in the cardiovascular system during the evolution from primitive tetrapods to mammals correlate for the most part with the increase in activity and metabolism. The heart becomes completely divided morphologically into a right side receiving depleted blood from the body and sending it out to the lungs, and a left side receiving aerated blood from the lungs and sending it to the body. This complete division has doubtless been one factor in greatly increasing arterial blood pressure and the general efficiency of circulation. For example, the mean blood pressure in the caudal artery of a mouse is 136 mm. Hg compared with a systolic aortic pressure in the frog of 30 mm. Hg.

The aortic arches are further reduced, for the fifth is lost on both sides, as well as the first and second. The branches of the dorsal aorta continue to follow the basic pattern established in lower vertebrates.

In the anterior part of the venous system, a single anterior vena cava, or a pair of them, evolves from the anterior and common cardinals.

More caudally, the renal portal system is lost, and the posterior vena cava continues to the iliac and caudal veins. Neither the presence nor the loss of the renal portal system is completely understood. If the renal portal system evolved to insure the passage of a large volume of blood through the kidneys for excretory or unknown purposes, its loss in mammals may be correlated with an increased blood pressure, for a large volume of blood now enters the kidneys directly from the aorta. It has also been suggested that the posterior migration in most mammals of a part of the embryonic kidneys with the testes (see Chapter 11) would necessitate the loss of the renal portal system. Much of the anterior portion of the posterior cardinals is also lost, but a part of them is transformed into an azygos system of veins.

The primitive lateral abdominal system of veins is represented embryonically by the umbilical veins, but these are lost in the adult. The veins from the appendages are not connected with the umbilicals, but enter the venae cavae directly.

The cardiovascular system should be studied on specimens that have been at least doubly injected. Triply injected specimens are necessary to see all of the hepatic portal system, and quadruply injected ones for seeing most of the lymphatic system.

Heart and Associated Vessels

Carefully cut away the pericardial sac and thymus of your cat or rabbit from around the heart and its great vessels. The **heart** (**cor**) is a large, compact organ having a pointed posterior end (its **apex**) and a somewhat flatter anterior surface (its **base**). The **right** and **left ventricles**, which are completely separated internally, form the posterior two-thirds or more of the organ. They are approximately conical in shape, and have thick, muscular walls. The **right** and **left atria**, also completely separated internally, lie anterior to the ventricles, and are set off from them by a deep, often fat-filled groove called the **coronary sulcus**. The atria are thinner-walled and darker than the ventricles. They are separated from each other on the ventral surface by the great arteries leaving the anterior end of the ventricles. That portion of each atrium lying lateral to these arteries is called the **auricle**, or **auricular appendage**. The auricular appendages are somewhat ear-shaped, and tend to have scalloped margins. The separation between the ventricles appears, on the ventral surface, as a shallow groove extending from the left auricular appendage diagonally and toward the right.

Pick away the fat from around the large arteries leaving the anterior end of the ventricles. The more ventral vessel is the **pulmonary trunk** (Fig. 10–14). It arises from the right ventricle and extends dorsally to the lungs. Trace it later. The more dorsal vessel is the **arch of the aorta**. It arises from the left ventricle deep to the pulmonary trunk, but one cannot see much of it until it emerges on the right side of the pulmonary artery. Two small **coronary arteries** leave the base of the arch of the aorta, and pass to the heart wall. One can be found deep between the pulmonary artery and the left auricular appendage; the other, deep between the pulmonary artery and the right auricular appendage.

Push the heart to the left side of the thorax, and you will see the **posterior vena cava** coming through the diaphragm and entering the right atrium. An anterior vena cava will also be seen entering this chamber from the right side of the neck. The adult cat normally has only the **right anterior vena cava**, but the rabbit also has a **left anterior vena cava** that comes down the left side of the neck, crosses the dorsal surface of the heart, and enters the right atrium. Lift up the apex of the heart and you will see this vessel. Notice that it receives the **coronary veins** from the heart wall. The coronary veins of the cat collect into a **coronary sinus** which has a position comparable to the proximal end of the rabbit's left anterior vena cava. Carefully pick away connective tissue dorsal to the anterior venae cavae and from the roots of the lungs. You will find the **pulmonary veins** coming from the lungs and entering the left atrium. There are several veins, but those of each side generally collect into two channels before entering the heart. What is the course of the blood through the heart, and why is a separate system of coronary vessels necessary?

It will be noted that the mammalian heart contains only two of the primitive four chambers, but that these chambers have become completely divided. The apparently missing sinus venosus and conus arteriosus are present embryonically, but disappear as such in the adult. The sinus venosus is absorbed into the right atrium, and forms that part of the atrium receiving the venae cavae. The conus arteriosus (bulbus cordis of mammalian embryology), together with the ventral aorta, splits and forms the very base of the pulmonary trunk and arch of the aorta.

Venous System

(A) *Hepatic Portal System*

The hepatic portal systems of the cat and rabbit are basically the same, differing only in lesser details. Only the major parts of the system need be considered. In both animals, the **hepatic portal vein** (often simply called the **portal vein**, since mammals lack a renal portal system) can be found in the lesser omentum. It lies dorsal to the bile duct, and forms the ventral border of the foramen epiploicum. Trace the vein caudad. As it passes dorsal to the pylorus, it receives small tributaries from the stomach, duodenum, and pancreas. The ones from the stomach (**left** and **right gastric veins**) usually enter independently, but the others usually unite to form a common trunk, the **anterior pancreaticoduodenal** vein (Fig. 10-12). Slightly posterior to the pyloric region, you will see that the hepatic portal is formed by the confluence of two tributaries— a **gastrosplenic vein** coming in from the left, and a much larger **superior mesenteric vein** coming in posteriorly. Dissect away pancreatic tissue, and trace the gastrosplenic toward the spleen. It receives two **gastroepiploic veins** from the stomach, and two **splenic veins** (**venae lienalis**) from the spleen. Now trace the superior mesenteric. It extends caudad, receiving an **inferior mesenteric vein** from the large intestine, a **posterior pancreaticoduodenal vein**, from the posterior portion of the pancreas and duodenum, and numerous **intestinal veins** from the small intestine.

(B) *Posterior Systemic Veins*

The **posterior vena cava**, or **postcaval**, was seen entering the heart. Trace it caudad. Its tributaries are substantially the same in the cat and rabbit (Fig. 10-12). As it passes through the diaphragm, it receives several small **phrenic veins**, and then it disappears in the liver. Scrape away tissue from the anterior surface of the right medial lobe of the liver, and find the entrance of several large **hepatic veins**. The major part of the posterior vena cava, however, passes through the right lateral lobe, and it should also be exposed by scraping away liver tissue. Other hepatics, most very small, will be seen entering.

Upon emerging from the liver, the vessel continues caudad along the right side of the dorsal aorta, and can be followed by carefully picking away surrounding connective tissue. Its most anterior posthepatic tributaries are the paired **suprarenolumbar** and **renal veins**. Those of the right side enter more anteriorly, since the right kidney is more cranially situated than the left. Carefully remove the fat between the right kidney and the posterior vena cava, and find the right renal vein. It may be multiple. That is, the vessel may be represented by two or more channels. Now lift up the lateral edge of the kidney and look on the muscles dorsal to it. The vein that you see is the suprarenolumbar. Trace it toward the postcaval. It ordinarily enters the posterior vena cava slightly anterior to the renal, but it may join the renal. Just before entering, the suprarenolumbar vein crosses a small, hard, oval-shaped nodule imbedded in the fat between the cranial end of the kidney and the postcaval. This is the **suprarenal** or **adrenal gland** — one of the endocrine glands. The renal and suprarenolumbar veins and suprarenal gland on the left side should be found in the same way.

The next most posterior, paired tributaries of the posterior vena cava are the small **testicular** or **ovarian veins**, depending upon the sex. They are most easily found by locating the gonads, and then tracing the vessels to their point of entrance into the posterior vena cava. The **ovaries** are small, oval bodies lying near the cranial ends of the Y-shaped uterus. The testes have descended to the scrotum, and, in doing so, each has made an apparent hole (the **inguinal canal**) through the body wall. The inguinal canals can be found lateral to the attachments on the body wall of the lateral ligaments of the urinary bladder. The testicular veins, accompanied by comparable arteries and the sperm ducts (**ductus deferentes**), can be seen passing through these canals. The right gonadial vein enters the posterior vena cava posterior to the right renal — just posterior in the cat, some distance posterior in the rabbit. The left gonadial may enter either the left renal, or the posterior vena cava directly. The former entrance is more common in the cat; the latter more common in the rabbit.

Caudad to the entrance of the gonadial vessels, the posterior vena cava receives a pair of **iliolumbar veins** coming in from the musculature lying on the inner surface of the ilium. The rest of the dorsal musculature in the lumbar region is drained by a series of small **lumbar veins**. These drain both sides of the back, but enter the dorsal surface of the

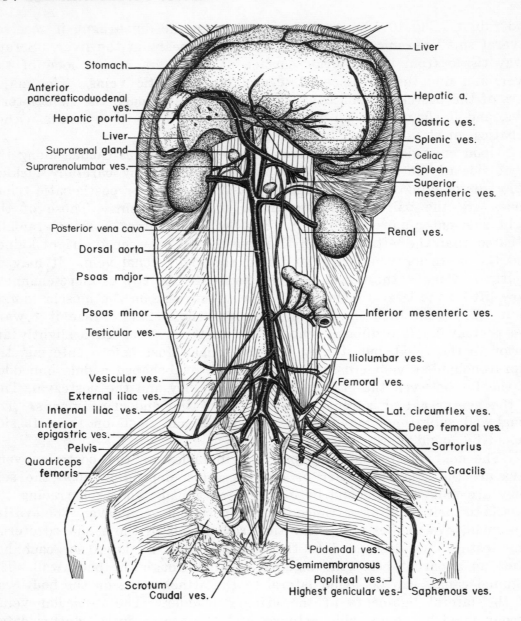

Figure 10–12. Ventral view of the posthepatic arteries and veins of the rabbit. The pelvic canal has been cut open. Arteries and veins running side by side and having the same name have been labeled only once. Most of the posthepatic arteries and veins of the cat are the same as these.

posterior vena cava as median vessels.

Just before entering the pelvic canal, the posterior vena cava passes deep to the arteries, and then receives the iliac veins from the pelvic region. To see this area clearly, one must open the pelvic canal. This is a simple procedure in the female. Cut the ventral ligament of the bladder and push it away from the anteroventral border of the pelvic girdle. Then take a scalpel, cut through the muscles on the ventral surface of the girdle, and continue to cut right through the midventral symphysis. Bone scissors may be used, but this is not usually necessary if you keep in the midventral line. Now take a firm grip on the thighs, and bend them as far dorsally as you can. The procedure for the male is the same, but one must use more caution in avoiding reproductive ducts. First locate the cremasteric pouches (narrow in the cat, wide in the rabbit) that extend from the inguinal canals, across the ventral surface of the girdle, and into the skin of the scrotum. They should be pushed aside before cutting. Also locate the penis emerging from the posterior end of the pelvic canal, and avoid cutting it. After the canal is opened, carefully pick away fat and connective tissue from around the vessels, bladder, and rectum. In so far as possible, confine your dissection to one side, and do not injure arteries or parts of the urogenital system.

In both cat and rabbit, an **external iliac vein** comes in from the leg and body wall, and a more medial **internal iliac vein** comes in from the pelvic region. These vessels enter the posterior vena cava independently in the rabbit, but those of each side first form a **common iliac vein** in the cat. The internal iliac collects from the rectum, bladder (in the cat), internal genital organs, and receives one or more vessels from the deeper muscles of the hip. A **caudal** (**median sacral**) **vein** from the sacrum and tail also enters one of the internal iliacs in the rabbit, usually the left. The caudal enters one of the common iliacs in the cat.

Next examine the tributaries of the external iliac. Near its entrance into the postcaval, it receives in the rabbit a **vesical vein** from the bladder. In both cat and rabbit it receives an **inferior epigastric vein** from the peritoneal surface of the rectus abdominis muscle just inside the body wall. The main continuation of the vessel in the leg is known as the **femoral vein**. Just before entering the body cavity, the femoral receives small tributaries from beneath the skin, from the external genitalia, and a large **deep femoral vein** from the deeper muscles of the thigh. The deep femoral enters the posteromedial surface of the femoral. Other tributaries of the femoral are shown in Figure 10–12.

(C) Anterior Systemic Veins

The **anterior vena cava**, or **precaval**, and its tributaries are sufficiently different in the cat and rabbit to warrant separate descriptions.

Cat

The cat has only a right anterior vena cava. Pick away fat and connective tissue from the front of the thorax and base of the neck, and trace the vessel forward. Do not injure any arteries. Its first tributary, which enters its dorsal surface very near the heart, is the **azygos vein**

(Fig. 10–13). The azygos curves around the anterior edge of the root of the right lung, and extends caudad along the right side of the vertebral column. It receives **intercostal veins** from between the ribs of both sides of the body. Most of the intercostals enter independently, but those of the first several intercostal spaces form a common trunk (the **highest intercostal**) before joining the azygos. The azygos also receives small vessels from the bronchi and esophagus, and helps to drain the anterior lumbar region.

As the anterior vena cava continues forward, it receives, on its ventral surface, a small vein from the thymus, and then a larger **internal thoracic** or **internal mammary vein**. Trace the internal thoracic. It passes to the ventral thoracic wall from which it receives a pair of internal thoracics lying deep to a layer of muscles (transversus thoracis). These continue caudad on either side of the midventral line of the thorax and abdomen. The abdominal portion of each vessel, which lies deep to the transversus abdominis, is called the **superior epigastric vein**. Each eventually anastomoses with one of the inferior epigastrics previously seen, but this union is hard to find.

About the level of the internal thoracic, the anterior vena cava receives on its dorsal surface the **right vertebral vein**. This vessel extends anteriorly, and, just anterior to the first rib, receives two vessels. The more anterior vessel is a continuation of the vertebral vein coming in from the transverse foramen of the cervical vertebrae. The more posterior vessel is the **right costocervical vein**. The costocervical receives a small tributary from the medial surface of the first rib, and then extends dorsally — one part going posterior to the vertebral attachment of the first rib toward the back muscles, and one part anterior to the first rib toward the deeper shoulder muscles. You will have to cut off the first rib close to the vertebral column, and dissect away parts of surrounding muscles (scalenus, serratus ventralis, intercostals) to see the deeper parts of the costocervical.

Anterior to the entrances of the internal thoracic and right vertebral, the precaval is formed by the union of two **brachiocephalic veins**. The left brachiocephalic is the longer, for it has to cross the base of the neck. The left brachiocephalic receives the **left vertebral vein** whose tributaries, being the same as those of the right, need not be dissected. Occasionally the right vertebral enters the right brachiocephalic. Each brachiocephalic is formed by the union of an **external jugular vein** descending from the side of the neck, and a **subclavian vein** coming in from the shoulder and arm.

Trace one of the subclavians. The vessel passes anterior to the first rib, and then is called the **axillary vein.** The portion of the vessel within the arm is known as the **brachial vein.** The brachial drains much of the arm, and the axillary receives a number of tributaries from the shoulder region, but these need not be considered. Several are shown in Figure 10–13.

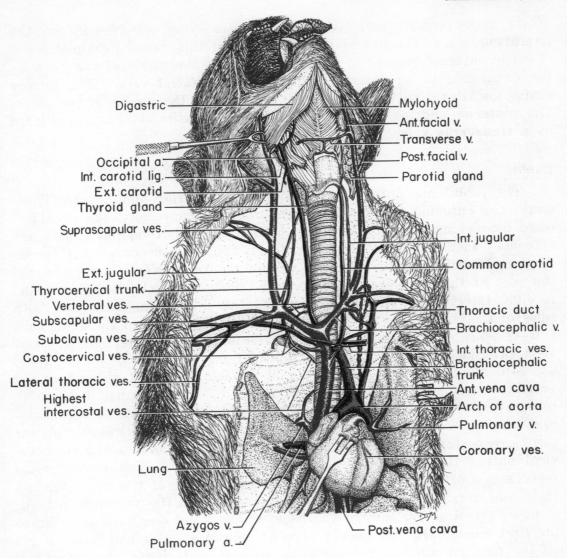

Digastric

Occipital a.
Int. carotid lig.
Ext. carotid
Thyroid gland
Suprascapular ves.

Ext. jugular
Thyrocervical trunk
Vertebral ves.
Subscapular ves.
Subclavian ves.
Costocervical ves.

Lateral thoracic ves.
Highest
intercostal ves.

Lung

Azygos v.
Pulmonary a.

Mylohyoid
Ant. facial v.
Transverse v.
Post. facial v.
Parotid gland

Int. jugular

Common carotid

Thoracic duct
Brachiocephalic v.

Int. thoracic ves.
Brachiocephalic
trunk
Ant. vena cava
Arch of aorta
Pulmonary v.

Coronary ves.

Post. vena cava

Figure 10–13. Ventrolateral view of the thoracic and posterior cervical arteries and veins of the cat.

Next trace one of the external jugulars. Very near its union with the subclavian, it receives on its medial surface a very small **internal jugular vein** that descends from the inside of the skull along the lateral surface of the trachea. **Lymphatic vessels** may also be seen entering the external jugular in this same region. They are not ordinarily injected, but in any event they can be recognized by their beadlike appearance. That is, the lymphatic vessels have expanded and contracted portions that give them the appearance of a string of beads.

More anteriorly, the external jugular receives, on its lateral surface, a large **suprascapular** or **transverse scapular vein** coming in from the front of the shoulder (Fig. 10–13.) If this vessel cannot be identified with certainty at its point of entrance, it can be located by first finding one of its major tributaries — the **cephalic vein** lying superficially on the lateral surface of the brachium — and tracing this vessel proximally.

The external jugular is formed caudad to the angle of the jaws by the confluence of an **anterior** and a **posterior facial vein**. The former drains the lateral portion of the face; the latter, the region of the pinna. The posterior ends of the anterior facials of opposite sides are connected by a **transverse vein**.

Rabbit

The rabbit has both a **left** and a **right** anterior vena cava. These were seen entering the heart. Trace the right anterior vena cava forward by picking away fat and connective tissue, but do not injure any arteries. Just before entering the heart, the right precaval receives on its dorsal surface an **azygos vein**. Follow the azygos. It curves around the root of the lung, extends posteriorly along the vertebral column, and receives **intercostal veins** from most of the intercostal spaces of both sides of the body. The more anterior intercostal spaces of the right side, however, are drained by the **highest intercostal**, which enters the anterior vena cava just anterior to the entrance of the azygos.

Still more anteriorly, on its dorsal surface, the precaval receives the **vertebral vein**. Trace it. It passes dorsally, and soon will be seen to receive two vessels. The anterior one is the continuation of the vertebral coming in from the transverse foramen of the cervical vertebrae; the more posterior is the **costocervical veins**. Occasionally, the costocervical enters the anterior vena cava independently. The costocervical receives a small tributary from the medial surface of the first rib, and then goes very deep, one part passing posterior to the vertebral attachment of the first rib toward the back muscles, and one part anterior to the first rib toward the deeper shoulder muscles. You will have to cut off the first rib close to the vertebral column, and dissect away parts of surrounding muscles (scalenus, serratus ventralis, intercostals) to see the deeper parts of the costocervical.

About the level of the vertebral, the precaval receives on its ventral surface one or more small veins from the thymus, and the **internal thoracic** or **internal mammary vein**. The internal thoracic comes in from the thoracic and abdominal wall where it will be found lateral to the midventral line and deep to a layer of muscles (transversus thoracis and abdominis). The abdominal portion of the vessel is called the **superior epigastric vein**. This vein anastomoses with the inferior epigastric previously seen, but it is difficult to trace the vessel this far caudad.

The anterior vena cava is formed anterior to the vertebral and internal mammary by the union of a lateral **subclavian vein**, and an anterior **external jugular vein**. After the subclavian passes lateral to the first rib, it is called the **axillary vein**. The portion of the vessel within the arm is called the **brachial vein**. The brachial drains most of the arm,

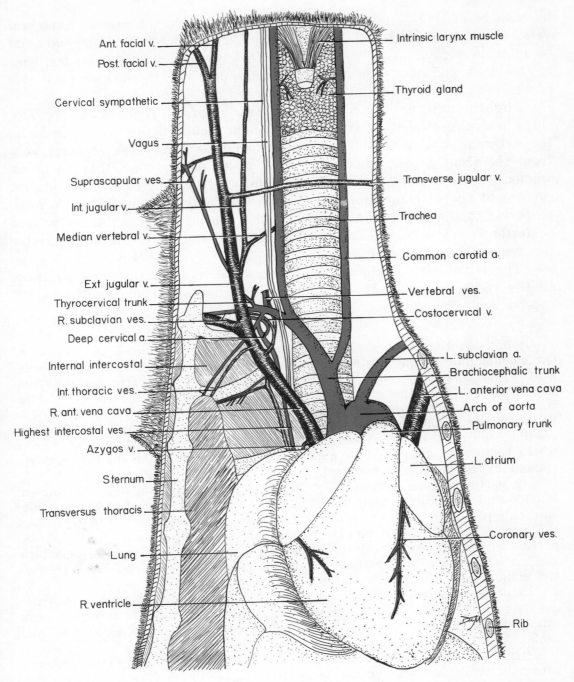

Figure 10–14. Ventral view of the thoracic and cervical vessels of the rabbit.

and the axillary receives several tributaries, which need not be traced, from the shoulder muscles.

Trace the external jugular. Near its origin with the subclavian, it receives on its medial side a very small **internal jugular vein** which descends from the inside of the skull along the lateral surface of the trachea. **Lymphatic vessels** may be seen entering the veins in the same region, but can be distinguished by their beaded appearance. A **median vertebral vein**, which courses dorsal to the esophagus, enters either the external or internal jugular near their union. This vessel descends from the base of the skull, and may either bifurcate and enter the jugulars on each side of the neck, or remain single and enter on one side only.

Slightly anterior to this region, the external jugulars of opposite sides are connected by a **transverse jugular vein.** Still more anteriorly, the external jugular receives a **suprascapular** or **transverse scapular vein** from the front of the shoulder. This vessel is the one that receives, among other tributaries, the **cephalic vein** lying superficially on the lateral surface of the shoulder and brachium. The external jugular is formed, posterior to the angle of the jaw, by the confluence of an **anterior** and **posterior facial vein**, the former draining the side of the face, the latter the region about the pinna.

The tributaries of the left anterior vena cava are the same as those on the right, except that it receives the coronary veins, and does not receive any azygos vein.

It has doubtless been noticed, during the above dissections, that parts of the mammalian venous system resemble parts of the system in other vertebrates, but that parts have changed considerably. The hepatic portal system is substantially the same as in lower tetrapods, and the primitive lateral abdominal veins are represented by the umbilical veins of the embryo (p. 264). The major change is the conversion of parts of the hepatic veins and the primitive cardinal and renal portal systems into a caval and azygos system. The way in which this comes about is best understood by recourse to the embryonic development of the veins in a mammal.

An early mammal embryo (Fig. 10–15, A) has a cardinal and an incipient renal portal system, for some of the blood in the posterior part of the posterior cardinals passes through the kidneys to a pair of subcardinals. In this stage the mammalian embryo is similar to the fish, except that in an adult fish the portion of the posterior cardinals situated just posterior to the anterior attachment of the subcardinals atrophies, and the flow of renal portal blood through the kidneys and into the subcardinals is mandatory.

Later in development (Fig. 10–15, B), the right hepatic enlarges, and a caudal extension of the vessel unites with the right subcardinal to form the proximal part of the posterior vena cava. The two subcardinals also unite with each other. This stage is not unlike the urodele.

Still later (Fig. 10–15, C), most of the anterior portion of the posterior cardinals atrophies, but the caudal portion on each side forms a large vessel connecting with the subcardinals. The essentially new feature, regarding the posterior veins of mammals, is the subsequent formation of a pair of **supracardinals** (Fig. 10–15, C), connecting anteriorly and posteriorly with the remnants of the posterior cardinals. The supracardinals also become connected with the subcardinals by a pair of **subsupracardinal anastomoses**, or the **renal collar**. This connection makes possible the elimination of most of the posterior portion of the posterior cardinals (the renal portal system of lower vertebrates).

During subsequent development, the supracardinals become divided into an anterior thoracic portion, and a posterior lumbar portion (Fig. 10–15, D). The

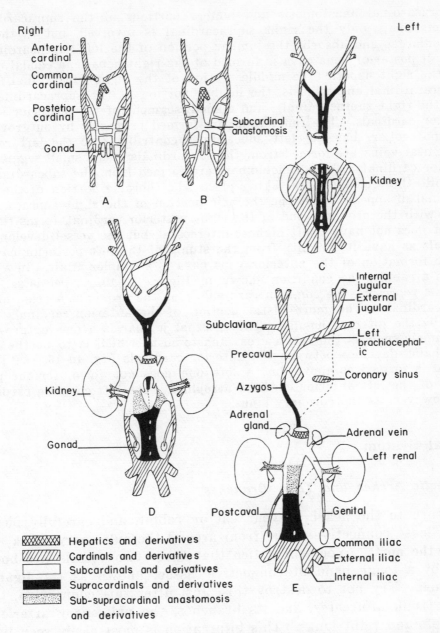

Right

- Anterior cardinal
- Common cardinal
- Postetior cardinal
- Gonad

A

- Subcardinal anastomosis

B

Left

- Kidney

C

- Kidney
- Gonad

D

- Subclavian
- Precaval
- Azygos
- Adrenal gland
- Postcaval

- Internal jugular
- External jugular
- Left brachiocephalic
- Coronary sinus
- Adrenal vein
- Left renal
- Genital
- Common iliac
- External iliac
- Internal iliac

E

- ⬛⬛ Hepatics and derivatives
- ⧄⧄ Cardinals and derivatives
- ▢ Subcardinals and derivatives
- ⬛ Supracardinals and derivatives
- ⣿ Sub-supracardinal anastomosis and derivatives

Figure 10–15. A series of diagrams ranging from a young embryo *A*, to an adult, *E*, to show the development of the major veins of the cat from the primitive cardinal and renal portal system. All are ventral views. For explanation, see text. (Slightly modified after Huntington and McClure, The development of the veins in the domestic cat, Anatomical Record, Vol. 20.)

right subsupracardinal anastomosis and lumbar portion of the supracardinal enlarge, while those of the left side do not. Renal veins grow out from the renal collar to the definitive kidneys, which have migrated cranially.

By the adult stage (Fig. 10–15, *E*), all but the most caudal segments of the posterior cardinals are lost, the left subsupracardinal anastomosis is lost, and the posterior vena cava is extended caudad by the enlargement of the right subsupracardinal anastomosis and lumbar portions of the supracardinals. In some mammals, only the right supracardinal is involved, but in the cat the right enlarges and absorbs the lumbar portion of the left supracardinal. Thus the adult posterior vena cava is formed of the right hepatic, a caudal outgrowth from the right hepatic, the middle section of the right subcardinal, the right subsupracardinal anastomosis, the lumbar portion of the supracardinals (especially the right supracardinal), and a small segment of the posterior end of the posterior cardinals. The renal veins are formed primarily by outgrowths from the renal collar, but the left subcardinal contributes to the left renal vein. The genital veins are formed from the subcardinals plus a small segment of the posterior cardinals; the adrenolumbars are formed from the subcardinals.

While these changes are taking place, the thoracic portion of the left supracardinal disappears. But the thoracic portion of the right supracardinal, together with the proximal end of the right posterior cardinal, forms the azygos. The cat does not have a left highest intercostal, but the vessel develops in such mammals as have it (rabbit) from the stump of the left posterior cardinal.

The formation of the anterior vena cava is a simpler affair. In a mammal such as a rabbit, the condition shown in Figure 10–15, C, persists. The two precavals represent the common cardinals plus the proximal portion of the anterior cardinals. The more distal portion of the anterior cardinals is represented by the internal jugular. The external jugular is a new outgrowth. But in mammals such as the cat, a cross anastomosis, which is to be the left brachiocephalic, develops between the anterior cardinals (Fig. 10–16, D). The right precaval is formed as above, but a left one does not form, for the proximal portion of the left anterior cardinal atrophies. The left common cardinal persists, however, as the coronary sinus.

Arterial System

(A) Aortic Arches and Their Branches

Return to the heart of your cat or rabbit, and carefully pick away fat and loose connective tissue from around the **pulmonary trunk** and the **arch of the aorta**. You will notice that these two vessels are bound together by a tough band of connective tissue, known as the **ligamentum arteriosum**. Try not to destroy this. Just after this connection, the pulmonary trunk bifurcates, and its branches, the **pulmonary arteries**, pass to the left and right lungs. This bifurcation is most easily seen by pushing the pulmonary trunk anteriorly, and dissecting between it and the anterodorsal portion of the heart.

The arch of the aorta, after giving off the **coronary arteries** previously described (p. 291), curves dorsally and to the left, disappearing dorsal to the root of the left lung. Two vessels arise from the front of this arch — a large **brachiocephalic trunk** nearest the heart, and then a smaller **left subclavian artery** (Figs. 10–13 and 10–14). The brachiocephalic passes forward, sends small branches to the thymus, and soon breaks up into three vessels — two **common carotid arteries** that ascend the neck on either side of the trachea and a **right subclavian artery**. In the cat and rabbit, these four major anterior arteries (two common carotids and two subclavians) usually spring from the arch of the aorta as described. But some variation is encountered in these animals, and other patterns are seen in other mammals. These may range from a common origin of all four vessels to an independent origin for each of the four.

Trace one of the subclavians forward, preferably the one on the side on which the veins were identified, if the arteries are still intact. Medial to the first rib, the subclavian gives rise to four or five arteries which are most accurately identified from their peripheral distribution. An **internal thoracic** or **internal mammary artery** arises from the ventral surface of the subclavian, and passes to the thoracic wall where it accompanies the corresponding vein. When the vessel enters the abdominal wall, it is referred to as the **superior epigastric artery**. A **vertebral artery** arises from the dorsal surface of the subclavian opposite, or nearly opposite, the internal thoracic. It enters the transverse foramen of the cervical vertebrae along with the corresponding vein.

The above two vessels are the same in the cat and rabbit. The others differ slightly. In both animals, a **highest intercostal artery** will be seen lying near the vertebral attachment of the anterior ribs. It sends out **intercostal arteries** to the first several intercostal spaces, and a deeper branch to the back muscles. Also, in both animals, a **deep cervical artery** will be seen passing anterior to the first rib and into the serratus ventralis muscle. It accompanies a part of the costocervical vein previously seen. These two arteries normally arise independently from the dorsal surface of the subclavian just posterior to the vertebral in the rabbit. But, in the cat, they arise in common from the dorsal surface of the subclavian anterior to the vertebral. The common trunk is called the **costocervical trunk**.

Anterior and lateral to all these vessels, the subclavian in both animals gives rise to a branch (the **thyrocervical trunk**) that passes forward near the external jugular vein, and then curves laterally with (cat), or near (rabbit), the suprascapular vein. The major peripheral branch of this trunk is the **suprascapular** or **transverse scapular artery**.

The subclavian then passes in front of the first rib and enters the armpit. This portion of the vessel is called the **axillary artery**. The axillary gives off a number of branches to shoulder muscles, which need not be considered, and then enters the arm as the **brachial artery**.

Next trace one of the common carotids. It ascends the neck lying between the internal jugular vein and the trachea. (If desired, a cervical extension of the sympathetic cord — the **cervical sympathetic** — and the **vagus nerve** can be found between the common carotid and internal jugular. These nerves form a common **vagosympathetic trunk** in the cat but are distinct in the rabbit, the vagus being the larger and the one that passes superficial to the brachiocephalic trunk.) As the common carotid extends toward the head, it gives off small branches to the trachea and adjacent structures, and a somewhat larger branch (the **superior thyroid artery**) to the anterior end of the thyroid gland. The cat also has a large, dorsal, muscular branch arising opposite the superior thyroid artery.

After this, the common carotid ascends for another 1/2 to 3/4 inch, and then it divides into an **external** and **internal carotid**. The former is distributed for the most part to the outside of the head; the latter passes through the carotid canal deep to the tympanic region, and enters the cranial cavity. The carotids are typical in the rabbit, and resemble those of the primitive carnivore shown in Figure 10–16, *A*. The internal carotid of the rabbit is a medium-sized vessel, and is the first dorsal branch of

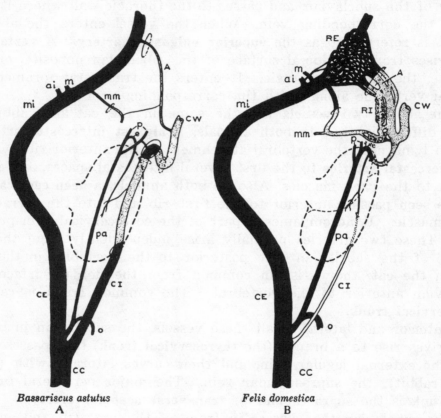

Bassariscus astutus
A

Felis domestica
B

Figure 10–16. Diagrams of the carotid circulation of a primitive carnivore, a member of the raccoon family, *A*, and of the domestic cat, *B*. Stippled parts are intracranial, or run through canals. The position of the tympanic bulla is shown by broken lines. Note the vestigial nature of the proximal part of the internal carotid of the cat (shown in outline), and the way in which other vessels have enlarged to carry blood to the brain. Some other carnivores have an intermediate condition. Abbreviations: *A*, anastomotic artery; *ai*, inferior alveolar; *CC*, common carotid; *CE*, external carotid; *CI*, internal carotid; *CW*, circle of Willis located on ventral surface of brain; *e*, eustachian; *m*, masseteric; *mi*, internal maxillary; *mm*, median meningeal; *o*, occipital; *p*, pharyngeal; *pa*, ascending pharyngeal; *RE*, external rete; *RI*, internal rete. (From Davis and Story, The carotid circulation in the domestic cat, Zoological Series, Field Museum of Natural History, vol. 28.)

the common carotid anterior to the superior thyroid. Just anterior to the internal carotid, an **occipital artery** arises from the external carotid and passes to the occipital musculature on the back of the neck. Near its origin, the occipital gives off an **ascending pharyngeal artery**.

The pattern of the carotids is more complicated in the cat (Figs. 10–13 and 10–16, *B*). The first dorsal vessel anterior to the superior thyroid is not the internal carotid, but the occipital and ascending pharyngeal. The proximal part of the internal carotid, although present as a vessel embryonically, is vestigial in the adult cat. It is represented by a small ligament that can be found by careful dissection just posterior to the occipital. The distal part of the internal carotid, however, is present, and the ascending pharyngeal anastomoses with it.

The major arteries described are derived from the aortic arches in the manner shown in Figure 10–17. All six of the primitive aortic arches appear during embryonic development, and connect the ventral aorta (paired anterior to the fourth aortic arch) with the dorsal aorta (paired in the region of the aortic arches and for a short distance caudad). The first, second, and fifth aortic arches, the dorsal part of the right sixth aortic arch, the paired dorsal aortae between arches three and four, and the right paired dorsal aorta caudad to the entrance of the right subclavian (an intersegmental artery) disappear during development. The dorsal part of the left sixth arch persists during embryonic life, as the **ductus arteriosus of Botallus** and shunts blood from the pulmonary arteries directly to the dorsal aorta. For a few hours after birth it shunts some blood in the opposite direction, thereby giving this portion of the blood a double aeration, but it soon becomes converted into the functionless **ligamentum arteriosum**.

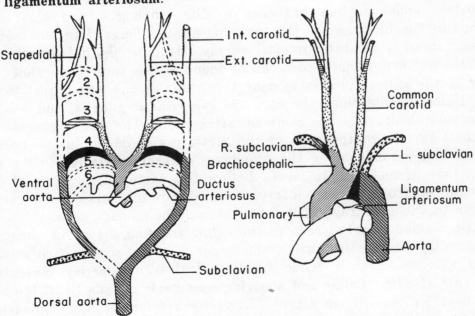

Figure 10–17. Diagrammatic ventral views of the mammalian aortic arches and their derivatives. *A*, embryonic condition; *B*, adult condition in man. (Slightly modified after Barry.)

The ventral portions of the sixth arches persist as the pulmonary arteries. The left fourth arch, together with part of the left dorsal aorta, form the arch of the aorta. (Differential growth has the effect of shortening this arch and the adjacent dorsal aorta, so that the left subclavian of the adult leaves the arch of the adult aorta much closer to the common carotids than it does in the embryo.) The right fourth arch, plus a segment of the right dorsal aorta, forms the proximal part of the right subclavian. A splitting of the caudal part of the ventral aorta, and of the conus arteriosus, results in the direct origin of the arch of the aorta and the pulmonary trunk from the ventricles.

The paired ventral aortae between the fourth and third arches form the common carotids. The ventral aortae anterior to the third arch become the external carotids; the third arches, plus the dorsal aortae anterior to them, the internal carotids. The internal carotid of the embryonic mammal not only supplies the intracranial part of the head, but also, by its stapedial branch passing through the stapes, much of the outside of the head. But, in the adults of most mammals, the external carotid taps into the stapedial and pirates most, or all, of its peripheral distribution. If the external carotid takes it all over, the stapedial disappears.

(B) Dorsal Aorta and Its Branches

After curving to the dorsal side of the body, the arch of the aorta is known as the **dorsal aorta**, or simply as the **aorta**, for adult mammals do not have a ventral aorta. Trace the aorta caudally. Its branches can be grouped into the categories shown in Figure 10-6. As it passes through the thorax along the left side of the vertebral column, it gives off paired **intercostal arteries** to those intercostal spaces not supplied by the highest intercostals, small median branches to the esophagus, and also small branches to the branches to the bronchi, for the lungs, like the wall of the heart, need a separate arterial supply. (If desired, the **thoracic portion** of the left sympathetic cord can be found at this time by carefully dissecting in the connective tissue near the heads of the ribs dorsal to the aorta. Enlargements along the cord are **sympathetic ganglia**, and delicate strands passing dorsally are **communicating rami**. The left vagus crosses the lateral surface of the arch of the aorta, passes dorsal to the root of the lung, and caudally along the esophagus.) **Phrenic arteries** to the diaphragm may arise from the aorta before the aorta passes through the diaphragm, or from the last intercostals, or they may arise from vessels posterior to the diaphragm (first lumbar, adrenolumbars, celiac).

Push the abdominal viscera to the right, and find the aorta emerging from the diaphragm. Just after emerging, it gives rise to two median visceral vessels — first a **celiac trunk** and then a **superior mesenteric artery** (Fig. 10-12). **Celiac and superior mesenteric ganglia** lie at the base of the superior mesenteric artery. They receive one or more **splanchnic nerves** from the sympathetic cord, and send out minute branches which travel along the vessels to the viscera. The superior mesenteric artery passes to most of the intestinal region; the celiac to the stomach, spleen, liver, pancreas, and anterior part of the duodenum. The peripheral branches of the mesenteric can be seen by spreading the mesentery, but one must carefully pick away surrounding connective tissue to trace the celiac. After a distance of about 3/4 inch, the celiac breaks up into three vessels — a **splenic (lienalis)**, **left gastric**, and **hepatic artery**. The hepatic and left gastric may arise in common by a short trunk. Follow each of these vessels. The left gastric passes to the anterior part of the stomach. The splenic goes to the spleen, but also sends branches to the stomach and pancreas. The hepatic passes deep to the stomach, and into the lesser omentum where it lies to the left of the hepatic portal vein (Fig. 9-17). As its name implies, the hepatic goes to the liver, but it also sends branches to the stomach, pancreas, and beginning of the duodenum. Most of the peripheral branches of the celiac and, superior mesenteric accompany tributaries of the hepatic portal system.

Slightly caudad to the superior mesenteric, the aorta gives rise to a pair of **suprarenolumbar arteries**, and to a pair of **renal arteries** in the cat. In the rabbit, the suprarenolumbars generally spring from the renals very near the aorta. Each renal passes to a kidney, but may send a small branch to the adrenal on its way. Each suprarenolumbar passes dorsal to a suprarenal gland, to which it sends a branch, and then dorsal to the kidney to help supply the lumbar musculature. The rest of the lumbar musculature is supplied by the iliolumbars (see below) and by **lumbar arteries**. The lumbar arteries arise as median vessels from the aorta, but soon bifurcate. All these vessels accompany corresponding veins.

Paired **testicular** or **ovarian arteries**, depending on the sex, arise from the aorta some distance posterior to the renals. They pass to the gonads in company with comparable veins. The ovarian also helps to supply the uterus, and is very large in pregnant specimens. A median **inferior mesenteric artery** arises from the aorta about the same level as the gonadial arteries (rabbit), or caudal to this region (cat). After a short distance, the inferior mesenteric bifurcates and passes fore and aft along the large intestine in company with the inferior mesenteric vein.

Caudad to the inferior mesenteric, the aorta gives rise to a pair of **iliolumbar arteries**, which pass laterally with comparable veins to the musculature and body wall ventral to the ilium, and to the paired **external** and **internal iliac arteries**. The iliacs of each side arise from a **common iliac** in the rabbit, but independently from the aorta in the cat. Occasionally the iliolumbars arise from the common iliac.

Trace one of the external iliacs. It extends diagonally laterally and posteriorly to the body wall with the external iliac vein. As it approaches (cat) or passes through (rabbit) the body wall, it gives off, from its posteromedial surface, a **deep femoral artery** that goes deep into the thigh. An **inferior epigastric artery** arises either from the external iliac near the deep femoral (rabbit), and ascends along the peritoneal side of the rectus abdominis muscle to anastomose with the superior epigastric previously seen. After going through the body wall, the external iliac is known as the **femoral artery**. The femoral passes down the medial side of the thigh, but gives off small branches to the groin and external genitals before doing so.

Next follow the internal iliac, very near its origin, it gives off a small **vesical artery** that passes to the urinary bladder and, in females, to the uterus as well. The vesical artery is the remnant of the large umbilical artery of the embryo. The proximal parts of the umbilicals persist in the adult to supply the bladder, but the portion from the bladder to the umbilicus atrophies. The internal iliac then passes into the pelvic canal and supplies the rectum and deeper hip muscles.

After giving off the iliacs, the dorsal aorta continues caudad as the small **caudal artery** (**median sacral artery**). It travels along the median surface of the sacrum, and enters the tail.

Bronchi and Internal Structure of the Heart

Cut the great vessels near the heart of your specimen, remove the heart, and examine the roots of the lungs. The bifurcation of the trachea into bronchi referred to earlier (p. 252) can now be exposed.

Review the external features and general structure of the heart (p. 291), and then examine its internal features by dissecting either the heart of your own specimen, or a separate sheep heart. The latter is preferable, if material is available, for the structures are larger and the chambers are not clogged with the injection mass. If a sheep heart is used, you will have to remove the pericardial sac, and clean and identify the great vessels. They are similar to those of the cat and rabbit, except that both of the subclavian and common carotid arteries leave the arch of the aorta by a common brachiocephalic. A small left anterior vena cava is also present in the sheep, and the ligamentum arteriosum is conspicuous.

Open the right atrium by making an incision that extends from the auricular appendage into the posterior vena cava; the left atrium, by an incision extending from its auricular appendage through one of the pulmonary veins. To open the ventricles, first cut off the apex of the heart in the transverse plane. Cut off a sufficient amount to expose the cavities of both ventricles. Then make a cut through the ventral wall of the right ventricle, and extend it from the cut surface made by removing the apex into the pulmonary artery. This will be a diagonal incision. Open the left ventricle by making an incision through its ventral wall that extends from the cut surface as far forward as the base of the arch of the aorta. This will be a longitudinal incision. Clean out the chambers of the heart if necessary.

Find the entrances of the **right anterior vena cava** and **posterior vena cava** into the **right atrium**. The entrance of the **coronary sinus** (cat), or **left anterior vena cava** (rabbit) lies just posterior to the entrance of the posterior vena cava. The extent of the coronary sinus can be determined by probing. Also find the entrances of the **pulmonary veins** into the **left atrium**. The atria have relatively thin muscular walls; however, the muscles in the **auricular appendages** form prominent bands known as **pectinate muscles** because they resemble a comb, or pecten.

The two atria are separated by an **interatrial septum**. Examine the septum from the right atrium, and you will find an oval-shaped depression, the **fossa ovalis**, beside the point where the postcaval enters. Put your thumb in one atrium and forefinger in the other, and palpate this region. You will feel that the septum is unusually thin here. During embryonic life, there is an opening, the **foramen ovale**, through the septum at this point, and much of the blood in the right atrium (mostly blood coming in by the postcaval) is sent directly to the left atrium and out to the body. This opening closes at birth.

The **atrioventricular openings** will be seen in the floor of the atria. The right one is guarded by the **right atrioventricular**, or **tricuspid valve** which consists of three flaps; the left one by the **left atrioventricular**, or **bicuspid valve**, which consists of two flaps (Fig. 10-18). Since these

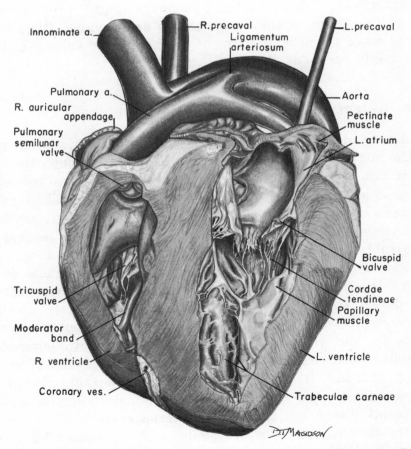

Figure 10-18. Ventral view of a dissection of the sheep heart.

flaps extend into the ventricles, they can be seen better from that aspect. Note that little tendinous cords (**chordae tendineae**) connect the margins of the flaps with the walls of the ventricles. Many of the chordae attach onto papilla-like extensions of the ventricular muscles (**papillary muscles**). The chordae tendineae may help to open the valves, but in any case they prevent the valves from everting into the atria during ventricular contraction.

Notice that the ventricles are separated from each other by an **interventricular septum**, and that the walls of the ventricles are much thicker than those of the atria. The left ventricular wall is also much thicker than the right one. Why? In addition to the papillary muscles, the inside of the ventricular walls bears irregular bands (**trabeculae carneae**), and sometimes bands that cross the lumen (**moderator bands**). There is a particularly prominent moderator band in the right ventricle of the sheep. Moderator bands are believed to prevent the overdistention of the ventricle.

Notice where the pulmonary trunk and arch of the aorta leave the ventricles. Three pocket-shaped, **semilunar valves** are located in the base of each vessel, for this part of each vessel developed from a splitting of the conus arteriosus. Those in the pulmonary artery are known as the **pulmonary valve**; those in the aorta, as the **aortic valve**. The two coronary arteries leave from behind two of the semilunar valves in the aorta. One has probably been cut through.

Lymphatic System

The relation of the lymphatic to the cardiovascular system was considered in the introduction to this chapter (p. 262). Although the lymphatic system is not conspicuous enough in most vertebrates to be studied easily, parts, at least, of the system can be seen in mammals even though it has not been specially injected. The following directions are based on the cat, inasmuch as the lymphatics are easier to find in that animal than in the rabbit, but are applicable to other mammals as well.

The major lymphatic vessel of the body is the **thoracic duct** (Fig. 10-19). This is a brownish vessel that can be found in the left pleural cavity just dorsal to the aorta. Sometimes the vessel is divided into two or more channels. Trace it forward. It passes deep to most of the anterior arteries and veins, and then curves around to enter the left external jugular beside the entrance of the internal jugular (Fig. 10-13). Now trace it caudad. It passes through the diaphragm dorsal to the aorta, and, dorsal to the origin of the celiac and superior mesenteric arteries, expands into a sac called the **cisterna chyli**.

Next stretch out a section of the mesentery supporting the small intestine, and hold it up to the light. Very small lymphatic vessels, in

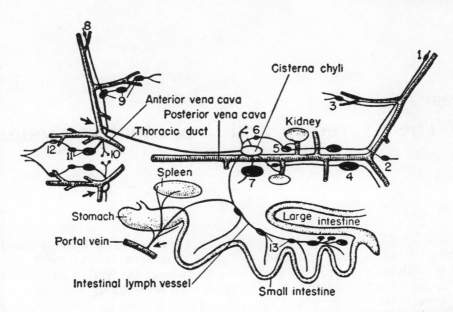

Figure 10–19. Diagrammatic ventral view of the deeper lymphatic vessels of the rat. Anterior is toward the left. The lymphatic vessels are shown in solid black beside the major veins that they accompany. Most of the lymphatics ultimately enter the subclavian veins (at two arrows on left) in the rat, but they enter the proximal ends of the external jugulars in the cat. Some lymphatics also enter the portal vein in the rat (lower arrow). The lymph nodes are named according to the region in which they lie: 1, knee; 2, tail; 3, inguinal; 4, lumbar; 5, kidney; 6, nodes near cisterna chyli; 6, intestinal; 8, elbow; 9, axilla; 10, thoracic; 11, cervical; 12, submaxillary nodes receiving lymphatics from tongue and lips (t); 13, mesenteric nodes (aggregate to form pancreas of Aselli in cat). (From Romer, The Vertebrate Body. After Job.)

this case called **lacteals**, for absorbed fat passes through them, can be seen outlined by little streaks of fat. These ultimately lead into an aggregation of lymph nodes (the **pancreas of Aselli**) located at the base of the mesentery. The pancreas of Aselli in turn is drained by one or more larger lymphatics that pass along the superior mesenteric artery to the cisterna chyli. These vessels have probably been destroyed. Lymphatic vessels from the stomach, liver, pelvic canal, and hind legs also pass to the cisterna chyli, but they are hard to see. Thus the cisterna chyli receives all the lymphatic drainage of the body posterior to the diaphragm, and passes it on the thoracic duct. The thoracic duct receives the lymphatic drainage of the thorax as it ascends through this region.

Other lymphatic vessels, which parallel the larger veins, drain the arms, neck, and head. Those of the left side enter the thoracic duct, or the left external jugular close to the entrance of the thoracic duct. Those of the right side enter the right external jugular near the entrance of the internal jugular, either independently, or by a short common trunk (the **right lymphatic duct**).

11 / The Excretory and Reproductive Systems

THE EXCRETORY system is the last of the group of systems concerned with metabolism. It plays an important role in eliminating the nitrogenous waste products of cellular metabolism, and helps to control the water balances of the body. The reproductive system is concerned with an entirely different function, namely, perpetuating the species. However, the two must be considered together morphologically, for in the males of most vertebrates excretory passages are utilized for the transport of the sperm, and in some cases the female genital ducts develop from excretory ducts. In view of this intimate morphological association, the two systems are sometimes referred to as the urogenital system.

History of the Kidney

A brief consideration of the history of the kidneys and their ducts is a prerequisite for an understanding of the urogenital system. The functional units of the kidneys are the **nephrons (renal tubules)**. Details of the structure of these tubules need not concern us at this time, but in all vertebrates they develop embryonically from a pair of bands of nephrogenic tissue (**nephric ridges**) located dorsal to the coelom between the somites and lateral plate mesoderm. In the ontogeny of an amniote, a pronephric kidney (**pronephros**) is succeeded by a **mesonephros**, which in turn is succeeded by the definitive **metanephros** (Fig. 11–1). These kidneys have a linear relationship from anterior to posterior along the nephric ridge. The pronephros forms the **pronephric**, or **archinephric**, **duct**, which extends caudad to the cloaca. The mesonephric tubules tap into this duct (sometimes called the **mesonephric** or **wolffian duct**), but the metanephros is drained by a **ureter** which develops as a craniad outgrowth from the caudal end of the archinephric duct.

A common statement is that the above ontogeny is a recapitulation of phylogeny, but this is an oversimplification. In anamniotes, the kidney tubules develop from all of the nephric ridge as in amniotes, but the adult kidney is one that occupies both the mesonephric and the potential metanephric portion of the ridge. The anterior part, at least, of such a kidney is drained by the archinephric duct, but the caudal part may be drained by one or more accessory, ureter-like ducts. Such a kidney has often been called a mesonephros, but is more appropriately called an **opisthonephros**, for it is obviously somewhat different from the mesonephros of an amniote embryo. A still more primitive kidney would be one that occupies the entire nephric ridge, and retains the

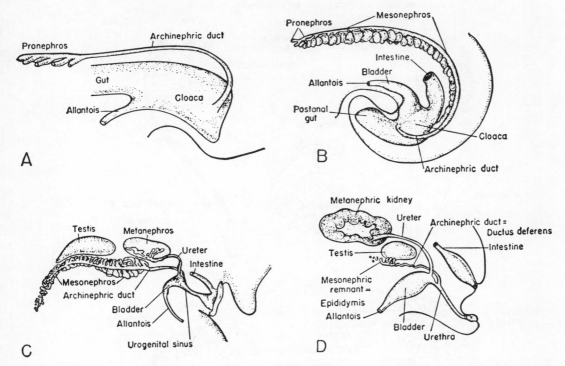

Figure 11–1. Diagrams to show the succession of kidneys in an amniote embryo. All are lateral views as seen from the left side. The genital details in these diagrams are those of a male. A, an early stage in which the pronephros and its duct are formed; B, mesonephric tubules tapping into the archinephric duct; C, pronephros degenerating, mesonephros functional, metanephros and ureter forming; D, definite stage in which the metanephros is functional and the remnants of the mesonephros and its duct are taken over by the male genital system. (From Romer, The Vertebrate Body.)

primitive segmental arrangement of the tubules that is seen in the pronephros. (Later kidneys have a multiplication of tubules so there is more than one per segment.) Such a kidney is called an **archinephros**, or **holonephros**. The holonephros is largely a theoretical kidney, but a close approach to it is seen in the larval hagfish in which the main kidney consists of segmentally arranged tubules and occupies all but the pronephric portion of the ridge. (The pronephric tubules form a specialized head kidney.) Thus the evolutionary history appears to be one of a progressive shortening of the kidney and posterior concentration of its functions, with a holonephros succeeded by an opisthonephros, and the opisthonephros succeeded in amniotes by a metanephros.

Genital Ducts and Their Relation to Excretory Ducts

In very primitive vertebrates, such as the cyclostomes, the gametes of both sexes are discharged into the coelom and pass to the outside through a pair of genital pores. This may have been the ancestral condition, but in other vertebrates ducts have evolved that transmit the gametes. In the embryo of either sex, primordia for the ducts of both the male and female are laid down during **a sexually indifferent stage**. In the dogfish, and some other primitive vertebrates,

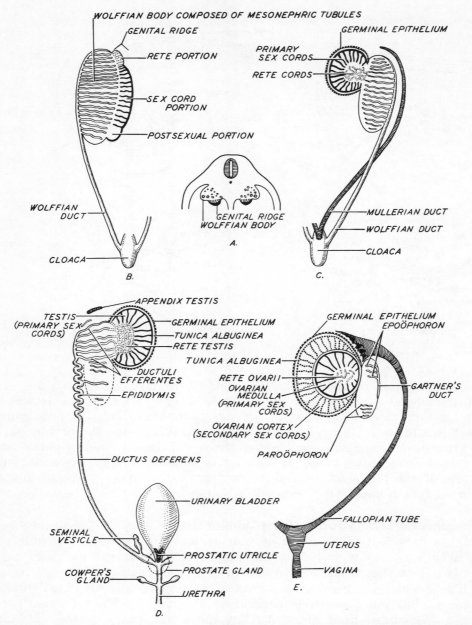

Figure 11–2. Diagrams of the development of the amniote genital system. *A*, a cross section of an early embryo showing the location of the mesonephros (wolffian body) and developing gonad (genital ridge). *B*, ventral view of an early embryo. *C*, the indifferent stage. *D*, differentiation of the male condition. The primary sex cords of the gonad of the indifferent stage become the seminiferous tubules. *E*, differentiation of the female conditions. The primary sex cords regress, and the follicles develop from secondary sex cords. (From Turner, General Endocrinology.)

the **oviduct** (embryonically known as the **müllerian duct**) develops from a splitting of the archinephric duct, and the coelomic opening of the duct (ostium) develops from the coelomic funnel (**nephrostome**) of a primitive kidney tubule. This mode of origin may be primitive, but in higher vertebrates the oviduct arises from a folding of the coelomic epithelium. In addition to the primordium of an oviduct, the same embryo acquires a series of cords, the **cords of the urogenital union** (**rete cords**), that connect the gonad with the anterior mesonephric tubules, and through them with the archinephric duct. Thus two potential routes are present for gamete transport.

During the subsequent differentiation of the female, the oviduct further develops, while the cords of the urogenital union and the adjacent parts of the mesonephros degenerate. In the adult female anamniote, this part of the embryonic mesonephros forms the slender, more or less functionless, anterior end of the opisthonephros. In adult female amniotes, which have a metanephros and ureter, the mesonephros is represented by minute, functionless groups of tubules called the **epoophoron** (more anterior mesonephric tubules) and **paroophoron** (more posterior tubules), and the archinephric duct degenerates, or forms a vestige known as **Gartner's duct** (**longitudinal duct of the epoophoron**) (Fig. 11–2).

During the subsequent differentiation of the male, the oviduct degenerates (although some vestiges may persist), and the route through the kidney becomes the functional pathway for the sperm. In adult anamniotes, the testis and the anterior end of the mesonephros (now the anterior end of the opisthonephros) are some distance apart. The sperm passes through the cords of the urogenital union (now called the **ductuli efferentes**, or **vasa efferentia**) located in the mesorchium, through the anterior kidney tubules, and into the archinephric duct. In adult amniotes, the testis and the anterior end of what was the mesonephros are close together, the pathway for the sperm is the same, but the terminology is different. The cords of the urogenital union constitute the **rete testis**; the anterior part of the mesonephros constitutes the head of the **epididymis**, and its tubules are called the **ductules of the epididymis** (**ductuli efferentes** is the Nomina Anatomica term, but it should be noted that a different structure, the cords of the urogenital union, is given this name in lower vertebrates) ; the highly coiled anterior part of the archinephric duct constitutes the body and tail of the epididymis, and is called the **ductus epididymis**; and the rest of the archinephric duct is called the **ductus deferens**, or **vas deferens**. The more caudal parts of the mesonephros sometimes form a functionless vestige called the **paradidymis**.

Study of the Excretory and Reproductive Systems

It is assumed that the major parts of the excretory and reproductive systems have been observed in previous dissections. In these exercises, the finer aspects will be examined and related to the more conspicuous parts. In studying these systems, you should not only dissect your own specimen, but also examine the dissection of a specimen of the opposite sex. Since someone else, in turn, will have to examine your specimen, make a particularly careful dissection. If possible, sexually mature specimens should be dissected.

FISHES

The excretory and reproductive systems of the cartilaginous fishes are a good example of a reasonably primitive vertebrate condition in most respects. The kidneys are opisthonephroi drained by archinephric ducts and, in the male at least, accessory urinary ducts. The gonads are situated far forward in the body cavity. The eggs are discharged through the coelom and a pair of oviducts; the sperm through the kidneys and archinephric ducts. A cloaca is present.

In certain more subtle features, however, the urogenital system is not entirely primitive. The kidney tubules of cartilaginous fishes have large glomeruli that remove a considerable volume of water from the blood. This type of tubule is generally regarded as primitive. It is advantageous in a fresh-water environment, which may have been the ancestral vertebrate environment, but its presence in these marine fishes, in which the osmotic problem is seemingly one of conserving water, poses problems. Cartilaginous fishes compensate for water loss through the kidneys by retaining considerable urea in their blood. As a consequence, the blood osmotic pressure slightly exceeds that of sea water, and water enters the body by osmosis. Retention of urea results from the reabsorption of urea by the kidney tubules and a reduction in the permeability of the gill membranes to urea so that little is excreted by this route. These features are peculiar to cartilaginous fishes.

Another specialized feature of many cartilaginous fishes, including *Squalus*, is the retention of their young within a uterus until embryonic development is complete. More primitive vertebrates are egg-laying.

Kidneys and Their Ducts

The kidneys of the dogfish are a pair of bandlike organs lying dorsal to the parietal peritoneum, a position called **retroperitoneal**, on either side of the dorsal mesentery. A conspicuous, white **caudal ligament** arises from the vertebral column between them and passes into the tail. They are **opisthonephric kidneys**, for they extend nearly the length of the pleuroperitoneal cavity; they are drained in part, at least, by **archinephric ducts** leading to the **cloaca**, and have a relatively primitive tubule structure. (Traces of microscopic nephrostomes are associated with some of the anterior tubules.) You may have to cut the parietal peritoneum along the lateral border of a kidney to trace it anteriorly, for the anterior two-thirds of each kidney is narrower and less conspicuous than the posterior one-third. It is in the posterior third that most of the urine production occurs. The anterior part is related to the reproductive system in the male, and is somewhat degenerate in the female.

The archinephric duct can easily be seen in the mature male, for it is a large, highly convoluted tube lying on the ventral surface of the opisthonephros. It is much smaller and straighter in the immature male. The duct of the female resembles that of an immature male, and cannot be seen until the oviduct is studied.

Further aspects of the excretory system must be considered separately in each sex.

Male Urogenital System

Notice that the paired **testes** are located near the the anterior end of the pleuroperitoneal cavity adjacent to the anterior end of the kidneys

(Fig. 11-3). Each is supported by a **mesorchium** through the front of which pass several, small, inconspicuous tubules called the **ductuli efferentes** or **vasa efferentia**. The ductuli efferentes carry the sperm from the testis to modified anterior kidney tubules. After passing through these tubules, the sperm enter the **archinephric duct**, and descend through it to the cloaca. The large size of the mature male archinephric duct is attributed to its role in sperm transport.

The part of the opisthonephros receiving the ductuli efferentes is homologous to the head of the epididymis of amniotes; the highly coiled portion of the archinephric duct adjacent to this region is homologous to the ductus epididymis, which comprises the body and tail of the epididymis; and the rest of the archinephric duct is comparable to the ductus deferens, or vas deferens.

The portion of the opisthonephros between the posterior end of the testis and the enlarged, posterior excretory region of the kidney is known as **Leydig's gland**. Most of the tubules in this region are modified to produce a secretion analogous to the seminal fluid of higher vertebrates. This secretion is of course discharged into the archinephric duct.

As the archinephric duct approaches the excretory portion of the kidney, it straightens and enlarges to form a **seminal vesicle**. Remove the parietal peritoneum from this portion of the kidney, and trace the seminal vesicle caudad. You will also have to open the cloaca on the side on which you are working by cutting through the side of the cloacal aperture and into the lateral wall of the intestine. The posterior end of the seminal vesicle passes dorsal to a **sperm sac**, whose blind anterior end should be freed from the seminal vesicle. The posterior end of the sperm sac, and of the seminal vesicle, unite to form a **urogenital sinus**. The sinus develops from the archinephric duct. Cut open the ventral surface of the sperm sac and sinus, clean them out, and notice the papilla that bears the opening of the seminal vesicle. The urogenital sinuses of opposite sides unite posterior to the entrances of the seminal vesicles, and extend into the **urogenital papilla** located dorsally in the cloaca. Probe posteriorly through the urogenital sinus, and notice the emergence of the probe through the tip of the urogenital papilla.

The sperm sacs appear to be diverticula from the urogenital sinus, but they develop from the posterior ends of the oviducts, which are present in the sexually indifferent stage of the embryo. Other remnants of the paired **oviducts** are tubular folds which can be found anteriorly on either side of the falciform and coronary ligaments. These portions of the oviducts even unite in the falciform ligament, and have a common entrance the **ostium**, into the coelom. The ostium can be found along the posterodorsal edge of the ligament. The intervening portions of the oviducts are lost in adult males.

The tubules of the excretory part of the kidney do not enter the seminal vesicle, but rather enter an **accessory urinary duct** which lies

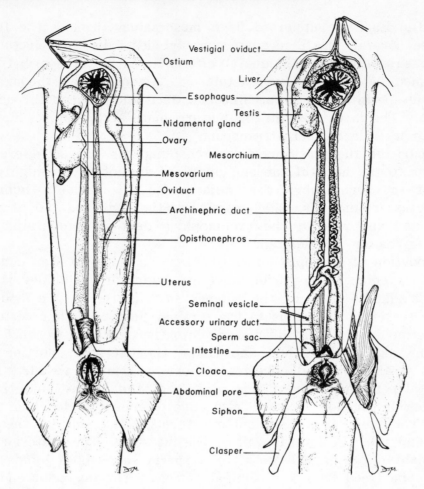

Figure 11-3. Ventral views of the urogenital system of *Squalus*. Left, female; right, male. The siphon has been dissected free on one side of the male.

against the dorsomedial edge of the seminal vesicle. This duct can be found by freeing the lateral edge of the seminal vesicle and dissecting dorsal to the vesicle. The accessory urinary duct, despite its name, carries virtually all the urine, for little if any urine is excreted by the anterior parts of the kidney in a mature male. Trace the accessory duct caudad. Its posterior end expands slightly to form a small **urinary bladder** which is analogous, but not homologous, to that of higher vertebrates, and then the duct enters the urogenital sinus posterior to the entrance of the seminal vesicle.

It can now be appreciated that the cloaca is a sort of sewer (Latin, cloaca = a sewer), for it receives the feces from the digestive system, the urine from the excretory system, and the gametes from the reproductive system. The cloaca is not divided in the male, but the excretory and genital products enter more dorsally and posteriorly than the feces.

Unlike most fishes, fertilization in the dogfish and other cartilaginous fishes is internal. During copulation one of the **claspers** on the pelvic fins of the male is turned forward and inserted into the cloaca and oviduct orifices of the female. The sperm proceed from the cloaca of the male into the groove on the dorsal surface of each clasper, and thence into the female. A sac-shaped muscular walled **siphon** is associated with each clasper. One can be found by skinning the ventral surface of the pelvic fin, if this has not already been done (p. 117). Cut open the siphon, and you can pass a probe through it into the groove on the clasper. It had long been thought that sea water enters the siphon and then is forcibly ejected during copulation, thus propelling the sperm along the clasper groove. But Heath (1956) finds no evidence that water is taken up. Rather, the siphons secrete copious amounts of a mucopolysaccharide which may lubricate the claspers and contribute to the seminal fluid.

Female Urogenital System

The **ovaries** are a pair of large organs located near the front of the pleuroperitoneal cavity adjacent to the anterior ends of the kidneys (Fig. 11-3). Each is supported by a **mesovarium**, and contains eggs in various stages of maturity. When the eggs are mature, they attain a diameter of nearly one inch and contain an enormous amount of yolk. Cut into one of the larger eggs to see the yolk. Each egg is surrounded by a sheath of follicular cells, but this cannot be seen grossly.

When the eggs are mature, they break through the wall of the follicle and ovary (a process called **ovulation**), pass into the coelom, and enter the front of the paired **oviducts**. In mature females, each oviduct is a prominent tube suspended by a **mesotubarium** from the ventral surface of the kidney. In immature specimens, the oviducts are small tubes lying against the kidneys, and mesotubaria are lacking. Trace an oviduct anteriorly. It passes dorsal to the ovary, and then curves ventrally and posteriorly in front of the liver to enter the falciform ligament. The oviducts of opposite sides unite within the falciform ligament, and have a common opening, the **ostium**, into the coelom. The ostium is located on the posterodorsal edge of the ligament, and can be opened by spreading its lips apart. Its location appears to be an adaptation for the reception of the unusually large, heavy eggs. In most lower vertebrates the ostia are separate and situated dorsally near the front of the pleuroperitoneal cavity.

The oviduct is narrow in diameter throughout much of its length, but enlarges in two regions. One enlargement lies dorsal to the ovary. This is the **nidamental gland** — a gland that secretes a thin, horny shell around groups of several eggs as they come down the oviduct. There is also evidence that sperm is stored in this region for some time before fertilization. The other enlargement, the **uterus**, occupies approximately the posterior one-third to one-half of the oviduct. It is very large in pregnant females, for the embryos develop here.

Open the cloaca by cutting through the side of the cloacal aperture, and into the lateral wall of the intestine. The two oviducts enter the posterodorsal part of the cloaca just ventral to a **urinary papilla**. The urogenital portion of the cloaca, known as the **urodeum**, is partially separated by a horizontal fold from the anteroventral, fecal portion of the cloaca (**coprodeum**).

The kidneys of the female *Squalus* are drained by the **archinephric ducts**, for there are no accessory urinary ducts as there are in the male. Some female elasmobranchs, however, have accessory ducts. An archinephric duct can be found by making an incision through the parietal peritoneum along the lateral border of the posterior part of the kidney, and very carefully reflecting the parietal peritoneum from the surface of the kidney. The archinephric duct lies on the ventral surface of the kidney directly dorsal to the attachment of the mesotubarium. If you do not see it on the kidney, it probably adhered to the dorsal surface of the parietal peritoneum, and can be picked off the peritoneum. The archinephric duct is much smaller than in the male, and is not convoluted. Trace it caudad. The posterior ends of the ducts of opposite sides enlarge slightly, and unite to form a small **urinary sinus** which opens through the tip of the urinary papilla. The urinary sinus constitutes a small urinary bladder that is analogous, but homologous, to the bladder of higher vertebrates. The sinus is too small to be easily dissected.

Reproduction and Embryos

As the eggs develop within the follicles in the ovary, they decrease in number but increase in size through the accumulation of yolk. At the time of ovulation, each ovary contains two or three ova averaging 3 cm. in diameter. After ovulation, the follicular cells are converted into a corpus luteum. This is a large body 2 cm. in diameter in an early pregnancy, but it gradually regresses. The eggs enter the oviduct which can stretch greatly in a living specimen. As they pass through the nidamental gland, they are fertilized, and a thin, horny shell (the "candle") is deposited around several of them. This mass then passes to the uterus in which it may be seen in specimens in an early stage of pregnancy.

After several months, the shell breaks down and the embryos develop within the uterus. Hisaw and Albert (1947) find that the gestation period lasts nearly two years. Pups that are slightly over a year old, which is a stage often seen in pregnant specimens obtained from biological supply houses, range in length from 12 to 20 cm. and much of the yolk is contained within an **external yolk sac** suspended from the underside of the embryo (Fig. 11-4). This is a **trilaminar yolk sac** (Fig. 11-6), for it contains all three germ layers. The rest of the yolk is carried in an **internal yolk sac**, which can be found within the pleuroperitoneal cavity. Pups just before parturition range in length from 23 to 29 cm. The yolk in the external sac has been consumed, but a small reserve remains in the internal sac.

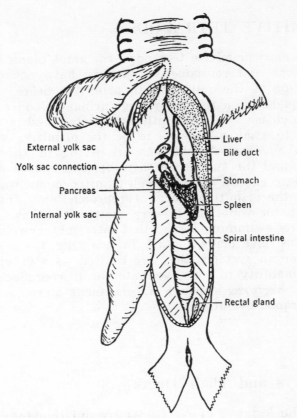

Figure 11–4. External and internal yolk sacs in a 220 mm. embryo of *Squalus suckleyi* (From Brown, Physiology of Fishes, Academic Press. After Hoar.)

External yolk sac
Yolk sac connection
Pancreas
Internal yolk sac

Liver
Bile duct
Stomach
Spleen
Spiral intestine
Rectal gland

Numerous vascular **uterine villi** line the uterus and are applied to the surface of the embryo, especially its thin walled, vascular, external yolk sac. The embryo doubtless obtains water from the mother by this pseudoplacental relationship, but most, if not all, of the other nutritional needs are supplied by the yolk, which is moved by ciliary action into the embryo's intestine where it is digested and absorbed. During the course of development, there is a steady decrease in the weight of organic matter in the embryo and yolk, but a 78 per cent gain in weight because of a great increase in water and minerals. This mode of development, in which the young are born as miniature adults, but do not receive much nutrition from the mother, is referred to as ovoviviparous development. In a few sharks, such as *Mustelus laevis,* there is an intimate union between the yolk sac and maternal tissues, and the embryos derive most of their nutritional requirements from the mother through this yolk sac placenta. Such animals are said to be **viviparous**. Sharks regarded as reproductively more primitive than *Squalus* are **oviparous**, for they lay eggs and the embryos develop in the surrounding water. The eggs of many oviparous vertebrates are provided with only a moderate amount of yolk, and the embryos soon hatch as free swimming larvae that feed for themselves. In the case of oviparous elasmobranchs, the eggs are provided with a large amount of yolk and the embryos develop within a horny egg case. The candle of the dogfish is regarded as a vestige of such a case.

PRIMITIVE TETRAPODS

The most primitive tetrapods, the amphibians, are still anamniotes, and their excretory and reproductive systems have not changed significantly from the condition of these systems in primitive fishes. The only new feature of any consequence is a relatively large urinary bladder formed as a ventral outgrowth of the cloaca. In many other respects, these systems in amphibians are even closer to what is believed to be the primitive vertebrate condition than they are in *Squalus*. Amphibians are living in the primitive environment, fresh water, and the tubules in their opisthonephros are of the primitive type that eliminate excess water as well as nitrogenous waste products. Amphibians also retain the primitive mode of reproduction. They are oviparous, laying their eggs in the water, or in very moist situations. In most cases, the eggs hatch into free-swimming larvae that later metamorphose to adults. Thus, while amphibians have made a start in adapting to terrestrial conditions in most of their organ systems, they are limited, as a group, to moist habitats because of their inability to conserve water or to reproduce under truly terrestrial conditions. *Necturus* although a permanent larva, is a good example of this level of tetrapod evolution.

Kidneys and Their Ducts

The kidneys of *Necturus* are **opisthonephric**. They lie in the posterior half of the pleuroperitoneal cavity on either side of the dorsal mesentery. They are easily seen in the male, but you will have to push the ovary and oviduct apart to see one in the female. Notice that the kidneys have bulged into the body cavity, so that both their dorsal and ventral surfaces are covered by visceral peritoneum. Also notice that the posterior part of a kidney is much larger than the anterior part. The anterior part of the male kidney is related to reproductive system, while that of the female is somewhat degenerate. The anterior part of the kidney is best seen by lifting up the lateral border of the organ and looking on its dorsolateral surface.

The kidneys are drained in both sexes exclusively by the **archinephric ducts**, for accessory urinary ducts are absent in *Necturus*. The archinephric duct of the male is a large, convoluted tube extending down the lateral border of the kidney to the cloaca. The archinephric duct of the female is similarly located, but is much smaller, and is not convoluted. Small **collecting tubules** may be seen entering the archinephric duct from the posterior, excretory portion of the kidney.

A **urinary bladder** lies ventral to the large intestine, and is connected to the midventral body wall by a mesentery known as the **median ligament of the bladder**. The bladder enters the anteroventral wall of the cloaca.

Urine reaches the bladder from the cloaca, for there is no direct connection between the archinephric ducts and bladder.

The **suprarenal** or **adrenal glands** of primitive tetrapods consist of cortical and medullary cells clustered together in irregular patches. You may see a number of small patches of adrenal tissue along the ventral surface of the kidney of *Necturus* (Fig. 11-5).

Male Urogenital System

The **testes** are a pair of oval organs located near the anterior ends of the kidneys. Each is supported by a **mesorchium**. Several inconspicuous **ductuli efferentes** pass through the front of the mesorchium carrying sperm from the testes to the modified anterior kidney tubules. This portion of the kidney thus functions as the head of the epididymis. From here, the sperm pass to the cloaca through the **archinephric duct**, which is therefore serving as a ductus deferens as well as an excretory duct. Notice that the anterior part of the archinephric duct, a part comparable to the ductus epididymis of amniotes, is much more convoluted than the rest of the duct. The small black line along the edge of the anterior part of the archinephric duct, and extending forward along the side of the posterior cardinal vein, is a remnant of the oviduct that has persisted from

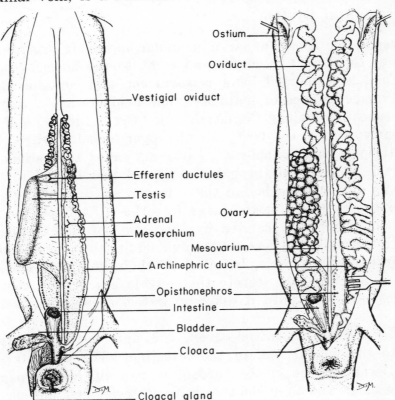

Figure 11-5. Ventral views of the urogenital system of *Necturus*. Left, male; right, female.

the sexually indifferent stage of the embryo. A shorter, but similar remnant of the oviduct can sometimes be seen along the very posterior end of the archinephric duct.

Cut through the body musculature lateral to the cloaca, and find the point at which the archinephric duct joins the dorsal wall of the cloaca just posterior to the large intestine. Open the cloaca by making an incision that extends from the anterior end of the cloacal aperture into the lateral wall of the intestine. The archinephric ducts (Fig. 11-5) enter the anterodorsal wall of the cloaca just posterior to a transverse ridge formed by the entrance of the intestine, but their openings probably will not be seen.

Skin one side and the ventral surface of cloaca, if this has not been done, and observe the large **cloacal gland** consisting of many small tubules. The secretions of this gland, together with the secretions of a less conspicuous **pelvic gland** in the dorsal wall of the cloaca, agglutinate the sperm into clumps called **spermatophores**. The spermatophores of salamanders generally are deposited in the water to be picked up by the female, but in a few species are transmitted directly to the cloaca of the female. Finally, notice the numerous **papillae** on the cloacal lips, which are a characteristic feature of the male.

Female Urogenital System

The **ovaries** are a pair of large, granular-appearing organs located on either side of the dorsal mesentery adjacent to the anterior part of the kidneys. Each is supported by a **mesovarium**, and contains eggs within follicles in various stages of maturity. The oviducts are a pair of large, convoluted tubes lying along the lateral side of each kidney, and extending forward nearly to the front of the pleuroperitoneal cavity. Each terminates anteriorly in a funnel-shaped opening called the **ostium**, and posteriorly in the cloaca. Cut through the musculature lateral to the cloaca to see that an oviduct attaches to the anterodorsal wall of the cloaca just posterior to the entrance of the large intestine. At ovulation, the eggs pass into the coelom, are carried by the action of cilia on the coelomic epithelium to the ostia, and descend the oviducts to the cloaca. The oviducts are not differentiated grossly, but part of their lining, at least, is glandular. The oviduct glands secrete the jelly layers around the eggs, but the jelly does not swell until it takes up water when the eggs are deposited.

Open the cloaca by cutting from the anterior end of the cloacal aperture into the lateral wall of the intestine. The oviducts enter the anterodorsal wall of the cloaca through a pair of **genital papillae**. If these are not clear, slit an oviduct near its posterior end, and pass a probe through it into the cloaca. The posterior end of each **archinephric duct** can be seen to leave the kidney and pass onto the wall of the oviduct, with which it becomes intimately united. However, the archinephric duct enters the cloaca independently beside the opening of the oviduct.

The **cloacal gland**, which helped to produce the spermatophores in the male, is present in a much reduced state in the female. It can be found by skinning one side and the ventral surface of the cloaca. The homologue of the pelvic gland of the male is transformed into a **spermatheca** within whose tubules the sperm are stored. In *Necturus,* this would be from the breeding season in the fall until egg laying in the spring. The spermatheca is located in the dorsal wall of the cloaca but cannot be seen glossly. In most salamanders, its point of entrance is marked by a dark pigment spot, but this is not apparent in *Necturus*. Finally, notice that the lips of the female cloaca bear smooth folds rather than the papillae characteristic of the male.

MAMMALS

In the evolution through reptiles to mammals, the problems concerned with terrestrial excretion and reproduction have been met, as have those concerned with the great increase in metabolic activity. A metanephric kidney and ureter evolve, and what is left of the mesonephros and archinephric duct is taken over by the male genital system. Thus there is a more complete separation of excretory and genital functions in amniotes than in anamniotes. There is also a division of the cloaca in mammals that separates the bladder and urogenital tract from the digestive tract. The problem of water concentration has been met in part by the evolution of a special segment, the loop of Henle, in the renal tubule. This loop helps to concentrate sodium ions in the intercellular fluid deep within the kidney, and thereby makes an osmotic gradient that causes water to be reabsorbed from the terminal portion of the tubule (collecting tubule). There has also been a tremendous increase in the number of kidney tubules to take care of the increase volume of excretory products. Whereas a kidney in *Necturus* contains approximately 50 tubules in its posterior excretory portion, the number in a small mammal such as a mouse is on the order of 20,000, and man has an estimated 1,000,000 to 4,000,000 per kidney.

In respect to reproduction, terrestrial life necessitates internal fertilization. In this connection, a **copulatory organ** evolves in the male, as do **accessory sex glands** which produce a fluid medium in which the sperm can be transported. The gonads of most male mammals also undergo a marked posterior migration (descent), and come to lodge outside the body cavity in a **scrotum**. This is apparently correlated with the inability of spermatogenesis to occur at the high temperatures within the body cavity.

Terrestrial life also necessitates a method of suppressing a free-swimming aquatic larva. Primitive reptiles remain oviparous, but suppress the larva by the evolution of a **cleidoic egg** (Fig. 11-6). This is an egg in which provision is made for all the requirements of the embryo, so that it can develop directly into a miniature adult capable of living on the land. The egg is supplied with a large store of yolk, which eventually becomes suspended in a **yolk sac**. As it descends the oviduct, **albumin**, or similar secretions which supply other metabolic needs, and a protective **shell** are added to the egg. The embryo itself

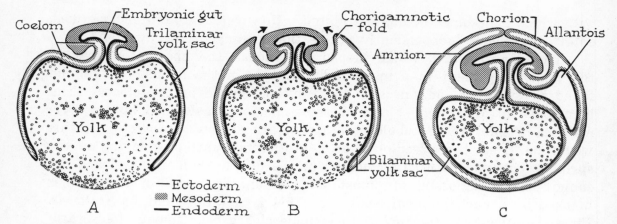

Figure 11–6. Diagrammatic sections of vertebrate embryos to show the extraembryonic membranes. *A*, trilaminar yolk sac of a large yolked fish embryo; *B*, hypothetical derivation of the chorioamniotic fold of the amniote embryo from the superficial layers of the trilaminar yolk sac; *C*, extraembryonic membranes of an amniote embryo. (From Villee, Walker, and Smith, General Zoology.)

early develops **extraembryonic membranes** that fulfill other needs. A protective **chorion**, and a fluid-filled **amnion** (which provides a local aquatic environment) evolve from ectodermal and mesodermal layers that covered the yolk sac in such an animal as the dogfish. The yolk sac of amniotes is therefore **bilaminar** with a wall of just mesoderm and endoderm. Finally, a respiratory and excretory allantois evolves from the urinary bladder of the Amphibia.

Primitive prototherian mammals lay this type of egg, but in therian mammals this egg, minus its shell and albumin, is retained in the female reproductive tract, and a **placenta** evolves. In eutherian mammals, the placenta is simply a union of the chorion and allantois (**chorioallantoic membrane**) on the one hand with tissues of the female on the other. As will be seen, this mode of reproduction necessitates changes in the female reproductive tract, notably the evolution of a **uterus**.

Excretory System

The **kidneys (renes)** of mammals, which are **metanephroi**, are located against the dorsal wall of the peritoneal cavity in a retroperitoneal position. Each is surrounded by a mass of fat (**adipose capsule**), which should be removed, and each is closely invested by a **fibrous capsule**. Notice that a kidney is bean shaped. The indentation on the medial border is called the **hilus**. Carefully remove connective tissue from the hilus, and you will see the **renal artery** and **vein** entering the kidney, and, posterior to them, the **ureter** that drains the kidney.

Remove one of the kidneys, and cut it longitudinally through the hilus. Study the half that includes the largest portion of the ureter. The hilus expands within the kidney into a chamber called the **renal sinus**. Pick away fat in the sinus in order to expose the blood vessels and the proximal end of the ureter. The sinus is largely filled with these structures, so its size is often not appreciated. If the vessels and ureter were removed, the space left would be the sinus.   The portion of the ureter within the sinus is expanded, and is termed the **renal pelvis** (Fig. 11-7). The substance of the kidney converges in the cat and rabbit to form a single, nipple-shaped **renal papilla**, which projects into the renal pelvis. (In some mammals, man among them, there are many renal papillae, and the proximal portion of the renal pelvis is subdivided into chambers for them, called **calyces**.) Notice that the substance of the kidney can be subdivided into a peripheral, light **cortex** and a deeper, darker, **medulla**. The medullary substance of the cat and rabbit constitutes one large **renal pyramid** whose apex is the renal papilla.   But in species with many papillae, there is a pyramid for each one.

Trace one of the ureters.   It extends posteriorly dorsal to the parietal

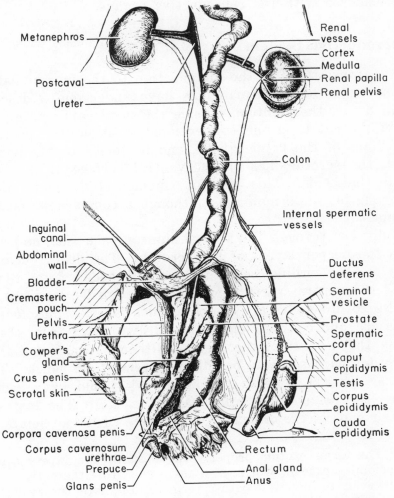

Figure 11-7.   Ventral view of the urogenital system of a male rabbit.   One kidney has been sectioned to show its internal structure. The pelvic canal has been opened and the urethra and penis twisted slightly to one side to show the accessory sex glands.   The scrotum is shown intact on the left side of the drawing, and opened on the right. A male cat differs in the size of the cremasteric pouch and in the accessory sex glands.

peritoneum, and then turns ventrally in the lateral ligament of the bladder to enter the posterior part of the **urinary bladder (vesica urinaria)**. As the ureter enters the lateral ligament of the bladder, it passes dorsal to the ductus deferens (male), or horn of the uterus (female). The urinary bladder itself is a pear-shaped organ with a broad, rounded, anterior end (its **vertex**) and a narrow, posterior end (its **fundus**). Cut open the bladder, and you may be able to see the points of entrance of the ureters in the dorsal wall of the fundus. Clean away connective tissue from around the bladder. It gradually narrows posterior to the entrance of the ureters, and passes into the pelvic canal. This narrow passage, which carries urine to the outside, is called the **urethra**. The urethra begins just posterior to the entrances of the ureters. Its more posterior parts will be considered with the reproductive organs.

Male Reproductive System

For reasons stated in the introduction to the mammalian urogenital system, the testes of most mammals have undergone a descent so that they lie in a sac, the **scrotum**, outside the body cavity (Fig. 11–7). The scrotum of the cat is in the conventional position just posterior to the ventral surface of the pelvic girdle, and posterior to the penis. But in the rabbit, the scrotum lies on the ventral surface of the pelvic girdle anterior to the penis. This position, incidentally, occurs only in lagomorphs among placental mammals, although a comparable location is seen in marsupials.

Carefully cut through the scrotal skin on each side, and separate the skin from the deeper layers of the scrotum. The deeper layers take the form of a pair of cordlike sacs that extend caudad from the abdominal wall, cross the ventral surface of the pelvic girdle, and enter the skin sacs. These cordlike sacs are composed of a layer of muscle (**cremasteric muscle**) fascia, and coelomic epithelium drawn down from the abdominal wall during the descent of the testes. Although considered to be a part of the scrotum, each sac may be called a **cremasteric pouch**.[17] The cremasteric pouches lie just beneath the skin, and should have been seen and saved when the pelvic canal was opened (p. 295). The testes lie within the posterior ends of the cremasteric pouches. This portion of each pouch

---

[17] The term scrotum properly includes all the layers separating the coelomic space in which each testis lies from the outside. It is useful, however, to have a separate term to designate the complex of tissue, apart from the skin, enfolding each testis and its duct and vessels, because this complex is clearly defined and must be separated from the skin in order to fully see the anatomical relationships that result from the descent of the testes. I am following Prof. Gerard (in P.–P. Grassé, Traité de Zoology, Masson et Cie., Paris, 1954, Vol. 12) in calling this part of the scrotum the cremasteric pouch.

is quite large in the cat, while the rest is a constricted tube, for the testes remain permanently in the scrotum. But all of the pouch is wide and of nearly uniform diameter in the rabbit, for the testes move back and forth — into the scrotum during the breeding season, back into the abdomen the rest of the time. The atrophy of the cremasteric muscle in the cat and its hypertrophy in the rabbit is also correlated with whether the testes remain permanently in the scrotum or are migratory.

Leave the cremasteric pouch intact on one side, but cut open the other one along its ventral surface. Extend the cut from the posterior end of the pouch to the body wall, but do not cut through the body wall. Notice that the cremasteric pouch contains a cavity, the lumen of the **processus vaginalis**. Pass a probe forward through this cavity near the body wall, and the probe will enter the peritoneal cavity. The processus vaginalis is a coelomic sac that descends with the testis. Thus the wall of the cremasteric pouch is lined with coelomic epithelium, and the structures within it (testis, epididymis, ductus deferens, testicular vessels and nerves) are covered with coelomic epithelium. The complex of ductus deferens and associated vessels and nerves, together with their covering of coelomic epithelium, is known as the **spermatic cord** (**funiculus spermaticus**). Notice that the cord is supported by a mesentery passing from the dorsal wall of the pouch, and that the most posterior structures in the pouch (testis and epididymis) and their mesentery are united to the posterior end of the pouch by a band of tissue called the **gubernaculum**. The most posterior part of the processus vaginalis (that surrounding the testis and epididymis) is known as the **tunica vaginalis**; the coelomic epithelium covering the testis being the **visceral layer** of the tunica vaginalis; that lining the adjacent cremasteric pouch, the **parietal layer**. In many mammals, man included, only the tunica vaginalis persists in the adult, the rest of the processus vaginalis atrophies. But this is not the case in the cat and rabbit.

The contents of the cremasteric pouch can now be examined in more detail. The testis is the relatively large, round (cat), or elongate (rabbit) body lying in the posterior part of the pouch. Testicular blood vessels and nerves attach to its cranial end. The **epididymis** is a band-shaped structure closely applied to the surface of the testis. It can be divided into three regions — a **head** (**caput epididymis**) at the cranial end of the testis, a **body** (**corpus epididymis**) on the lateral surface of the testis, and a **tail** (**cauda epididymis**) at the caudal end of the testis. The head of the epididymis is functionally connected with the testis, and is made up of modified kidney tubules homologous to those of the anterior end of the opisthonephros of anamniotes. The rest of the epididymis consists of the highly convoluted **ductus epididymis** (former anterior end of the archinephric duct) imbedded in connective tissue.

The **ductus deferens**, the first part of which is somewhat convoluted, leaves the tail of the epididymis and, in company with the testicular vessels and nerves, ascends the cremasteric pouch and passes through the abdominal wall. The passage through the body wall is known as the **inguinal canal**. The canal is not long in the cat or rabbit, its anterior end (**internal inguinal ring**) being the entrance into the peritoneal cavity; its posterior end (**external inguinal ring**), the attachment of the outermost layer of the wall of the cremasteric pouch to the external surface of the abdominal muscles. In man the inguinal canal passes diagonally through the abdominal wall, hence is much longer.

Continue to follow the ductus deferens. It passes forward in the peritoneal cavity for a short distance, then loops over the ureter, and extends posteriorly into the pelvic canal between the urethra and large intestine. The ductus deferentes of opposite sides then converge, and soon enter the urethra. The portion of the urethra distal to this union carries both sperm and urine. Various accessory sex glands, which secrete the seminal fluid, are associated with the ends of the ductus deferentes and adjacent parts of the urethra; however, these glands, and the details of the union of the ductus deferentes with the urethra, are different in the cat and rabbit.

In the cat, the two ductus deferentes enter the urethra independently, and a small **prostate** surrounds their point of entrance and the adjacent urethra. At the caudal end of the pelvic canal, a pair of **bulbourethral**, or **Cowper's**, **glands** enter the urethral canal. These glands lie dorsal to a pair of processes (crura of the penis) that extend from the base of the penis to the ischia.

In the rabbit, the two ductus deferentes pass between the urethra and a heart-shaped **seminal vesicle**. Carefully separate these structures from each other. The ductus deferentes enter the narrow posterior end of the seminal vesicle, which in turn enters the urethra. The dorsal wall of the seminal vesicle is rather thick, and includes the **prostate**. It is possible, by very careful dissection, to free the anterior end of the prostate and turn it caudad. Further dissection will reveal overlapping anterior and posterior lobes of the prostate. Both enter the urogenital canal just posterior to the entrance of the seminal vesicle. A bilobed **bulbourethral**, or **Cowper's**, **gland** enters the dorsal surface of the urethral canal posterior to the prostate. It lies anterodorsal to a pair of processes (crura of the penis) that extend from the base of the penis to the ischia.

The **crura of the penis** extend from the base of the penis to the ischia, and each is covered by muscular tissue (**ischiocavernosus muscle**). If the crura was not torn when the pelvic canal was opened, you may have to cut one now to see the paired (cat) or bilobed (rabbit) Cowper's gland clearly.

The **penis** of both the cat and rabbit encloses the part of the urethra lying outside the pelvic canal. The free end of the penis, **glans penis**, lies in a pocket of skin called the **prepuce**. Cut open the prepuce to better see the glans and the opening of the urethral canal. A number of small spines are borne on the glans of the cat. The rest of the penis is a firm, cylindrical structure which should be exposed by removing the skin and surrounding loose connective tissue. Make a cross section of this portion of the penis, and examine it with a hand lens. The urethral canal lies along the dorsal surface of the penis (if the organ is flaccid) imbedded in a column of spongy tissue called the **corpus cavernosum urethrae**. A pair of columns of spongy tissue separated by a septum, which is often indistinct, lie along the opposite surface, and are surrounded by a ring of dense connective tissue. These columns are the **corpora cavernosa penis**. The glans penis is simply a caplike fold of the corpus cavernosum urethrae that covers the distal ends of the corpora cavernosa penis. The crura of the penis are the diverging proximal ends of the corpora cavernosa penis. The spongy tissue of which all of the corpora cavernosa consist is known as **erectile tissue**, and the spaces within it become filled with blood during erection of the penis.

Before leaving the urogenital system, dissect beneath the skin on either side of the rectum near the anus, and find a pair of **anal glands**. These are elongate in the rabbit, round in the cat. They produce an odoriferous secretion that enters the anus, and presumably is for sexual attraction, or stimulation.

Female Reproductive System

The **ovaries** are a pair of small, oval bodies (Fig. 11-8). In the adult they lie slightly posterior to the kidneys, for they have undergone a partial descent, and the metanephroi shift cranially during development. The small size of the mammalian ovaries is correlated with the uterine development of the embryos. Fewer eggs are produced, and the eggs do not contain much yolk. The eggs are microscopic, but you may see small vesicles, the **graafian follicles** (**folliculi vesiculosi**), each of which contains an egg, protruding on the surface of the ovary. **Corpora lutea** may also be seen, especially in pregnant specimens. Each ovary is supported by a **mesovarium** — a mesenteric fold extending from a point near the caudal end of the kidney to the ovary and uterus.

Typical oviducts are present in early mammalian embryos, but they differentiate into several regions during development, and their posterior ends fuse in varying degrees (Fig. 11-9). Thus the adult reproductive tract is more or less Y-shaped. The anterior part of each wing of the Y forms a narrow, convoluted **uterine**, or **fallopian**, **tube** lying lateral to the ovary. Notice that a uterine tube curves over the front of the ovary and forms a hoodlike expansion (**infundibulum**) with fringed (fimbriated) lips. Spread open the lips, and you will see the coelomic opening of the tube, the **ostium**. The mesentery supporting the uterine tube is called the **mesosalpinx**.

The rest of each wing of the Y lies posterior to the ovary, and forms

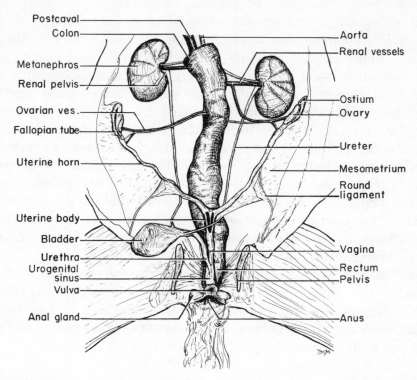

Figure 11-8. Ventral view of the urogenital system of a female cat. The pelvic canal has been cut open. The uterine body has been sectioned to show its bipartite nature.

a much wider tube — the **horn of the uterus** (cat), or **uterus** (rabbit). This section is straight in the cat, but somewhat convoluted in the rabbit. It is very large in pregnant specimens, for the embryos develop within it. The mesentery supporting this part of the reproductive tract is known as the **mesometrium**. The mesosalpinx and mesometrium of mammals are homologous to the mesotubarium of lower vertebrates. These two mesenteries and the mesovarium more of less merge with each other, and the combined mesentery is known as the **broad ligament**. Pull the uterine horn (cat) or uterus (rabbit) toward the midline, thereby stretching the the broad ligament. The mesenteric fold extending diagonally across the broad ligament from a point near the anterior end of the uterine horn (or uterus) to the body wall, and lying perpendicular to the broad ligament, is the **round ligament** (**ligamentum teres uteri**). Notice that the round ligament attaches to the body wall at a point comparable to the location of the inguinal canal in the male. The round ligament is the female counterpart of the male gubernaculum — a strand concerned with the descent of the testis.

The two uterine horns (cat) or uteri (rabbit) converge anterior to the pelvic canal and enter a common median passage. This would be the stem of the Y, and it is formed in part by the fusion of the lower ends of the oviducts and in part by the division of the cloaca (see page 336). In the cat, the anterior part of this median passage is the **body of the uterus**. The body is not long, and soon leads into the **vagina**. The demarcation between these two will be seen presently when the tract is cut open. In the rabbit, the two uteri enter the **vagina** directly, for there is no median body to the uterus. The vagina of both animals proceeds posteriorly through the pelvic canal, lying between the urethra and large intestine. Carefully separate these structures from each other, and find the point where the vagina and urethra unite. The common passage from here on is the **vestibule**. Because of its greater length in these mammals compared with human beings, it is often called the **urogenital sinus**, but it is not homologous with a similarly named structure in the dogfish. The lateral lips of the opening of the urogenital sinus on the body surface are the **labia majora**. The labia, together with the opening, constitute the **vulva** (**pudendum femininum**).

Cut through the skin around the vulva, and completely free the urogenital sinus and vagina from the rectum. A pair of **anal glands**, which are elongate in the rabbit and round in the cat, can be found by dissecting beneath the skin on the lateral surface of the vulva (rabbit), or of the rectum near the anus (cat). These glands produce an odoriferous secretion that enters the anus, and presumably is for sexual attraction, or stimulation. Now open the median portion of the genital tract by making a longitudinal incision through its dorsal wall that extends from the vulva to the horns of the uterus (cat), or uteri (rabbit). Veer away from the middorsal line toward one of the uterine horns as you open the body of the uterus in the cat. A small bump will be seen in a pocket of tissue in the midventral line of the urogenital sinus near the orifice of the sinus. This is the **glans clitoridis**. The clitoris consists of only the small glans in the cat, but is large and continues anteriorly in the rabbit. In the rabbit, the rest of it can be exposed by removing the skin from the ventral surface of the urogenital sinus. The clitoris and its glans develop from a phallus-like structure which is present in the sexually indifferent stage of the embryo. Make a cross section through the organ, and it will be seen to consist of a pair of columns of spongy tissue, the **corpora cavernosa clitoridis**, homologous to the corpora cavernosa penis.

More anteriorly in the urogenital sinus, you will see the entrance of the urethra. The genital passage anterior to this union is the vagina, and it continues forward to the neck, or **cervix of the uterus**. In the cat, the cervix lies about halfway between the urethral orifice and the horns of the uterus, and appears as a pair of folds constricting the lumen of the reproductive tract. The body of the uterus lies between the cervix and the horns. Notice that the anterior part of the body is subdivided into right and left sides by a vertical partition. The uterus of the cat is therefore bipartite (Fig 11-9). In the rabbit, the vagina extends forward to the two uteri, and each uterus has a papilla-like **cervix** extending into the vagina. This type of uterus is called **duplex** (Fig. 11-9).

The origin of most of the mammalian male and female urogenital tracts from those of the lower vertebrates has been indicated. However, a consideration of the cloacal region and its fate in mammals has been deferred until the terminal portions of the urogenital passages could be studied in the mammal.

A cloaca was seen in the lower vertebrates, and one is still present in monotremes and in the embryos of the higher mammals. But it becomes divided and contributes to the intestinal and urogenital passages in the adults of the higher mammals. In an early, sexually indifferent, eutherian embryo, the cloaca consists of a chamber derived from the enlargement of the posterior end of the hindgut (Fig. 11-10, *A*). At first this endodermal cloaca is separated from the ectodermal proctodeum by a plate of tissue, but this plate soon breaks down and the proctodeum contributes to the cloaca.

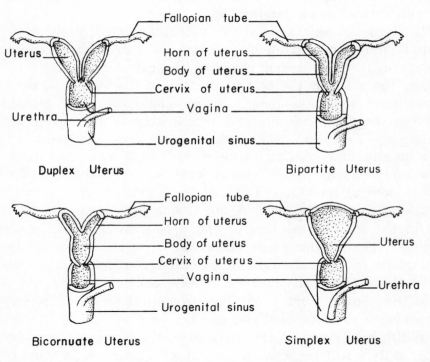

Figure 11-9. Diagrams to show the progressive fusion of the posterior ends of the oviducts in placental mammals. The uterus and part of the vagina have been cut open. The duplex type of uterus, in which the lower ends of the oviducts have united to form a vagina but the uteri remain distinct, is found in rodents and lagomorphs. In the bipartite uterus of carnivores, the lower ends of the uteri also have fused to form a median body from which uterine horns extend, but a partition is present in the body of the uterus. This partition is lost in the bicornuate uterus of ungulates. In the simplex uterus of primates, the uteri have completely united to form a large median body from which the fallopian tubes arise. In primates, the urogenital sinus also divides, so that the vagina and urethra open independently on the body surface. (Modified after Wiedersheim, Comparative Anatomy of Vertebrates, Macmillan Company.)

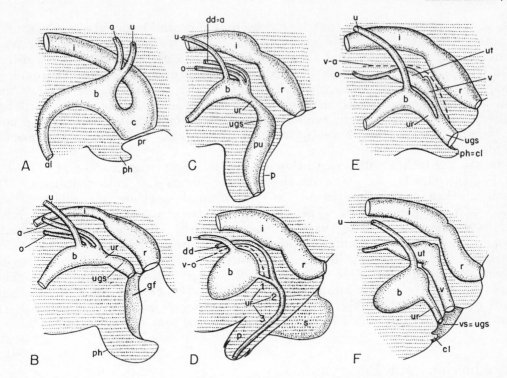

Figure 11-10. A series of diagrams in lateral view showing the division of the cloaca that occurs during the embryonic development of a placental mammal. Anterior is toward the left, and only one member of the paired ducts is shown. *A*, an early sexually indifferent stage in which the cloaca is undivided; *B*, a later sexually indifferent stage in which the cloaca has become divided into a dorsal rectum and a ventral urogenital sinus; *C*, early male condition; *D*, adult male; *E*, early female condition; *F*, adult female. See text for further explanation. Abbreviations: *a*, archinephric duct; *al*, stalk of the allantois; *b*, urinary bladder or proximal part of the allantois; *c*, cloaca; *cl*, clitoris; *dd*, ductus deferens; *gf*, genital fold; *i*, intestine; *o*, oviduct; *p*, penis; *ph*, phallus; *pr*, proctodeum; *pu*, penile urethra; *r*, rectum; *s*, scrotum; *u*, ureter; *ugs*, urogenital sinus; *ur*, urethra; *ut*, uterus; *v*, vagina; *v-a*, vestige of archinephric duct; *v-o*, vestige of oviduct; *vs*, vestibule. (From Romer, The Vertebrate Body.)

The cloaca receives the intestine dorsally and the allantois ventrally. Even at an early stage (Fig. 11-10, *A*), the anterior portion of the cloaca is partly divided into a dorsal **coprodeum** receiving the intestine, and a ventral **urodeum** receiving the allantois, the ureters, and the archinephric ducts. Finally, a small **phallus** is present on the ventral surface of the body anterior to the cloaca. This stage is similar to the cloaca of lower vertebrates except for the more ventral entrance of the urogenital ducts.

Later in the sexually indifferent period (Fig. 11-10, *B*), the urodeum and coprodeum become completely separated from each other by a fold of tissue (**urorectal fold**) and form the urogenital sinus and rectum respectively. Oviducts now enter the front of the urogenital sinus beside the archinephric ducts, but the attachments of the ureters shift onto the allantois and developing urinary bladder.

In the subsequent differentiation of a male (Fig. 11-10, *C* and *D*), the constricted neck of the bladder (allantois) and the urogenital sinus form that portion of the urethra that is not included in the penis (segments 1 and 2). The archinephric ducts form the ductus deferens and enter the urethra. Their point of entrance is a landmark that separates the portion of the urethra derived from the allantois from the portion derived from the urogenital sinus. The oviducts disappear in the male, although their point of entrance into the urethra may form a small sac (**prostatic utricle**) within the prostate. The phallus enlarges to form the penis, and a groove on its ventral surface closes over, by the coming together of the **genital folds**, to form the penile portion of the urethra (segment 3).

In the subsequent differentiation of most female mammals (Fig. 11-10, *E* and *F*), the constricted neck of the bladder forms the entire urethra. The female urethra is thus comparable to only a small portion (the allantoic segment) of the male urethra. The urethra and the two oviducts, whose lower ends have fused to form the vagina and uterus (Fig. 11-9), enter the urogenital sinus. In most female mammals, the urogenital sinus remains undivided. But in primates, it too becomes divided and continues the urethra and vagina to the surface as separate passages. The archinephric ducts disappear, and the phallus forms the small remnant known as the clitoris. The genital folds of the indifferent stage form the labia minora in such mammals as have these labia. The labia majora are skin folds comparable to the scrotum of the male.

Reproduction and Embryos

As stated earlier, eutherian mammals are viviparous, and the embryos develop within the uterus. If any of the specimens are pregnant, cut open the uterus and examine the embryos. Mammalian embryos produce the various extraembryonic membranes characteristic of amniotes (Fig. 11-6). Since the outermost membrane is the chorion, the whole complex of embryo and extraembryonic membranes is often called the **chorionic sac**. **Villi** arise from the surface of the chorioallantoic membrane in eutherian mammals, and penetrate or unite in various ways with the uterine lining. This combination of uterine lining and villi constitutes the **placenta**. In many mammals, only certain villi establish a definitive union with the uterine lining, and the gross shape of the placenta depends on the distribution of these villi. In the cat, a belt-shaped band of villi unites with the uterine lining to form a definitive placenta of the **zonary** type. This band is easily seen on the surface of the chorionic sac. In the rabbit, a disc-shaped patch of villi unites, and the placenta is said to be **discoidal**.

In many mammals, including the two considered, the union of villi and uterine lining is intimate, and some of the uterine lining is discharged at birth. Such a placenta is said to be **deciduous**, in contrast to a **nondeciduous** placenta in which the union is not intimate, and maternal tissue is not discharged at birth.

Still other terms describe the microscopic details of the union. In the cat, the surface epithelium lining the uterine portion of the placenta disappears. The villi penetrate the substance of the uterine lining and come in contact with the endothelial walls of the maternal capillaries (**endotheliochorial placenta**). In the rabbit this happens, but in addition the maternal blood vessels break open, and the epithelium on the surface of the villi disappear. The circulation of fetus and mother is then separated only by the endothelial walls of the fetal capillaries (**hemoendothelial placenta**). To summarize, the cat has a chorioallantoic, zonary, deciduous, endotheliochorial placenta; the rabbit has a chorioallantoic, discoidal, deciduous, hemoendothelial placenta.

Cut open the chorionic sac and placenta, and you will see the embryo enclosed by the **amnion**. Open the amnion. The cord of tissue extending from the underside of the embryo to the chorionic sac and placenta is the **umbilical cord**. It contains the allantoic stalk, umbilical, or allantoic, blood vessels, and a vestige of the yolk sac stalk. Its surface is covered by the amnion.

I / Terms for Directions, Planes, and Sections

It is necessary when discussing the anatomy of animals to have a set of terms that can be used in describing the location of parts. The following are the ones in most common usage in comparative anatomy. Most of them are illustrated in the accompanying diagram (Fig. A–1).

1. *Terms for Direction*

The terms for direction can be arranged in contrasting sets. **Dorsal** refers to the upper surface, or back of the body of an animal; **ventral**, to the under-surface, or belly.

Anterior, cephalic, cranial, or **superior** pertains to the head end; **posterior, caudal**, or **inferior** pertains to the tail end.

Lateral refers to the side of the body; **medial** to a position toward the midline. **Median** is used for a structure in the midline.

Distal pertains to the part of some organ, such as an appendage or blood vessel, that is furthest removed from some major point of reference, such as the origin of the organ, or the center of the body; **proximal** to the opposite end of the organ, i.e., the part nearest the point of reference.

Left and **right** are self-evident, but it should be emphasized that in anatom-

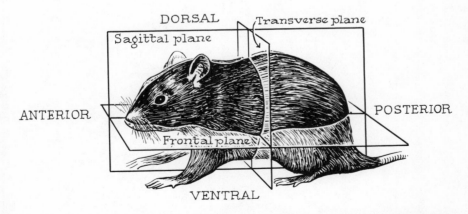

Figure A–1. A diagram to illustrate the planes of the body. (From Villee, Walker, and Smith, General Zoology.)

ical directions they always pertain to the specimen's left or right, regardless of the way the specimen is viewed by the observer.

Adverbs may be formed from the above adjectives by adding the suffix **ly** or **ad** to the root, in which case the term implies motion in a given direction. Thus caudally or caudad would mean movement in the direction of the tail.

It should also be remarked at this time that some of these terms, unfortunately, are used in a slightly different way in human anatomy. Anterior of human anatomy is equivalent to ventral of comparative anatomy, and posterior is equivalent to dorsal. The terms dorsal and ventral are seldom used in human anatomy.

2. *Planes and Sections of the Body*

The body is frequently cut in various planes to obtain views of internal organs. A longitudinal, vertical section from dorsal to ventral that passes through the median longitudinal axis of the body is a **sagittal** section. Such a section lies in the sagittal plane. Sections or planes parallel with, but lateral to, the sagittal plane are often referred to as **parasagittal**.

A section cut across the body from dorsal to ventral, and at right angles to the longitudinal axis, is a **transverse**, or **cross**, section, and it lies in the transverse plane.

A **frontal**, or **coronal**, section or plane is one lying in the longitudinal axis, and passing horizontally from side to side.

II / The Preparation of Specimens

It is common practice to purchase specimens from biological supply houses, but students are frequently interested in the way they are prepared, and it is sometimes necessary to prepare an occasional specimen. The following directions are presented with these needs in mind. Only the simpler and more common procedures are described. Further details can be found in such a book as Wilder and Gage, *Anatomical Technology*, or can be obtained by writing to a biological supply house.[18]

1. *Killing the Specimen*

After a specimen is caught, it must of course be killed. And incidentally, some specimens, cats for example, should be obtained from the proper sources. Long years have passed since Wilder and Gage could write, "There is usually no difficulty in taking a cat when it is wanted. Such as will not come when called may be secured by means of a strong net......"!

(A) Aquatic Vertebrates. Aquatic specimens may be killed by direct immersion either in a preserving solution (formalin or alcohol), or in a solution of chloretone. One ounce of a saturated chloretone solution added to one quart of water will kill most specimens, and it has the advantage of leaving them relaxed.

(B) Terrestrial Vertebrates. Terrestrial specimens should be killed by chloroforming or etherizing in a tight container, injecting ether into the body cavity, or by gassing. The last is preferable for such a mammal as a cat. To do it, simply put the cat in a resonably tight container, and insert a hose from a gas line. Turn the gas off when the specimen stops struggling, but leave the specimen in the container for 5 to 10 minutes.

2. *Preserving Specimens*

(A) Simple Preserving. Many animals can be preserved simply by cutting open the body cavity and immersing the specimen in a solution of alcohol or

---

[18] I am indebted for much of the following information to various "Turtox Service Leaflets" published by the General Biological Supply House, Chicago, Illinois.

formalin. The preserving solution can also be injected into the body cavity, and into the larger muscle masses. For most vertebrates, use a 70 per cent solution of alcohol, or an 8 to 10 per cent solution of formalin. Commercial formalin is a saturated, aqueous solution (40 per cent) of formaldehyde gas. An 8 per cent formalin is solution is a mixture of 8 parts of commercial formalin and 92 parts of water. It is best to fix the specimen in 10 per cent formalin for a day or two, and then to transfer it to an 8 per cent solution.

Formalin-preserved specimens are generally unpleasant to work with for any length of time. Several things may be done to alleviate the situation. Lanolin may be applied to one's hands before and after working on such specimens. Pure lanolin is better than the "toilet" variety, as it makes a more lasting coating on one's hands. The specimens may be soaked for several hours in water before using them, or the specimens may be soaked for a while (the longer the better) in a deformalizing solution. Such solutions remove some, but not all, of the formalin. One formula suggested by Fort, Wilson, and Goldberg (*Science*, vol. 94, pp. 169-170, 1941) is an aqueous solution containing 5.7 per cent by weight of sodium bisulfate and 3.8 per cent by weight of sodium sulfite.

(B) Embalming. Better results can be obtained on large specimens, however, by injecting the preserving fluid into the circulatory system — a process called embalming. After a cat, for example, has been killed, it should be tied out on a spreading board in the position in which one desires it to harden. Then make an incision through the skin on the medial surface of the thigh to expose the femoral artery. The artery is smaller and lighter in color than the accompanying vein. Mobilize the artery and slip a thread beneath it. Then make a V-shaped cut half way through the vessel with a pair of scissors. Insert a 16- or 18- gauge injection needle into the vessel, and tie it in. Force a wad of cotton into the animal's mouth to prevent the embalming solution from leaking.

An embalming solution that is often used is made according to the following formula:

Carbolic acid (melted crystals)	5 parts
Formalin (40%)	5 parts
Glycerin	5 parts
Water	85 parts

The formalin is the chief preservative; the carbolic acid is a disinfectant and also helps to preserve the color of the tissues, and gives a sanitary odor to the specimen; the glycerin prevents the specimen from drying out too quickly.

The solution can be injected with a gravity bottle, or with a large hand syringe. Inject sufficient fluid so that the specimen appears bloated, and the fleshy parts are firm. It should not be possible to move the appendages or head easily if enough fluid has been used. An average-sized cat takes about 1 liter. If a syringe is used, it must obviously be refilled. Be sure to cap the base of the needle with your finger when the syringe is taken off, or the fluid will escape. After injecting, remove the needle, and tie the artery.

If embalmed specimens become unusually dry while being dissected they may be moistened with a wetting fluid made as follows:

Carbolic acid crystals	30 g.
Glycerin	250 cc.
Water	1000 cc.

3. *Injection of the Cardiovascular System*

The blood vessels of an animal can be rendered more conspicuous by injecting a colored solution into them that will later harden. Colored latex is usually used. If specimens are to be simply preserved, the injection should be made before preservation. But embalmed specimens should be left for several days before injecting the vessels in order to let the embalming solution penetrate the tissues. Injection is made with an injecting syringe and with the needle inserted and tied into a cut in one of the vessels. Avoid injecting air. After injecting, the vessel should be tied off with a piece of string, or clamped with small "Tip Top" paper clips, on both sides of the cut. The vessel through which the injection is made varies with the animal, as described below. Ordinarily the arteries are injected before the veins. The success of the injection should be checked by seeing whether or not the smaller peripheral vessels are filled. It must be remembered, however, that veins contain valves, and these generally prevent some of the veins from filling—notably veins in the appendages.

(A) Dogfish. To inject the arteries of the dogfish, cut across the tail and inject forward through the caudal artery. Inject the hepatic portal system through the hepatic portal vein, or one of the intestinal veins.

(B) *Necturus*. In *Necturus*, the systemic veins should be injected first. Insert the needle into the posterior part of the postcaval and inject forward until the vessels in the liver are filled (8 to 10 cc.). Just before finishing the injection, press against the postcaval near the liver and thus force some injection mass into the posterior cardinals. The arteries are injected by inserting the needle into the dorsal aorta about 1/2 inch anterior to the celiacomesenteric artery. Inject forward until the gills assume the color of the mass (2 to 3 cc.), and posteriorly until the smaller intestinal arteries are filled (2 to 3 cc.). The hepatic portal system is injected through the mesenteric vein (8 to 10 cc.).

(C) Cat or Rabbit. Inject the arteries of the cat or rabbit caudally through one of the common carotids until the vessels in the intestine, or those beneath the skin, fill (30 cc.). The systemic veins are injected caudally through the external jugular (30 to 50 cc.), and the hepatic portal system through one of the intestinal veins in the mesentery (12 to 18 cc.). If only the arteries are to be injected, the blood may be allowed to accumulate in the veins, but if the veins are to be injected as well, the blood should be drained off during the arterial injection through a cut in the external jugular vein.

4. *Injection of the Lymphatic Vessels*

If the lymphatics of the mammal are to be studied, they can be injected in a freshly killed specimen. A saturated solution of Berlin blue should be injected subcutaneously in various parts of the body, for this substance will enter the lymphatic capillaries, but not those of the cardiovascular system. Injections should be made in all the foot pads, in the pad at the end of the nose, and in the lips. It is necessary to inject slowly, and to maintain a pressure on the syringe for 15 to 30 minutes at each site. Also massage the limbs and neck, working toward the center of the body, to aid the flow of the injection fluid. To inject the lymphatics at the base of the mesentery, it is necessary to open the body, and to inject into the peripheral parts of the mesenteric lymph nodes (pancreas of Aselli) The efferent lymphatic from any lymph node may be injected in the same way.

If a more permanent preparation is desired, a weak solution of warm gelatin may be mixed with the Berlin blue. In this case, the injection must be made while the body and the gelatin are still warm. Injecting with gelatin solution is more difficult.

5. *Preparation of Skeletons*

Skeletons should be prepared from mature specimens that have been freshly killed, or from specimens preserved in brine. Fresh specimens are better. One should first skin and dismember the specimen, removing the head and legs. Then cut off the larger masses of flesh from the bones. The rest of the flesh can easily be scraped off after it has been loosened in one of several ways. (1) Let it macerate (decay) in a closed container of water at room temperature for several months. (2) Allow insects to remove most of the soft parts. Dermestid beetles are particularly good, and some institutions maintain a colony for this purpose. (3) Simmer the specimen in water, or in a soap solution made by diluting 1 part of the following stock solution with 3 or 4 parts of water:

Ammonia (strong)	150 cc.
Hard soap	75 g.
Potassium nitrate (saltpeter)	12 g.
Water	2000 cc.

The brain can be removed through the foramen magnum with a wire having a flattened loop at one end.

If disarticulated bones are desired, one can let the bones simmer for a considerable time (an hour or two). But if a skeleton in which the bones are held together by ligaments is the objective, one must look at the specimen frequently to be sure that the muscles are soft enough to scrape off, but that the ligaments are not so soft that they detach easily. It is also necessary to cut some of the larger ligaments and tendons, especially those on the underside of the paws of mammals, or they will shorten and distort the specimen. When such a preparation is cleaned satisfactorily, the parts of the skeleton must be nailed out in the desired position before the preparation dries.

After the flesh has been removed, it is generally desirable to degrease the bones by leaving them for a day or so in turpentine, benzine, or carbon tetrachloride. The bones should then be bleached for a day or two in a solution of hydrogen peroxide.

To get a disarticulated skull, simply continue the boiling process until the sutures are soft enough to permit pulling the bones apart, Another method is to fill the cranial cavity with dried peas, tightly cork the foramen magnum, and put the skull in water. The peas will swell and loosen the bones. Young adults should be used for disarticulated skulls, as some of the skull bones grow together in old specimens.

III / References

The references given below include those cited in the text, those of particular value for laboratory studies in comparative anatomy and certain key references on the functional significance and interrelationships of the various organs. More inclusive bibliographies can be found in standard textbooks, and in many of the works cited below.

GENERAL

Bolk, L., and others: Handbuch der vergleichenden Anatomie der Wirbeltiere. 6 vols. Berlin and Vienna, Urban und Schwarzenberg, 1931-1938.

Bronn, H. G., and others: Klassen und Ordnungen des Thier-Reichs. Abteilung VI (Vertebrates). Leipzig and Heidelberg, C. F. Winter'sche Verlagshandlung, 1874-1938.

DeBeer, G. R.: The Vertebrate Skull. Oxford, Clarendon Press, 1937.

DeBeer, G. R.: Vertebrate Zoology. Revised ed. New York, The Macmillan Company, 1953.

Edgeworth, F. H.: The Cranial Muscles of Vertebrates. London, Cambridge University Press, 1935.

Fort, W. B., Wilson, H. C., and Goldberg, H. G.: The daily removal of formalin from preserved biological specimens used in class work. Science, vol. 94, pp. 169–70, 1941.

Foxon, G. E. H.: Problems of the double circulation in vertebrates. Biological Reviews of the Cambridge Philosophical Society, vol. 30, pp. 196-228, 1955.

Goodrich, E. S.: Studies on the Structure and Development of Vertebrates. New York, Dover Publications, 1958.

Grassé, P.-P., editor: Traité de Zoologie. Vols. 11-17 deal with protochordates and vertebrates. Paris, Masson et Cie., 1948-1958.

Gray, J.: Studies in the mechanics of the tetrapod skeleton. Journal of Experimental Biology, vol. 20, pp. 88-116, 1944.

Hughes, G. M.: Comparative Physiology of Vertebrate Respiration. Cambridge, Harvard University Press, 1963.

Hyman, L. H.: Comparative Vertebrate Anatomy. 2nd ed. Chicago, University of Chicago Press, 1942.

Jaeger, E. C.: A Source-Book of Biological Names and Terms. 3rd ed. Springfield, Charles C Thomas, 1955.

Kappers, C. U. A., Huber, G. C., and Crosby, E. Ç.: The Comparative Anatomy of the Nervous System of Vertebrates, Including Man. New York, Macmillan Company, 2 vols., 1936.

Kopsch, K., and Knese, K. H.: Nomina Anatomica. Vergleichende Übersicht der Basler, Jenaer und Pariser Nomenklotur. Stuttgart, Georg Thieme Verlag, 1957.

Nelsen, O. E.: Comparative Embryology of the Vertebrates. New York, The Blakiston Company, 1953.

Parker, T. J., and Haswell, W. A.: A Text-Book of Zoology. 6th ed., revised by Forster-Cooper, C. London, Macmillan Company, 1940, vol. 2.

Prosser, C. L., and Brown, F. A., Jr.: Comparative Animal Physiology, 2nd ed. Philadelphia, W. B. Saunders Company, 1961.

Rand, H. W.: The Chordates. Philadelphia, The Blakiston Company, 1950.

Romer, A. S.: The Vertebrate Body. 3rd ed. Philadelphia, W. B. Saunders Company, 1962. ,

Romer, A. S.: Vertebrate Paleontology. 2nd ed. Chicago, University of Chicago Press, 1945.

Romer, A. S.: The Vertebrate Story. Chicago, University of Chicago Press, 1958.

Walls, G. L.: The Vertebrate Eye and Its Adaptive Radiation. Bloomfield Hills, Cranbrook Institute of Science, Bull. No. 19, 1942.

Wiedersheim, R.: Comparative Anatomy of Vertebrates. 3rd English ed., adapted by W. N. Parker. London, The Macmillan Company, 1907.

Wilder, B. G., and Gage, S. H.: Anatomical Technology as Applied to the Domestic Cat. New York, A. S. Barnes and Company, 1882.

Williams, E. E.: Gadow's arcualia and the development of tetrapod vertebrae. Quarterly Review of Biology, vol. 34, pp. 1-32, 1959.

Young, J. Z.: The Life of Vertebrates. 2nd ed. London, Oxford University Press, 1962.

FISHES

Barrington, E. J. W.: The supposed pancreatic organs of *Petromyzon fluviatilis* and *Myxine glutinosa*. Quarterly Journal of Microscopical Science, vol. 85, pp. 391-417, 1945.

Bigelow, H. B., Schroeder. W. C., and Farfante. I P.: Fishes of the Western North Atlantic. Part I. Lancelets, Cyclostomes and Sharks. New Haven, Sears Foundation for Marine Research, Yale University, 1948.

Brown, M. E., editor: The Physiology of Fishes. 2 vols. New York, Academic Press Inc., 1957.

Burger, J. W., and Hess, W. N.: Function of the rectal gland in the spiny dogfish. Science, vol. 131, pp. 670-671, 1960.

Daniel, J. F.: The Elasmobranch Fishes. 3rd. ed. Berkeley, University of California Press, 1934.

Dean, B.: Fishes, Living and Fossil. New York, Macmillan, 1895.

Gans, C., and Parsons, T. S.: A Photographic Atlas of Shark Anatomy. New York, Academic Press, 1964.

Gibbs, S. P.: The anatomy and development of the buccal glands of the lake lamprey (*Petromyzon marinus* L.) and the histochemistry of their secretion. Journal of Morphology, vol. 98, pp. 429-470, 1956.

Goodrich, E. S.: A Treatise on Zoology, edited by E. Ray Lankester. Part IX.

Vertebrata Craniata, Fasc. I. "Cyclostomes and Fishes." London, Adam and Charles Black, 1909.

Goodrich, E. S.: On the development of the segments of the head of Scyllium. Quarterly Journal of Microscopical Science, vol. 63, pp. 1-30, 1918.

Heath, G. W.: The siphon sacs of the smooth dogfish and spiny dogfish. Anatomical Record, vol. 125, p. 562, 1956.

Hisaw, F. L., and Albert, A.: Observations on the reproduction of the spiny dogfish, *Squalus acanthias*. Biological Bulletin, vol. 92, pp. 187-119, 1947.

Marinelli, W., and Strenger, A.: Vergleichende Anatomie und Morphologie der Wirbeltiere. I. Lieferung. Lamperta fluviatilis L., III. Lieferung. Squalus acanthias L. Vienna, Franz Deuticke Verlag, 1954 and 1959.

Murray, R. W.: Evidence for a mechanoreceptive function of the ampullae of Lorenzini. Nature, vol. 179, pp. 106-107, 1957.

Norris, H. W., and Hughes, S. P.: The cranial, occipital, and anterior spinal nerves of the dogfish, *Squalus acanthias*. Journal of Comparative Neurology, vol. 31, pp. 293-402, 1920.

Nursall, J. R.: Swimming and the origin of paired appendages. American Zoologist, vol. 2, pp. 127-141, 1962.

O'Donoghue, C. H., and Abbot, E. B.: The blood vascular system of the spiny dogfish, *Squalus acanthias* Linne, and *Squalus sucklii* Gill. Transactions of the Royal Society of Edinburgh, vol. 55, pp. 823-894, 1928.

Oguri, M.: Rectal glands of marine and fresh water sharks, comparative histology. Science, vol. 144, pp. 1151-1152, 1964.

Sand, A.: The function of the ampullae of Lorenzini, with some observations on the effect of temperature on sensory rhythms. Proceedings of the Royal Society, Series B, vol. 125, pp. 524-553, 1938.

Satchell, G. H.: The reflex coordination of the heart beat with respiration in the dogfish. Journal of Experimental Biology, vol. 37, pp. 719-731, 1960.

Willemse, J. J.: The way by which flexures of the body are caused by muscular contractions. Koninklijk Nederland Akademie Wentschap. Proceedings Series C, vol. 62, pp. 589-593, 1959.

Young, J. Z.: The autonomic nervous system of selachians. Quarterly Journal of Microscopical Science, vol. 75, pp. 571-624, 1933.

AMPHIBIANS AND REPTILES

Barclay, O. C.: The mechanics of amphibian locomotion. Journal of Experimental Biology, vol. 23, pp. 177-203, 1946.

Chase, S. W.: The mesonephros and urogenital ducts of *Necturus maculosus*, Rafinesque. Journal of Morphology, vol. 37, pp. 457-532, 1923.

DeLong, K. T.: Quantitative analysis of blood circulation through the frog heart. Science, vol. 138, pp. 693-694, 1962.

Figge, F. H.: A morphological explanation for the failure of Necturus to metamorphose. Journal of Experimental Zoology, vol. 56, pp. 241-265, 1930.

Francis, E. B.: The Anatomy of the Salamander. London, Oxford University Press, 1943.

Harris, J. P., Jr.: Necturus papers: The skeleton of the arm. The pelvic musculature. The muscles of the forearm. The levator anguli scapulae. The musculus depressor mandibulae. Natural history. Field and Laboratory, vols. 20, 21, 22, 25, 27, 1952-1959.

Herrick, C. J.: The Brain of the Tiger Salamander, *Ambystoma tigrinum*. Chicago, University of Chicago Press, 1948.

Kingsbury, B. F.: On the brain of *Necturus maculatus*. Journal of Comparative Neurology, vol. 5, pp. 139-205, 1895.

Miller, W. S.: The vascular system of *Necturus maculatus*. University of Wisconsin Science Series, vol. 2, pp. 211-226, 1900.

Noble, G. K.: The Biology of the Amphibia. New York, Dover Publications, 1954.

Romer, A. S.: Osteology of Reptiles. Chicago, Chicago University Press, 1956.

Reed, H. D.: The morphology of the sound-transmitting apparatus in caudate amphibia and its phylogenetic significance. Journal of Morphology, vol. 33, pp. 325-287, 1920.

Wilder, H. H.: The skeletal system of Necturus maculatus Rafinesque. Memoirs of the Boston Society of Natural History, vol. 5, pp. 387-439, 1903.

Wilder, H. H.: The appendicular muscles of *Necturus maculosus*. Zoologische Jahrbucher, sup. 15, part 2, pp. 383-424, 1912.

MAMMALS

Abell, N. B.: A comparative study of the variations of the postrenal vena cava of the cat and rat and a description of two new variations. Denison University Journal of the Science Laboratory, vol. 40, pp. 87-117, 1947.

Arey, L. B.: Developmental Anatomy. 6th ed. Philadelphia, W. B. Saunders Company, 1954.

Barry, A.: The aortic arch derivatives in the human adult. Anatomical Record, vol. III, pp. 221-238, 1951.

Bensley, B. A.: Practical Anatomy of the Rabbit. 7th ed. revised by Craigie, E. H. Philadelphia, The Blakiston Company, 1946.

Bloom, W., and Fawcett, D. W. : A Textbook of Histology. 8th ed. Philadelphia, W. B. Saunders Company, 1962.

Davis, D. D., and Story, H. E.: The carotid circulation in the domestic cat. Zoological Series Field Museum of Natural History, vol. 28, pp. 1-47, 1943.

Field, H. E., and Taylor, M. E.: An Atlas of Cat Anatomy. Chicago, Chicago University Press, 1954.

Gross, C. M., editor: Gray's Anatomy of the Human Body. 27th ed. Philadelphia, Lea and Febiger, 1960.

Hamilton, W. J., Boyd, J. D., and Mossman, H. W.: Human Embryology. London, W. Heffer & Sons, Ltd., 1945.

Huntington, G. S., and McClure, C. F. W.: The development of the veins in the domestic cat. Anatomical Record, vol. 20, pp. 1-31, 1920.

Jayne, H.: Mammalian Anatomy. Part I. The Skeleton of the Cat. Philadelphia, J. B. Lippincott Company, 1898.

Kerr, N. S.: The homologies and nomenclature of the thigh muscles of the opossum, cat, rabbit, and rhesus monkey. Anatomical Record, vol. 121, pp. 481-493, 1955.

Ranson, S. W.: The Anatomy of the Nervous System. 10th ed. revised by Clark, S. L. Philadelphia, W. B. Saunders Company, 1959.

Rasmussen, A. T.: The Principal Nervous Pathways, 4th ed. New York, Macmillan Company, 1952.

Reighard, J. E., and Jennings, H. S.: Anatomy of the Cat. 3rd ed. revised by Elliott, R. New York, Henry Holt and Company, Inc., 1935.

Voge, M., and Bern, H. A.: Cecal villi in *Dicrostonyx torquatus* (Rodentia: Mircrotinae). Anatomical Record, vol. 123, pp. 125-132, 1955.

Weber, M., Burlet, H. M. de, and Abel, O.: Die Saugetiere. 2 vols., 2nd ed. Jena, G. Fischer, 1927-1928.

Young, J. Z.: The Life of Mammals. New York and Oxford, Oxford University Press, 1957.

INDEX

Abdomen, 37
Abdominal muscle layers, *131*
Abdominal pores, 230
Abdominal veins, lateral. See *Lateral abdominal veins*.
Abdominal viscera, 254, *260*
Abducens nerve, in *Necturus*, 205
  in sheep, 213
  in *Squalus*, 190
Abduction, 111, 117
Accessory nerve, 207
  in sheep, 214
Accessory process, 68
Accessory sex glands, 325
Accessory urinary duct, 317
Acelous, 68
Acetabular surface, 88
Acetabulum, in cat, 100, 102
  in *Chelydra*, 94
  in *Necturus*, 90
Acoustic meatus, external, 38, 74, 174, 178
  internal, 79, 80, 179
Acousticolateral area, 187
Acromiocoracoid ligament, 92
Acromiodeltoid muscle, 138
Acromion, in cat, 98
  in *Chelydra*, 92
Acromiotrapezius muscle, 137
Adduction, 111, 117
Adductor brevis muscle, 150
Adductor femoris muscle, in mammals, 144
  in *Necturus*, 126
Adductor longus muscle, 144, 150
Adductor mandibulae muscle, 118
Adipose capsule, 326
Adrenal glands, 293
  in *Necturus*, 323
Afferent branchial arteries, in *Necturus*, 286, *286*, 287
  in *Squalus*, 238, 271
Afferent neurons, 182
  internuncial, 183

Afferent renal veins, 267
Afferent spiracular artery, 273
Agnatha, 26
Air sinuses, 79
Albumin, 325
Alisphenoid bone, 73, 78
Allantoic arteries, 264
Allantois, 229
Alveoli, 246, 252
Amia, 28, 50
  fin of, 87, *87*
  pectoral girdle of, *87*
  pelvic girdle of, 87
  skull of, *50*
Ammocoetes larva, 22-25, *23*
Amnion, 326, 337
Amniote, 28
  embryo of, blood vessels of, *263*
    succession of kidneys in, *313*
  genital system of, *314*
Amphibia, 28
Amphicelous, 42
Amphioxus, 2, 6, *7, 8*
  cross sections of, 11-13
  external features of, 7-8
  whole mount slide, 8-11
Ampulla(e), 177
  of Lorenzini, 162
    in *Squalus*, 33, 163
Anal glands, 331, 333
Anamniotes, 28
Anapophysis, 68
Anapsid condition, in *Chelydra*, 62
  in primitive tetrapods, 56
Angular process, in cat, 82
  in *Chelydra*, 66
  in *Necturus,* 61
Ankle. See *Tarsus*.
Anterior (defined), 339
Anterior cardinal sinus, 193, 269
Anterior cardinal veins, 264
  in lamprey, 21, 22

Anterior cardinal veins (*Continued*)
  in *Necturus*, 285
Anterior chamber, 167
Anterior chorioid plexus, in *Necturus*, 205
  in *Squalus*, 186
Anterior commissure, 216
Anterior coracoid, in *Chelydra*, 92
  in primitive tetrapods, 88
Anterior cranial fossa, 79
Anterior facial vein, 298, 300
Anterior intestinal artery, 277
Anterior intestinal vein, 267
Anterior lacerate foramen, 80
Anterior lobe of brain, 200
Anterior pancreaticoduodenal vein, 292
Anterior peduncle, 213
Anterior perforated substance, 212
Anterior systemic veins, in mammals, 295-298
  in *Necturus*, 284
Anterior trapezius muscle, 137
Anterior trunk muscles, 152-156
Anterior utriculus, 177
Anterior vena cava, 291, 295, 308
Anterior ventrolateral artery, 277
Anterior zygapophyses, in cat, 68
  in *Necturus*, 53
Antibrachium, in mammals, 37
  in *Necturus*, 34
  in primitive tetrapod, 88
Anticlinal vertebra, 71
Antorbital process, in *Necturus*, 60
  in *Squalus*, 46
Anura, 28
Anus, in *Amphioxus*, 8, 11, 13
  in mammals, 38, 258
Aorta, dorsal. See *Dorsal aorta*.
  radices of, 287
  ventral. See *Ventral aorta*.
Aortic arch, 291
  adult, *274*
  embryonic, *274*
  in mammal, 291, 302, *305*
  in *Necturus,* 285
Aortic valve, 310
Apoda, 28
Aponeurosis, 111
Appendicular muscles, 103
  in primitive tetrapods, 123-127
  in *Squalus*, 116-117
Appendicular skeleton, 40, 84-102
  in fishes, 84-88
  in mammals, 96-102
  in *Necturus*, 90
  in primitive tetrapods, 88-96
  in turtle, 91-96
Appendix, vermiform, 260
Aquatic vertebrate, preparation of, 341
Aqueous humor, 167
Arachnoid membrane, 207
Arachnoid space, 220
Arbor vitae, 217
Archenteron, 226
Archinephric duct, 312
  in lamprey, 22
  in *Necturus*, 242, 322, 323, 324
  in *Squalus*, 231, 316, 317, 320
Archinephros, 313
Archipallium, 206, 210
Archipterygium fin, 84

Arcualia, 44
  in lamprey, 18
Area acoustica, 214
Area vestibularis, 214
  in cat, 302-308
  in *Necturus*, 285-290
  in *Squalus*, 270
Arteries, 262. See also names of arteries.
Articular bone, in *Chelydra*, 64
  in primitive tetrapod, 58
Articular facet, 68
Arytenoid cartilage, 251
Ascending colon, 259
Ascending pharyngeal artery, 305
Ascidiacea, 3
Aselli, pancreas of, 311
Astragalus, in cat, 102
  in *Chelydra*, 96
Atlantal foramen, 69
Atlas, 52
  in cat, 69
Atriopore, 8, 10
Atrioventricular aperture, in mammals, 309
  in *Squalus*, 278
Atrium, 266. See also *Auricle*.
  in *Amphioxus*, 10, 12
  in lamprey, 21
  in mammals, 308, 309
  in *Molgula*, 5
  in *Necturus*, 281
  in *Squalus*, 278
Auditory ossicles, *175*
Auditory tube, 227
  in mammal, 81, 178, 250
  in primitive tetrapod, 177
Auricle, of cerebellum, in *Necturus*, 205
    in *Squalus*, 186
  of ear, 38, 178
  of heart, 266, 291, 309
Autonomic nervous system, 181
  parasympathetic subdivision of, 182
Autostylic suspension of jaw, 49
Aves, 28
Axial muscles, 103
  in fishes, 113-116
  in *Necturus*, 121-123
Axial skeleton, 40-83
  postcranial, in fishes, 41-44
    in mammals, 66-72
    in primitive tetrapods, 51-56
Axillary artery, 303
Axillary nerve, 222
Axillary vein, 296, 298
Axis, 69
Azygos vein, 295, 298

Balanoglossus, 1
Basal cartilage, 44
  in *Squalus*, 87, 88
Basal plate, 46
Basapophyses, 44
Basibranchial arch, 49
  in *Necturus*, 61
Basiclavicularis muscle, 137
Basihyal arch, 49
  in cat, 83
Basioccipital arch, 60
Basioccipital bone, 64
Basipterygoid processes, 47

Basisphenoid bone, in cat, 78
  in *Chelydra*, 64
Basitrabecular process, 47
Batoidea, 26
Biceps brachii, 141
Biceps femoris, 142, 147, 148
Bicipital groove, 98
Bicuspid valve, 309
Bilaminar, 326
Bile duct, common. See *Common bile duct.*
  in *Salamandra*, *243*
  in *Squalus*, 230, 233
Bladder, urinary. See *Urinary bladder.*
Blind spot, 167
Blood capillaries, 262
Blood vessels, 265
  cutaneous, 129
  in amniote embryo, *263*
Bones. See names of bones.
Bony palate, 75, 249
Botallus, ductus arteriosus of, 305
Brachial artery, in mammals, 303
  in *Necturus*, 289
  in *Squalus*, 277
Brachial plexus, 221-223, *222*
Brachial vein, in mammals, 296, 298
  in *Necturus*, 285
  in *Squalus*, 270
Brachialis, 141
Brachiocephalic trunk, 302
Brachiocephalic vein, 296
Brachium, in mammals, 37
  in *Necturus*, 34
  in primitive tetrapod, 88
  muscles of, 140
Brachium conjunctivum, 213
Brachium pontis, 212
Brain, 180
  cortex of, 181
  development of, *183*, 184
  hemispheres of, 206, 212
  intermediate lobe of, 200
  in *Ammocoetes* larva, 23
  in lamprey, 18
  in *Necturus*, 203-206, *204*
  in sheep. See *Sheep brain.*
  in *Squalus*, 184-187, *185*, *189*, 200-202, *201*
  ventricles of. See *Ventricles of brain.*
Brainstem, nuclei of, 180
  sheep, *211*, *213*
Branchial adductor muscle, 119
Branchial arches, 48
Branchial arteries, in *Necturus*, 286, *286*, 287
  in Squalus, 238, 270, *271*
Branchial bars, 227
Branchial basket, 18
Branchial muscles, 110
Branchial nerve, *199*
Branchial pouch, in lamprey, 21
  in *Squalus*, 237
Branchial vessels, *288*
Branchiohyoideus, 128
Branchiomeric muscles, 110
  in fishes, 117-119
  in mammals, 158-160
  in *Necturus*, 127
Branchiostoma, 6
Brevis muscle, 125

Broad ligament, 332
Bronchi, 252, 308-310
Buccal branch of facial nerve, 191
Buccal cavity, in *Ammocoetes* larva, 23
  in lamprey, 20
Buccal funnel, 15
Buccal glands, in lamprey, 20
  in mammals, 248
Bulbourethral glands, 330
Bulbus arteriosus, 281
Bulbus oculi, 165
Bursa, omental, 255

Calcaneus, in cat, 102
  in *Chelydra*, 96
Calcaneus tendon of Achilles, 152
Calyces, 327
Canine (teeth), 82
Canthi, 169
Capillus, 39
Capitate, 100
Capitulum, 99
Capsularis, 146
Caput epididymis, 329
Carapace, 92
Cardiac region, in mammals, 255
  in *Necturus*, 242
  in *Squalus*, 233
Cardinal veins, common. See *Common cardinal veins.*
  posterior. See *Posterior cardinal veins.*
Cardiovascular system, 262
  development of, 263
  injection of, for study, 343
Carnassial teeth, 83
Carnivora, 36
Carnivore, primitive carotid circulation in, *304*
Carotid artery, 302
  external. See *External carotid artery.*
  in lamprey, 21
  internal. See *Internal carotid artery.*
Carotid canal, 81
Carotid circulation, *304*
Carotid duct, in *Necturus*, 287
Carotid foramen, 46
Carpals. See also *Carpus.*
  in cat, 99
  in *Chelydra*, 93
  in *Necturus*, 90
  in primitive tetrapod, 88
Carpus. See also *Carpals.*
  in cat, 99
  in *Necturus*, 34
Cartilage replacement bone, 40
Cat. See also *Mammals.*
  caudal vertebrae in, 69
  cervical vertebrae in, 69
  deep muscles of pelvic girdle in, 144
  deep muscles of shoulder in, 138
  deep thigh muscles in, *145*
  epaxial muscles in, 132, *154*, 155
  hyoid apparatus in, 83, 160
  hypaxial muscles in, 130-132, 152
  iliopsoas complex in, 151
  jaw in, 82, *159*
  lateral thigh muscles in, 142

Cat (*Continued*)
  lumbar vertebrae in, 68
  mandibular muscles in, 158
  muscles of brachium in, 140
  muscles of forearm in, 141
  neck muscles in, *135*
  oral cavity in, 249
  pectoral girdle in, 98-100, *97*
  pectoral muscles in, 132, *133, 135*
  pelvic bones in, *101*
  pelvic girdle muscles in, 43
  posterior branchiomeric muscles in, 160
  posteromedial thigh muscles in, 144
  posthyoid muscles in, 156
  prehyoid muscles in, 157
  quadriceps femoris complex in, 144
  ribs in, 70
  sacral vertebrae in, 69
  shank muscles in, 146, 152
  skull in, *76*
    composition, 76-78
    foramina, 80-81
    general features, 74-75
    sagittal section, 79
  sternocleidomastoid muscles in, 135
  sternum in, 70
  superficial muscles of shoulder in, 137
  superficial muscles of pelvic girdle in, 143
  teeth in, 82
  thoracic vertebrae in, 67
  trapezius muscles in, 135
  vertebrae in, *67*
  vertebral skeleton in, 66
Cauda, 37
Cauda epididymis, 329
Caudal (defined), 339
Caudal artery, 267
  in lamprey, 21
  in mammals, 308
  in *Necturus*, 290
Caudal fin, in *Ammocoetes* larva, 22
  in *Amphioxus*, 7, 13
  in lamprey, 15
  Squalus, 32
Caudal ligament, 316
Caudal region, 15
Caudal vein, 267
  in lamprey, 21
  in mammals, 295
  in *Necturus*, 283
Caudal vertebrae, 52
  in cat, 67, 69
Caudate nucleus, 218
Caudocruralis, 126
Caudofemoralis, in mammals, 143
  in *Necturus*, 126
Caudopuboischiotibialis, 126
Caval fold, 252
Cecum, in *Amphioxus*, 11, 12
  in mammals, 259, 260
Celiac trunk, in mammals, 306
  in *Squalus*, 277
Celiacomesenteric artery, 289
Central canal of nervous system, 180, 220
Central nervous system, 180
Centralia, 93
Centrum, in cat, 68
  in *Necturus*, 53
  in *Squalus*, 42

Cephalic (defined), 339
Cephalic vein, 298, 300
Cephalobrachial muscle, 137
Cephalochordata, 1, 6-13
Ceratobranchial arch, 49
  in *Necturus, 61*
Ceratohyal arch, 49
  in *Necturus,* 61
Ceratohyal process, 83
Ceratotrichia, 32, 44
  in *Squalus,* 87
Cerebellum, auricle of. See *Auricle, of cerebellum.*
  in *Necturus,* 205
  in sheep, 212
Cerebral aqueduct of Sylvius, 202
  in sheep, 217
Cerebral fissure, longitudinal, 209
Cerebral hemisphere(s), left, *208*
  in *Necturus,* 204
  in sheep, 209
  in *Squalus,* 185
Cerebral peduncles, 212
Cerebrum, sheep, dissection of, 217-219, *218*
Cervical artery, 303
Cervical plexus, 221
Cervical sympathetic nerve, 303
Cervical vertebra, 52
  in cat, 66, 69
Cervical vessels, *299*
Cervix, of uterus, 37, 333, 334
Chamber of vitreous body, 167
Chelonia, 28
Chelydra. See also *Turtle.*
  hyoid apparatus in, 66
  lower jaw in, 64
  pectoral girdle in, 91-94, *91, 93*
  pelvic girdle in, 94-96, *95, 96*
  plastron in, *92*
  skull in, 62-63, *65*
Chevron bones, 52
  in cat, 69
Chimaerae, 28
Choanae, 75
Choanichthyes, 28
Choledochal veins, 266
Chondrichthyes, 26
Chondrochranium, in fishes, 45
  in lamprey, 18
  in *Necturus,* 60
  in *Squalus,* 45
  sagittal section of, 47
Chondrostei, 28
Chorda tympani, 179
Chordae tendineae, 310
Chordata, 1
Chordates, lower, 1-13
Chorioallantoic membrane, 326
Chorioid coat, 167
Chorioid plexus, 214, 217
  posterior. See *Posterior chorioid plexus.*
Chorion, 326
Chorionic sac, 336
Chromatophores, in *Ammocoetes* larva, 22
  in *Squalus,* 33
Cilia, 39
Ciliary body, 167
Ciliary nerve, 188
Circulatory system, 262-310
  course of, 262

Circulatory system (*Continued*)
  functions of, 262
  in fishes, 265-280
  in lamprey, 21
  in mammals, 290-312
  in primitive tetrapods, 280-290
  parts of, 262
Circumvallate papillae, 250
Cirri, 8, 9, 12
Cisterna chyli, 310
Claspers, 32, 88, 318
Clava, 215
Clavicle, in cat, 98
  in *Chelydra*, 92
Clavodeltoid muscle, 137, 138
Claws, in mammals, 37, 39
  in primitive tetrapods, 36
Cleidoic egg, 325
Cleidomastoid muscle, 136, 137
Cleithrum, 87
Cloaca, division of, *335*
  in *Ammocoetes* larva, 25
  in lamprey, 17
  in *Necturus*, 242
  in *Squalus*, 230
Cloacal aperture, in lamprey, 17
  in *Necturus*, 36
  in *Squalus*, 31
Cloacal artery, 290
Cloacal gland, 125, 324, 325
Cloacal vein, 270
Coccygeal vertebrae, 67
Coelom, 224-226
  in fishes, 229-239
  in lamprey, 20
  in mammals, 246
  in primitive tetrapods, 239-245
  subdivisions of, *225*
  subendostylar, in *Amphioxus*, 12
  visceral organs and, *225*
Coelomic canals, dorsal, 12
Collateral sulcus, 210
Collecting tubules, 322
Collector loops, 272
Colliculi, 212
Colon, 258-260
Columella, 57
Commissure of fornix, 218
Common bile duct, in mammals, 257
  in *Necturus*, 243
Common cardinal veins, 264
  in lamprey, 21
  in *Necturus*, 285
  in *Squalus*, 269
Common carotid artery, 302
Common iliac arteries, 307
Common iliac vein, in mammals, 295
  in *Necturus*, 282
Common jugular vein, 285
Communicating rami, 181, 220, 306
Conchae, 79, 173
Condyles, 98, 102
  occipital. See *Occipital condyles.*
Condyloid canal, 81
Condyloid process, 82
Conjunctiva, 169
  in *Squalus*, 165, 166
Connective tissue, 230
Constrictors, 118, 119, 238

Conus arteriosus, 266
  in *Necturus*, 281
  in *Squalus*, 278
Coprodeum, in mammals, 335
  in *Squalus*, 320
Copula, 49
Copulatory organ, 325
Cor, 291. See also *Heart.*
Coracoarcuals, 115
Coracobrachialis, in mammals, 141
  in *Necturus*, 125
Coracobranchials, 116
Coracohyoid muscle, 115
Coracoid bar, 86
Coracoid foramen, in *Necturus*, 90
  in *Squalus*, 86
Coracoid head, 125
Coracoid plate, 90
Coracoid process, 98
Coracomandibular muscle, 115
Coracoradialis, 124, 125
Cords of urogenital union, 315
Corium, 30
Cornea, 167, 169
Cornua, 66
Coronal section (defined), 340
Coronary artery, in mammals, 291, 302
  in *Squalus*, 273
Coronary ligament, 228
  in mammals, 255
  in *Necturus*, 244
  in *Squalus*, 235
Coronary sinus, 308
Coronary sulcus, 291
Coronary veins, in mammals, 291
  in *Squalus*, 270
Coronoid bone, 66
Coronoid fossa, 82
Coronoid process, 82, 99
Corpora cavernosa clitoridis, 333
Corpora cavernosa penis, 331
Corpora lutea, 331
Corpora quadrigemina, 212
Corpus callosum, 209, 215
Corpus cavernosum urethrae, 331
Corpus epididymis, 329
Corpus striatum, 218
Cortex, of brain, 181, 217
  of kidney, 327
Cosmoid scales, 33
Costae, 70
Costal cartilage, 70
Costocervical trunk, 303
Costocervical veins, 296, 298
Cotylosauria, 28
Countershading, 31
Cowper's glands, 330
Cranial (defined), 339
Cranial cavity, in cat, 79
  in *Squalus*, 46, 47
Cranial fossae, 79
Cranial nerves. See also names of nerves.
  in *Necturus*, *204*, 205
  in *Squalus*, 187-200
  posttrematic branch of, 239
  pretrematic branch of, 239
  stumps of, 209-215
Cranial region, 38, 74
Craniata, 1

Cremasteric muscles, 328
Cremasteric pouch, 129, 328
Crest of ilium, 100
Crest of tibia, 102
Cribriform plate, 79
Cricoid cartilage, 251
Crista, 177
Crocodilia, 28
Cross section (defined), 340
Crossopterygii, 28
Crura, of penis, 330
Crus, 34, 37
Crystalline lens, 166
Ctenoid scales, 33
Cuboid bone, 102
Cucullaris, in *Necturus*, 124, 128
    in *Squalus*, 119
Cuneate nucleus, tubercle of, 215
Cuneiform bone, 99, 102
Cupula, 162
Cutaneous artery, 289
Cutaneous blood vessels, 129
Cutaneous nerves, 223
Cutaneous vein, 285
Cutis, 30
Cuvier, ducts of. See *Ducts of Cuvier.*
Cycloid scales, 33
Cyclostomata, 26
Cystic duct, in mammals, 256
    in *Necturus*, 243

Deciduous placenta, 336
Dens, 70
Dentary bones, in cat, 82
    in *Chelydra,* 66
    in *Necturus,* 61
Denticles, dermal, 33
Dentine, 33
Depressor conchae posterior, 137
Depressor mandibulae, 128
Depressores arcuum, 128
Dermal bones, 40
    in *Amia,* 50
Dermal denticles, 33
Dermal girdle, 84
Dermal plates, 92
Dermal roof, 50
    in *Chelydra,* 62
Dermal skeleton, 40
Dermis, 11, 30
Descending colon, 259
Diaphragm, 226, *226,* 252
Diapophysis, 51
    in cat, 68
    in *Necturus,* 53
Diastema, 82
Diencephalon, 184
    in sheep, 210
    in *Squalus,* 185
Digastric muscle, 157, 158, 160
Digestive system, development of, 226-229
    in lamprey, 20
    in mammals, 247, 258-261
    in *Necturus, 241,* 242
    in *Squalus,* 229-239
Digitiform gland, 231
Digitigrade locomotion, 37, 96

Dilatator laryngis, 128
Dipnoi, 28
Disc, intervertebral, 51, 68
Discoidal placenta, 336
Distal (defined), 339
Distal centrale, 96
Distal tarsals, 96
"Distant touch," 162
Dogfish. See also *Fishes* and *Squalus.*
    adult, major veins of, *268*
    adult aortic arches in, *274*
    brain of, *185, 189*
    cranial nerves of, *189*
    embryonic aortic arches in, *274*
    head of, *110*
    injection of, *343*
    respiration of, *237*
Dolichoglossus, 1
Dorsal (defined), 339
Dorsal aorta, 264
    in lamprey, 21
    in mammals, 253, 306
    in *Necturus,* 287, 289
    in *Squalus,* 272, 273, 276, *277*
Dorsal coelomic canals, 12
Dorsal fins, in *Ammocoetes* larva, 22
    in *Amphioxus,* 7, 9, 11
    in lamprey, 15
    in *Squalus,* 31
Dorsal funiculus, 220
Dorsal hyoid constrictor muscle, 118
Dorsal interarcual muscle, 119
Dorsal intercalary plate, 42
Dorsal intermediate sulcus, 215
Dorsal lamina, 5
Dorsal lateral sulcus, 215, 220
Dorsal median sulcus, 214, 220
Dorsal mesentery, 224
    in *Squalus,* 232
Dorsal pancreas, 228
Dorsal ramus, 181, 220, 276
Dorsal ribs, 44
Dorsal root ganglion, 181, 203, 220
Dorsal skeletogenous septum, 41
Dorsalis trunci, 121
Ducts of Cuvier, 264
    in lamprey, 21
    in *Squalus,* 269
Ductuli efferentes, 315
    in *Necturus,* 323
    in *Squalus,* 317
Ductus arteriosus of Botallus, 305
Ductus choledochus, 257
Ductus deferens, 293, 315, 329
Ductus epididymis, 315, 329
Duodenum, in cat, 257, 258
    in *Necturus,* 241
    in rabbit, 257, 259
Duodenorenal ligament, 259
Duplex uterus, 334
Dura mater, 203
    in mammals, 207, 219, 220
    in *Necturus,* 203

Ear, 174-179
    auricle of, 38, 178
    in mammals, 178-179

Ear (*Continued*)
  in primitive tetrapods, 177
  in *Squalus*, 175-177
  inner, in *Ammocoetes* larva, 23
  left, in *Heptanchus maculatus*, *176*
  middle, See *Middle ear*.
Ectepicondyle, 99
Ectotympanic bone, 73
Effectors, 161
Efferent branchial arteries, in *Necturus*, *287*, 287,
    288
  in *Squalus*, 238, 272
Efferent neurons, 182, 184
Efferent renal veins, 267
  in *Necturus*, 284
  in *Squalus*, 269
Efferent spiracular artery, 273
Egg, cleidoic, 325
Elasmobranchii, 26
Elbow joint, 34
Embalming, 342
Embryo, mammalian. See *Mammalian embryo*.
Embryonic aortic arches, *274*
Endochondrium, 185
Endolymph, 174
Endolymphatic ducts, 46, 176
Endolymphatic fossa, 46
Endolymphatic pores, 32
Endoskeletal girdle, 84
  in *Amia*, 87
Endoskeleton, 40
Endostyle, in *Amphioxus*, 10, 12
  in *Molgula*, 5
Endotheliochorial placenta, 337
Entepicondylar foramen, 99
Entepicondyle, 99
Entoplastral plate, 92
Entoplastron, 92
Entotympanic bone, 73, 77
Epaxial muscles, in mammals, 132, *154*, 155
  in *Necturus*, 121
  in *Squalus*, 113
Ependymal epithelium, 216
Epibranchial arch, 49
  in *Necturus*, 61
Epibranchial muscles, 115
Epicondyles, 102
Epidermis, 11, 30
Epididymis, 315, 329
Epigastric artery, 303, 307
  in *Necturus*, 290
Epigastric vein, inferior, 295
  superior, 296, 298
Epiglottic cartilage, 251
Epiglottic hamuli, 251
Epiglottis, 246, 250
Epihyal process, 83
Epipharyngeal groove, 10, 12
Epiphyseal foramen, 47
Epiphysis, in mammals, *164*
  in *Necturus*, 205
  in *Squalus*, 184, 186
Epiplastra, 92
Epiploic foramen of Winslow, 256
Epipterygoid bone, in *Chelydra*, 64
  in primitive tetrapods, 56
Epipubic cartilage, 94
Epithalamic region, 204
Epithalamus, in sheep, 210

Epithalamus (*Continued*)
  in *Squalus*, 186
Epithelium, 230
  ependymal, 216
Epitrochlearis, 140
Epoophoron, 315
Erectile tissue, 331
Erector spinae, 132
Esophagus, in *Ammocoetes* larva, 25
  in *Amphioxus*, 10
  in lamprey, 20
  in mammals, 250, 251, 253, 255
  in *Molgula*, 5
  in *Necturus*, 240, 244
  in *Squalus*, 230, 233
Ethmoid bone, 78
Ethmoid cells, 173
Ethmoid foramen, 80
Ethmoid plate, 59, 60
Ethmoturbinal bone, 79, 173
Eustachian tube. See *Auditory tube*.
Eutheria, 29
Excretory ducts, genital ducts and, 313-315
Excretory system, 312-337
  in fishes, 315-322
  in lamprey, 22
  in mammals, 325-337
  in primitive tetrapods, 322-325
Excurrent siphon, 3
Exoccipital bones, in *Chelydra*, 64
  in *Necturus*, 59, 60
Extension, 111
Extensors, in mammals, 141
  in *Necturus*, 125, 127
External acoustic meatus, 38, 74, 174, 178
External carotid artery, in *Necturus*, 286
  in rabbit, 304
  in *Squalus*, 273
External cuneiform bone, 102
External gills, in *Necturus*, 35, 244
  in primitive tetrapods, 239
External gill slits, in *Ammocoetes* larva, 25
  in lamprey, 16, 21
  in *Squalus*, 32
External iliac arteries, 307
External iliac vein, 295
External inguinal ring, 330
External intercostals, 153
External jugular vein, in mammals, 296, 298
  in *Necturus*, 285
External nares, in *Chelydra*, 62
  in mammals, 38, 74, 172
  in *Necturus*, 35, 172
  in primitive tetrapods, 56
  in *Squalus*, 32, 171, *172*
External oblique muscle, in mammals, 130
  in *Necturus*, 121
External yolk sac, 320
Exteroceptors, 161
Extraembryonic membrane, 326
Extrinsic muscles of eye, 116
Eye, 170
  extrinsic muscles of, 116
  in lamprey, 16
  in mammals, 38, 168-170
  in *Necturus*, 35
  in *Squalus*, 32, 116, 165-168, *166*
  lateral. See *Lateral eye*.
  median, 164

Eye (*Continued*)
  parapineal, 164
  pineal. See *Pineal eye.*
Eyeball, 165
Eyelashes, 39
Eyelids, in mammals, 38, 169
  in *Squalus*, 32, 165

Facial muscles, in mammals, 129, 160
  in primitive tetrapods, 130
Facial nerve, hyomandibular branch of, 191
  in *Necturus*, 205
  in sheep, 213
  palatine branch of, 191
  superficial ophthalmic branch of, 191
Facial region, 38, 74
Facial veins, 298, 300
Falciform ligament, in mammals, 254
  in *Necturus*, 240, 242
  in *Squalus*, 232
Fallopian tube, 331
False ribs, 70
False vocal cords, 251
Falx cerebri, 207
Fascia, 111
Fascia cruris, 152
Fascia lata, 142, 147
Fauces, 249
Femoral artery, in mammals, 307, 308
  in *Necturus*, 290
  in *Squalus*, 278
Femoral vein, in mammals, 295
  in *Necturus*, 282
  in *Squalus*, 270
Femur, in cat, 102
  in *Chelydra*, 95
  in *Necturus*, 91
  in primitive tetrapod, 88
Fenestra cochlea, 174
  in mammals, 81, 178
Fenestra ovalis, in cat, 81
  in *Necturus*, 59
Fenestra vestibuli, 174, 179
  in cat, 81
Fibrous capsule, 326
Fibrous tunic, 166, 167
Fibula, in cat, 102
  in *Chelydra*, 96
  in *Necturus*, 91
  in primitive tetrapod, 88
Fifth ventricle of brain, 216
Filiform papillae, 250
Fimbria, 217
Fin, caudal. See *Caudal fin.*
  in *Amia*, 87, *87*
  in *Squalus*, 85-87
  pelvic. See *Pelvic fin.*
  ventral, 7, 9, 13
Fin ray box, 9, 11
Fishes. See also *Dogfish* and *Squalus.*
  appendicular skeleton of, 84-88
  axial muscles of, 113-116
  caudal vertebrae of, 42-43
  circulatory system of, 265-280
  coelom of, 229-239
  digestive system of, 229-239
  excretory system of, 315-322
  integumentary derivatives of, 33-34
  nervous system of, 184-203
  postcranial axial skeleton of, 41-44

Fishes (*Continued*)
  respiratory system of, 229-239
  vertebral column of, 41
Flagellated pit, 9
Flexion, 111
Flexors, in mammals, 141
  in *Necturus*, 125, 127
Flocculus, 212
Folia, 212
Foliate papillae, 250
Folliculus vesiculosus, 331
Foot pad, 39
Foramen magnum, in cat, 74
  in *Chelydra*, 64
  in *Necturus,* 58
  in *Squalus*, 46
Foramen ovale, 309
Foramen rotundum, 80
Forearm, muscles of, 141
Foregut, 227
Fornix, body of, 216
  commissure of, 218
Fossa ovalis, 309
Fossa rhomboidea, 186
Fourth head of humerus, 141
Fourth ventricle of brain, in *Necturus*, 205
  in sheep, 214
  in *Squalus*, 186
Friction ridges, 39
Frontal bones, in cat, 77
  in *Chelydra*, 63
  in *Necturus*, 58
Frontal section (defined), 340
Frontal sinus, 79, 173
Fundus, 328
Fungiform papillae, 250
Funiculus, dorsal, 220
Funiculus cuneatus, 215
Funiculus gracilis, 215
Funiculus spermaticus, 329
Fusiform shape, 31

Gall bladder, in *Ammocoetes* larva, 24
  in mammals, 255
  in *Necturus*, 243
Ganoid scales, 33
Gartner's duct, 315
Gasserian ganglion, 191
Gastralia, 92
Gastric artery, in mammals, 307
  in *Squalus*, 277
Gastric vein, in fishes, 266
  in mammals, 292
  in *Necturus*, 282
Gastrocnemius muscle, 152
Gastrocolic ligament, 256
Gastroepiploic veins, 292
Gastrohepatic artery, 277
Gastrohepatic ligament, in *Necturus,* 242
  in *Squalus*, 232
Gastrohepatoduodenal ligament, in mammals, 255
  in *Squalus*, 232
Gastrolienic ligament, 256
Gastrosplenic artery, 289
Gastrosplenic ligament, in *Necturus*, 242
  in *Squalus*, 232
Gastrosplenic vein, in *Necturus*, 282
  in rabbit, 292
Gemellus inferior, 151
Gemellus superior, 145, 151

Geniculate body, lateral, 211
Geniculate ganglion, 191
Genioglossus, in mammals, 158
  in *Necturus*, 122
Geniohyoid, in mammals, 158
  in *Necturus*, 122
Genital ducts, excretory ducts and, 313-315
Genital folds, 336
Genital papillae, 324
Genital pores, 22
Genital ridges, 3
Genital system, in amniote, *314*
  in lamprey, 22
Genu, 215
Gill(s), external. See *External gills.*
  in *Squalus*, 237
Gill bars, 10, 12
Gill filaments, 244
Gill lamellae, 25
Gill pouch, 25
Gill rakers, in *Necturus*, 244
  in *Squalus*, 237, 238
Gill rays, 49, 238
Gill slits, 227
  external. See *External gill slits.*
  in *Amphioxus*, 10, 12
  in Chordata, 1
  in *Molgula*, 5
  in *Necturus*, 35, 244
  internal. See *Internal gill slits.*
Glands. See names of glands.
Glans clitoridis, 333
Glans penis, 330
Glenoid cavity, in cat, 98
  in *Chelydra*, 92
  in *Necturus*, 90
Glenoid surface, 86
Glossopalatine arches, 249
Glossopharyngeal foramen, 46
Glossopharyngeal nerve, in *Necturus*, 205
  in sheep, 214
  in *Squalus*, 192
Glottis, 250
  in mammals, 251
  in *Necturus*, 245
Gluteus maximus, 143, 148
Gluteus medius, 143, 148
Gluteus minimus, 145, 150
Gnathostomes, 26
Gonad(s), in *Amphioxus*, 8, 9, 12
  in *Molgula*, 6
  in *Squalus*, 231
Gonadial arteries, in *Necturus*, 289
  in *Squalus*, 278
Gonadial sinuses, 269
Gonadial veins, 284
Graafian follicles, 331
Gracilis muscle, 144, 150
Gray matter of central nervous system, 180, 217, 220
Greater curvature of stomach, 233
Greater omentum, in mammals, 255
  in *Necturus*, 242
  in *Squalus*, 232
Greater trochanter, 102
Guanine crystals, 167
Gubernaculum, 329
Gular fold, in *Necturus*, 35
Gular series, 51

Gymnophiona, 28
Gyri, 210

Habenula, in *Necturus*, 205
  in sheep, 210, 217
Habenular commissure, 210, 217
Habenular region, 186
Hamate process, 100
Hamulus, 75
  epiglottic, 251
Harderian gland, 170
Hatschek's groove, 9
Hatschek's pit, 9
Hair, 36, 39
Haustra coli, 261
Head, in lamprey, 14
  in mammals, 37
  in *Necturus*, 34
  in *Squalus*, 31, *110*
  of humerus, tubercle of, 98, 138, 139
  of muscles, 111
Head skeleton, 45
  in mammals, 72-83, *75*
  in *Necturus*, 56-61
  in primitive tetrapods, 56-61
  in reptiles, 62-66
Heart, 262, 263. See also *Cor.*
  auricle of, 266, 291, 309
  in *Ammocoetes* larva, 24
  in lamprey, 21
  in mammals, 254, 291, 308-310
  in *Necturus*, 244, 281
  in *Squalus*, 235, 266, 278, *279*
  internal structure of, 308-310
  sheep, dissection of, *309*
  ventricle of. See *Ventricle of heart.*
Hemal arch, 52
  in *Squalus*, 42
Hemal canal, 42
Hemal plate, 42
Hemal processes, 69
Hemal spine, 42
Hemibranch, 237
Hemichordata, 1-3
Hemispheres of brain, 206, 212
Hemoendothelial placenta, 337
Hepar, 254. See also *Liver.*
Hepatic. See also *Liver.*
Hepatic artery, in mammals, 307
  in *Necturus*, 289
  in *Squalus*, 277
Hepatic duct, in mammals, 256
  in *Necturus*, 243
  in *Squalus*, 233
Hepatic portal system, 263
  in mammals, 292
  in *Necturus*, 282
  in *Squalus*, 266
Hepatic portal vein, 266
  in mammals, 292
  in *Necturus*, 282
Hepatic ridges, 3
Hepatic sinus, 267
  in *Necturus*, 284
Hepatic veins, 263, 267
  in lamprey, 22
  in mammals, 293
  in *Necturus*, 284
  in *Squalus*, 266

Hepatoduodenal ligament, in *Necturus*, 242
    in *Squalus*, 232
Hepatopancreatic ampulla, 259
Hepatorenal ligament, 255
Heptanchus maculatus, left ear of, *176*
Heterocercal, 32
Heterodont, 82
Hilus, 326
Hindgut, 227. See also *Intestine*.
Hip girdle, 84
Hippocampus, 217
Holobranch, 237
Holocephali, 26
Holonephros, 313
Holostei, 28
Homodont, 49, 61
Homoiothermic, 246
Hoofs, 39
Horn, of uterus, 332
    posterior, 83
    trabecular, 60
Humeroantibrachialis, 124
Humerus, in cat, 98, 99
    in *Chelydra*, 93
    in *Necturus*, 90, 125
    in primitive tetrapod, 88
Hyoglossus, 158
Hyoid apparatus, in cat, 83
    in *Chelydra*, 66
    in *Necturus*, 61
    in primitive tetrapod, 58
Hyoid arch, 48
Hyoid bone, 251
Hyoid constrictor muscle, dorsal, 118
Hyoid muscles, 160
Hyoidean efferent artery, 272
Hyoidean sinus, 270
Hyomandibular arch, 49
Hyomandibular branch of facial nerve, 191
Hyomandibular canal, 164
Hyostylic suspension of jaw, 49
Hypaxial muscles, in mammals, 130-132, 152
    in *Necturus*, 121
    in *Squalus*, 113, 230
Hypobranchial arch, 49
Hypobranchial artery, 273
Hypobranchial muscles, in mammals, 156-158
    in *Necturus*, 122
    in *Squalus*, 115
Hypobranchial nerve, 193
Hypogastric artery, 290
Hypoglossal canal, 80
Hypoglossal nerve, 207, 214, 247
Hypohyal arch, 61
Hypopharyngeal groove, in *Amphioxus*, 10
    in *Molgula*, 5
Hypophyseal pouch, in *Ammocoetes* larva, 23
    in lamprey, 19
Hypophysis, 200
    in sheep, 210
Hypoplastra, 92
Hypothalamus, in sheep, 210
    in *Squalus*, 186

Ileocolic ring, 11
Ileum, 257
Iliac arteries, 307
    in *Necturus*, 290
    in *Squalus*, 278

Iliac process, 88
Iliac vein. See *Common iliac vein*.
Iliacus, 151
Iliocostalis, 132, 155
Ilioextensorius, 127
Iliofemoralis, 127
Iliofibularis, 127
Iliolumbar veins, 293
Iliopsoas complex, 146, 151
Iliotibialis, 127
Ilium, crest of, 100
    in cat, 100
    in *Chelydra*, 94
    in primitive tetrapod, 89
Incisive bone, 76
Incisive foramen, 81
Incisors, 82
Incurrent siphon, 3
Incus, 174
    in mammals, 73, 77, 179
Inferior (defined), 339
Inferior colliculi, 212
Inferior concha, 173
Inferior epigastric artery, 307
Inferior epigastric vein, 295
Inferior jugular vein, in lamprey, 22
    in *Squalus*, 269
Inferior lobe of lacrimal gland, 170
Inferior meatus, 173
Inferior mesenteric artery, in mammals, 307
    in *Squalus*, 278
Inferior mesenteric vein, 292
Inferior nasal meatus, 79
Inferior oblique muscles, in mammals, 170
    in *Squalus*, 165, 188
Inferior rectus muscle, 165, 188
Infraorbital canal, in cat, 80
    in *Squalus*, 164
Infraorbital fenestrae, 62
Infraorbital salivary gland, 170, 248
Infraorbital trunk, 191
Infraspinatus muscle, 138
Infraspinous fossa, 98
Infundibulum, 200
    in mammals, 210, 331
    in *Necturus*, 206
Inguinal canal, 293, 330
Inguinal ligament, 130
Inguinal ring, 330
Inner ear, 23
Innominate bone, 100
Inscription, of muscles, 111
Insectivora, 29
Insertion, 111
Integument, 30
Integumentary derivatives, in mammals, 38-39
    in *Squalus*, 33
Integumentary muscles, 129
Interarcual muscle, 117, 119
Interatrial septum, in mammals, 309
    in *Necturus*, 281
Interbranchial constrictor muscle, 117, 119, 238
Interbranchial septum, 237
Intercalary plate, dorsal, 42
Intercentrum, 51
Interclavicle, in *Chelydra*, 92
    in primitive tetrapod, 89
Intercondyloid fossa, 102
Intercostal artery, 303, 306

Intercostal muscles, 153
Intercostal veins, 296, 298
Interhyoideus, in *Necturus*, 127
  in *Squalus*, 118
Intermandibularis, in *Necturus*, 127
  in *Squalus*, 118
Intermedium, 93
Internal acoustic meatus, 79, 80, 179
Internal capsule, 219
Internal carotid artery, in *Necturus*, 287
  in rabbit, 304
  in *Squalus*, 273
Internal cuneiform bone, 102
Internal gill slits, in lamprey, 21
  in *Squalus*, 237
Internal gills, 239
Internal inguinal ring, 330
Internal iliac arteries, 307
Internal iliac vein, 295
Internal intercostal muscles, 153
Internal jugular vein, in mammals, 297, 299
  in *Necturus*, 285
Internal mammary artery, 303
Internal mammary vein, 296, 298
Internal nares, in cat, 75
  in *Chelydra*, 62
  in *Necturus*, 172, 245
  in primitive tetrapods, 56
Internal oblique muscle, in mammals, 130
  in *Necturus*, 122
Internal structure of heart, 308-310
Internal thoracic artery, 303
Internal thoracic vein, 296, 298
Internal yolk sac, 320
Internuncial neurons, 183
Interoceptors, 161
Interparietal bone, 77
Interpeduncular fossa, 212
Interpterygoid vacuities, 56
Intersegmental arteries, 276
  in *Necturus*, 290
  in *Squalus*, 278
Intersegmental veins, in *Necturus*, 284
  in *Squalus*, 269
Interspinalis muscle, 121
Interthalamic adhesion, 216
Intertubercular groove, 98
Interventricular foramen of Monro, 201
  in sheep, 217
Interventricular septum, 310
Intervertebral disc, 51
  in cat, 68
Intervertebral foramina, 68
Intestinal veins, in mammals, 292
  in *Necturus*, 282
  posterior, 267
Intestine. See also *Hindgut.*
  in *Ammocoetes* larva, 25
  in *Amphioxus*, 11, 13
  in lamprey, 17, 20
  in mammals, 257
  in *Molgula*, 5
  in *Necturus*, 240, 242
  in primitive tetrapods, 240
  in *Squalus*, 230
  valvular. See *Valvular intestine.*
Intrinsic muscles of larynx, 160
Iris, in mammals, 169
  in *Squalus*, 166, 167

Irritability, 161
Ischiocaudalis, 126
Ischiocavernosus, 330
Ischiofemoralis, 126
Ischioflexorius, 126
Ischium, in cat, 100, 102
  in *Chelydra*, 94
  in *Necturus*, 90
  in primitive tetrapod, 89
Islets of Langerhans, 229
Isthmus, 251
Ileocecal valve, 259, 260

Jacobson, vomeronasal organs of, 173
Jaw, in cat, *159*
  in rabbit, *159*
  lower, in cat, 82
    in *Chelydra*, 64, *65*
    in *Necturus*, 60, *60*
Jejunum, 257
Jugal bone, in cat, 77
  in *Chelydra*, 63
Jugular foramen, 80
Jugular process, 74
  common, 285
  external. See *External jugular vein.*
  inferior. See *Inferior jugular vein.*
  internal. See *Internal jugular vein.*
  transverse, 300

Keratin, 30
Kidney. See also *Renal.*
  history of, 312-313
  in mammals, 255, 326
  in *Necturus*, 242, 322
  in *Squalus*, 231, 316
  opisthonephic. See *Opisthonephric kidneys.*
  pronephric, 24
  succession of, in amniote embryo, *313*
Knee, in *Necturus*, 34

Labia, 37
Labia majora, 333
Labial cartilage, 49
Labial fold, 32
Labial pockets, 32
Labyrinthodonts, 28
Lacerate foramen, middle, 81
  posterior, 80
Lacrimal bone, 76, 169
Lacrimal gland, 169, 170
Lacrimal puncta, 169
Lacteals, 311
Lagena, 177
Lagomorpha, 36
Lamina, dorsal, 5
Lamina terminalis, 216
Lamprey, 14-25
  circulatory system in, 21
  coelom in, 20
  cross section of, 17-22, *19*
  digestive system in, 20
  excretory system in, 22
  external features in, 14-22
  genital system in, 22
  muscular system in, 18
  nervous system in, 18-20
  pineal eye of, *164*

Lamprey (*Continued*)
  respiratory system in, 20
  sagittal section of, 17-22, *17*
  sense organs in, 18-20
  skeletal system in, 18
Langerhans, islets of, 229
Laryngeopharynx, 250
Laryngotracheal chamber, 245
Larynx, in mammals, 246, 250, 251
  intrinsic muscles of, 160
Lateral (defined), 339
Lateral abdominal veins, 264
  in *Squalus*, 270
Lateral cartilages, 245
Lateral condyle, 102
Lateral eye, 164
  in *Ammocoetes* larva, 23
Lateral funiculus, 220
Lateral geniculate body, 211
Lateral interarcuals, 119
Lateral ligament, 258
Lateral line canals, 162
  in *Squalus*, 163
Lateral line system, 162-164
  in lamprey, 16
  in *Necturus*, 35
  in *Squalus*, 33
Lateral lingual swellings, 246
Lateral malleolus, 102
Lateral rectus muscle, 165, 190
Lateral styloid process, 99
Lateral teeth, 61
Lateral thigh muscles, 142, 147
Lateral vein, 285
Lateral visceral arteries, 276
Lateralis neurons, 162
Latissimus dorsi, in mammals, 138, 222
  in *Necturus*, 124
Lens, crystalline, 166
Lentiform nucleus, 219
Lepidotrichia, 34
Lesser curvature of stomach, 233
Lesser omentum, in mammals, 255, *257*
  in *Squalus*, 232
Lesser peritoneal cavity, 256
Lesser trochanter, 102
Levator hyomandibulae, 118
Levator mandibulae, 128
Levator palatoquadrati, 118
Levator palpebrae superioris, 170
Levator scapulae, in mammals, 138, 152
  in *Necturus*, 124
Levatores arcuum, 128
Leydig's gland, 317
Lien, 256. See also *Spleen*.
Lienalis, 307
Lienogastric artery, 277
Lienogastric vein, posterior, 267
Lienomesenteric vein, 266
Ligamentum arteriosum, 302, 305
Ligamentum hepatocavopulmonale, 242
Ligamentum teres hepatis, 255
Ligamentum teres uteri, 332
Linea alba, in mammals, 129
  in *Squalus*, 113
Lingua, 249
Lingual cartilage, 18
Lingual fenulum, 248
Lingual process, 66

Lingual swellings, lateral, 246
Lingual vein, 285
Lip, in *Ammocoetes* larva, 23
  in mammals, 37
  in *Necturus*, 35
Liver, 228. See also *Hepar* and *Hepatic*.
  in *Ammocoetes* larva, 24
  in lamprey, 20
  in mammals, 254
  in *Necturus*, 240
  in *Squalus*, 230, *231*
Longissimus muscle, 132, 155
Longitudinal bars, 5
Longitudinal bundles, 113
Longitudinal cerebral fissure, 209
Longus colli muscle, 154
Lorenzini, ampullae of. See *Ampullae of Lorenzini*.
Lower chordates, 1-13
Lumbar arteries, 307
Lumbar region, 37
Lumbar veins, 293
Lumbar vertebrae, 66, 68
Lumbosacral plexus, 221
Lung, 228. See also *Pulmonary*.
  in mammals, 252
  in *Necturus*, 240
  root of, 252
Lymph, 262
Lymph nodes, 247, 262
Lymph space, 12
Lymphatic capillaries, 262
Lymphatic duct, right, 311
Lymphatic system, 262, 310
Lymphatic vessels, 262
  deeper, in rat, *311*
  in mammals, 297, 299
  injection of, for study, 343
Lymphocytes, 262

Maculae, 177
Magnum, 100
Malleolus, 102
Malleus, 174
  in mammals, 73, 77, 178
Mamillary bodies, 211
Mamillary process, 69
Mammal(s). See also *Cat* and *Rabbit*.
  anterior trunk muscles in, 152-156
  appendicular skeleton of, 96-102
  branchiomeric muscles of, 158-160
  circulatory system of, 290-312
  coelom of, 246-261
  digestive organs of, 246-252
  ear of, 178-179
  excretory system of, 325-337
  external features of, 36-38
  eye of, 168-170
  head skeleton of, 72-83
  hypobranchial muscles of, 156-158
  iliopsoas complex of, 146
  integumentary derivatives of, 38-39
  muscular system of, 128-160
  nervous system of, 206-223
  nose of, 172-173
  pectoral muscles of, 132-141
  postcranial axial skeleton of, 66-72
  reproductive system of, 325-337
  respiratory organs of, 246-261

Mammalia, 29
Mammalian aortic arches, *305*
Mammalian embryo, diaphragm of, *226*
  mesenteries, of, *256*
  sagittal section, *227*
Mammary artery, internal, 303
Mammary glands, 38, 129
Man, development of brain in, *183*
Mandible, 82
Mandibular arch, 48
Mandibular canal, 164
Mandibular foramen, 82
Mandibular fossa, 74
Mandibular muscles, 158
Mandibular symphysis, 82
Mantle, 5
Manubrium, 70
Manus, in mammals, 37, 99, *100*
  in *Necturus*, 34
  in primitive tetrapod, 88
Marsupialia, 29
Massa intermedia, 216
Masseter, 158
Mastoid process, 74
Maxilla, in cat, 76
  in *Chelydra*, 63
Maxilloturbinal, 79, 173
Meckel's cartilage, in *Necturus*, 61
  in *Squalus*, 49
Medial (defined), 339
Medial condyles, 102
Medial cutaneous nerve, 223
Medial geniculate body, 211
Medial malleolus, 102
Medial rectus muscle, 165, 188
Medial styloid process, 99
Medial teeth, 61
Median (defined), 339
Median eye, 164
Median ligament of bladder, in mammals, 258
  in *Necturus*, 242, 322
Median nerve, 223
Median nostril, in *Ammocoetes* larva, 23
  in lamprey, 15
Median sacral artery, 308
Median vertebral vein, 300
Mediastinal septum, 253
Mediastinum, 252
Medulla, of kidney, 327
Medulla oblongata, in *Necturus*, 205
  in sheep, 214
  in *Squalus*, 186
Medulla spinalis, 219
Medullary velum, 214
Membrane bone, 40
Membranous labyrinth, 174
  in *Squalus*, 176
Meninges, 207-209
Meninx primitiva, 185
Mental foramina, 82
Mesencephalon, 184
  in *Ammocoetes* larva, 23
  in sheep, 212
  tectum of, in *Squalus*, 186
Mesenteric artery, 289
  inferior. See *Inferior mesenteric artery*.
  superior. See *Superior mesenteric artery*.
Mesenteric veins, 292
  in *Necturus*, 282

Mesentery, dorsal. See *Dorsal mesentery*.
  in lamprey, 22
  in mammalian embryo, *256*
  in mammals, 254, 258
  in *Necturus*, 242
  in *Squalus*, 232
  ventral, 224
Mesethmoid bone, 72, 78, 79
  perpendicular plate of, 173
Mesocolon, in mammals, 258
  in *Necturus*, 242
  in *Squalus*, 232
Mesoduodenum, 258
Mesogaster. See *Greater omentum*.
Mesometrium, 332
Mesonephric duct, 312
Mesonephros, 312
Mesopterygium, 87
Mesorchium, in *Necturus*, 242, 323
  in *Squalus*, 232, 317
Mesosalpinx, 331
Mesotarsal joint, 96
Mesotubarium, in *Necturus*, 242
  in *Squalus*, 232, 319
Mesovarium, in mammals, 331
  in *Necturus*, 242, 324
  in *Squalus*, 232, 319
Metacarpals, in cat, 100
  in *Chelydra*, 93
  in *Necturus*, 90
  in primitive tetrapod, 88
Metacromion, 98
Metanephroi, 326
Metanephros, 312
Metapleural folds, 7, 9, 12
Metapophysis, 69
Metapterygium, 87, 88
Metatarsals, in cat, 102
  in *Chelydra*, 96
  in *Necturus*, 91
  in primitive tetrapod, 88
Metatheria, 29
Metencephalon, 212
Middle concha, 173
Middle cranial fossa, 79
Middle ear, evolution of, *175*
  in *Chelydra*, 63
Middle ear cavity, 177
Middle lacerate foramen, 81
Middle meatus, of nose, 173
Middle peduncle, 212
Midgut, 11
Midgut cecum, 11, 12
Moderator bands, 310
Molar gland, 82, 248
Molgula, 3-6, *4*
  dissection of, 4-6
  external features of, 3
Monotremata, 29
Monro, interventricular foramen of. See *Interventricular foramen of Monro*.
Motor column, 182
Motor neurons, 182
Mouth. See also *Oral cavity*.
  in mammals, 37
  in *Necturus*, 35
  Squalus, 32
Mouth opening, in *Ammocoetes* larva, 23
  in lamprey, 15

Müllerian duct, 314
Multifidus muscle, 132, 155
Muscle slips, 111
Muscles. See also names of muscles.
  action of, 111
  anterior trunk, 152-156
  appendicular, 116-117
  axial. See *Axial muscles*.
  branchiomeric. See *Branchiomeric muscles*.
  epaxial. See *Epaxial muscles*.
  facial. See *Facial muscles*.
  groups of, 103-110
  hypaxial. See *Hypaxial muscles*.
  hypobranchial. See *Hypobranchial muscles*.
  integumentary, 129
  of expression, 129
  origin of, 111
  parietal, 103
  pelvic, in mammals, 141-152
    in *Necturus*, 125-127
  posterior trunk, 130
  somatic, 103
  study of, 112
  terminology of, 110
Muscular system, 103-160
  in lamprey, 18
  in mammals, 128-160
  in primitive tetrapods, 119-128
Musculi linguae, 158
Musculocutaneous nerve, 223
Myelencephalon, 184
  in sheep, 214
  in *Squalus*, 186
Mylohyoid, 158
Myocommata, in *Amphioxus*, 7
  in lamprey, 18
  in *Squalus*, 113
Myomeres, in *Amphioxus*, 7, 9, 11
  in lamprey, 17
  in *Necturus*, 121
  in *Squalus*, 113-115
Myosepta, in *Amphioxus*, 7, 11
  in *Necturus*, 121
  in *Squalus*, 41, 113
Myotomes, 23

Nails, 39
Nares. See also *Nose*.
  external. See *External nares*.
  internal. See *Internal nares*.
Nasal bones, 77
Nasal capsules, 45, 47
Nasal cavities, 172
Nasal meatus, inferior, 79
Nasal passage, 58
Nasal region, 47
Nasal septum, 173
Nasolacrimal canal, 81
Nasopalatine duct, 173, 249
Nasopharynx, 250
Nasoturbinal bones, 79, 173
Navicular bone, 102
Neck, in mammals, 37
  in *Necturus*, 34
  muscles of, in cat, *135*
    in rabbit, *136*
  of rib, 70

Necturus, 34
  appendicular skeleton of, 90
  arterial system of, 285-290
  digestive organs of, *241*, 242
  external features of, 34-36
  head skeleton of, 56-61
  injection of, for study, 343
  integumentary derivatives of, 36
  pectoral girdle of, 90
  pectoral muscles of, 123-125
  pelvic appendage of, 90
  pelvic muscles of, 125-127, *127*
  pleuroperitoneal cavity of, 240-243
  urogenital system of, 323, 324
  venous system in, 282-285, *283*
  visceral organs of, 240
Neopallium, 206, 210
Neoteny, 34
Nephric ridge, 312
Nephron, 312
Nephrostome, 314
Nerve(s). See also names of nerves.
  cranial. See *Cranial nerves* and names of individual nerves.
  spinal. See *Spinal nerves*.
Nerve cord, in *Amphioxus*, 9, 11
  in Chordata, 1
Nerve strand, 3
Nervous system, 180-223, *181*
  autonomic, 182
  central, 180
  divisions of, 180-182
  functional components of, 182-184
  in fishes, 184-203
  in lamprey, 18
  in mammals, 206-223
  in primitive tetrapods, 203-206
  peripheral, 181
  somatic, 182
  visceral, 182
Nervus terminalis, 187
Neural arch, 53
  in cat, 68
  in *Squalus*, 42
Neural gland, 6
Neural plate, 42
Neural spine, in cat, 68
  in *Necturus*, 53
  in *Squalus*, 42
Neurocoel, 180
  in *Amphioxus*, 11
Neuromasts, 162
Neurons, afferent, 182
  defined, 161
  efferent, 182
  internuncial, 183
  lateralis, 162
  motor, 182
  preganglionic, 182
  sensory, 182
  somatic, 182
  visceral, 182
Nictitating membrane, 169
  in mammals, 38
Nidamental gland, 319
Nodulus, 212
Nondeciduous placenta, 336
Nose, 171-173. See also *Nares*.
  in mammals, 38, 172-173

Nose (*Continued*)
  in primitive tetrapods, 172
  in *Squalus*, 171-173
Nostril, median. See *Median nostril*.
Notochord, in *Ammocoetes* larva, 23
  in *Amphioxus*, 9, 11
  in Chordata, 1
  in lamprey, 18
Notochordal tissue, 42
Nuchal crest, 75
Nucleus gracilis, tubercle of, 215
Nucleus (nuclei), cuneate, tubercle of, 215
  lentiform, 219
  of brainstem, 180
  red, 206

Oblique muscles, in mammals, 130
  in *Necturus*, 121, 122
  in *Squalus*, 165, 189, 190
  inferior. See *Inferior oblique muscles*.
  superior. See *Superior oblique muscles*.
Obturator externus, 151
Obturator foramen, in cat, 100
  in *Necturus*, 90
Obturator internus, 143, 146, 151
Occipital arch, 45, 60
Occipital artery, 305
Occipital bone, 77
Occipital condyles, in cat, 74
  in *Chelydra*, 64
  in *Necturus*, 59
  in *Squalus*, 46
Occipital nerves, in *Necturus*, 205
  in *Squalus*, 193
Occipital region, 46
Oculi, 38
Oculomotor nerve, in *Necturus*, 205
  in sheep, 212
  in *Squalus*, 188
Odontoid process, 70
Olecranon, 99
Olecranon fossa, 99
Olfactory bulb, in *Necturus*, 204
  in sheep, 210
  in *Squalus*, 171, 185
Olfactory foramina, 80
Olfactory lamellae, in *Necturus*, 172
  in *Squalus*, 171
Olfactory nerves, in cat, 173
  in *Necturus*, 204, 205
  in sheep, 210
  in *Squalus*, 188
Olfactory passage, 172
Olfactory sac, in lamprey, 19
  in *Squalus*, 32, 171
Olfactory tract, in sheep, 210
  in *Squalus*, 171, 185
Olive, 215
Omental bursa, 255
Omentum, greater. See *Greater omentum*.
  lesser. See *Lesser omentum*.
Omoarcuals, 122, 124
Opercular series, 51
Operculum, in *Amia*, 87
  in *Necturus*, 59
  in primitive tetrapod, 178
Ophthalmic foramina, superficial, 47
Ophthalmic trunk, superficial, 184

Opisthonephric, 322
Opisthonephric kidneys, in lamprey, 22
  in *Squalus*, 316
Opisthonephros, 312
Opisthotic bone, in *Chelydra*, 64
  in *Necturus*, 59, 60
Optic canal, 80
Optic chiasma, 200
  in *Necturus*, 206
  in sheep, 210
Optic foramen, 47
Optic lobes, 186
Optic nerve, in *Necturus*, 205
  in sheep, 210
  in *Squalus*, 166, 188
Optic pedicel, 166
Optic region, 46
Optic tract, 200
  in sheep, 211
Oral cavity, 226
  in mammals, 249
  in *Necturus*, 244, *245*
  in *Squalus*, 235, 236
Oral glands, 239
Oral hood, 7, 9, 12
Oral tentacles, 23
Orbicularis oculi, 169
Orbital fissure, 80
Orbital plate, 45
Orbital process, 49
Orbital sinus, 269
Orbits, in cat, 74
  in *Chelydra*, 62
  in *Necturus*, 58
  in *Squalus*, 45, 46, 165
Orbitosphenoid bone, 78
Origin of muscles, 111
Ornithorhynchus, 29
Oropharynx, 250
Os, 37
Os coxae, 100
Osseous labyrinth, 174
Ossicles, auditory, *175*
Osteichthyes, 28
Ostium, in mammals, 331
  in *Necturus*, 324
  in *Squalus*, 317, 319
Ostracoderms, 26
Otic capsule, in *Necturus*, 60
  in primitive tetrapod, 177
  in *Squalus*, 46
Otic notch, 56
Otolith, 177
Oval window, 174
  in mammals, 179
  in *Necturus*, 56
  in primitive tetrapods, 56, 178
Ovarian arteries, in mammals, 307
  in *Squalus*, 278
Ovarian veins, in mammals, 293
  in *Necturus*, 284
  in *Squalus*, 269
Ovaries, in lamprey, 22
  in mammals, 293, 331
  in *Necturus*, 242, 324
  in *Squalus*, 231, 319
Oviducts, 314
  in *Necturus*, 242
  in *Squalus*, 317, 319

Oviparous, 321
Ovulation, 319

Palatal series, 50
Palatine bones, in cat, 78
 in *Chelydra*, 64
Palatine branch of facial nerve, 191
Palatine canal, posterior, 81
Palatine tonsils, 250
Palatoquadrate cartilage, 49
Paleopallium, 206
 in sheep, 210
Palpebrae, 38, 169
Palpebral fissure, 169
Pancreas, dorsal, 228
 in lamprey, 20
 in mammals, 259
 in *Necturus*, 241
 in *Squalus*, 230, *231*
 of Aselli, 311
 ventral, 228
Pancreatic duct, in mammals, 259
 in *Salamandra*, *243*
 in *Squalus*, 235
Pancreaticoduodenal artery, 289
Pancreaticoduodenal vein, posterior, 292
Pancreaticomesenteric artery, 277
Pancreaticomesenteric vein, 266
Panniculus carnosus, 129
Pantotheria, 29
Papilla(e), in lamprey, 15
 mammae, 38
 of esophagus, 233
 urinary, 31
 urogenital. See *Urogenital papilla.*
Papillary muscles, 310
Parabranchial chamber, 237
Parachordals, 45
 in *Necturus*, 60
Paradidymis, 315
Parahippocampal gyrus, 210
Paraoccipital process, 74
Paraphysis, in *Necturus*, 204
 in *Squalus*, 186
Parapineal eye, 164
Parapophysis, 52, 53
Parasagittal section, 340
Parasphenoid bone, in *Amia*, 51
 in *Chelydra*, 64
 in *Necturus*, 59
Parathyroid glands, 227, 251
Parietal arteries. See *Intersegmental arteries.*
Parietal bones, in cat, 77
 in *Chelydra*, 63
 in *Necturus*, 58
Parietal muscles, 103
Parietal pericardium, in mammals, 254
 in *Necturus*, 244
 in *Squalus*, 235
Parietal peritoneum, in mammals, 132, 254
 in *Necturus*, 240
 in *Squalus*, 230
Parietal pleura, 252
Parietal veins. See *Intersegmental veins.*
Paroophoron, 315
Parotid duct, 247
Parotid gland, 247
Passeriformes, 29
Patella, 102

Pectineus, 144, 150
Pectoantibrachialis, 133
Pectoral appendage, muscles of, in *Squalus*, *116*
 in cat, 98-100
 in *Chelydra*, 91-94, *93*
Pectoral bones, *97*
Pectoral fin, *85*
 in *Squalus*, 31
 muscles of, 116
Pectoral girdle, 84
 in *Amia*, *87*
 in cat, 98-100
 in *Chelydra*, 91-94, *91*
 in *Necturus*, 90
 in *Squalus*, 85-87, *86*
Pectoral limb bud, *264*
Pectoral muscles, in cat, 133, *133*, *135*
 in mammals, 132-141
 in *Necturus*, 123-125
 in rabbit, 134, *136*
Pectoralis major, 134
Pectoralis minor, 134
Pectoralis primus, 134
Pectoriscapularis, 123, 124
Pedicle, 68
Peduncle, middle, 212
 posterior, 213
Pelvic appendage, in cat, 100-102
 in *Chelydra*, 94-96, *96*
 in *Necturus*, 90
Pelvic bones, *101*
Pelvic canal, in cat, 102
 in *Necturus*, 91
Pelvic fins, 31
 muscles of, 117
Pelvic girdle, 84
 in *Amia*, 87
 in cat, 100-102
 in *Chelydra*, 94-96, *95*
 in *Necturus*, 90
 in rabbit, 148, 150
Pelvic gland, 324
Pelvic muscles, of cat, 141-146, *142, 145*
 in *Necturus*, 125-127, *127*
 in rabbit, 146-152, *147*
Pelvic veins, 282
Pelvis, renal, 327
Penis, 38, 330
 crura of, 330
Perforated substance, 212
Pericardial artery, 274
Pericardial cavity, 224
 in lamprey, 20
 in mammals, 254
 in *Necturus*, 243
 in *Squalus*, 235
Pericardioperitoneal canal, 235, 279, 280
Pericardiopleural membranes, 226
Pericardium, parietal. See *Parietal pericardium.*
 visceral. See *Visceral pericardium.*
Perilymph, 174
Perilymphatic ducts, 46
Perineum, 38
Periotic bone, 77
Peripharyngeal bands, in *Amphioxus*, 10
 in *Molgula*, 5
Peripheral nervous system, 181
Peritoneal cavity, 224, 246, 254-261
Peritoneum, parietal. See *Parietal peritoneum.*

Peritoneum, visceral. See *Visceral peritoneum*.
Perivisceral cavity, 229
Perycosauria, 29
Pes, in mammals, 37, *100*, 102
   in *Necturus*, 34
   in primitive tetrapod, 88
Petromyzon marinus, 15, 16
Petrosal ganglion, 192
Petrous bone, 77
Phalangeal formula, 88
Phalanges, in cat, 100, 102
   in *Chelydra*, 93, 96
   in *Necturus*, 90, 91
   in primitive tetrapod, 88
Pharyngeal artery, 305
Pharyngeal pouch, 227
Pharyngobranchial arch, 49
Pharyngoesophageal artery, 272
Pharynx, in *Ammocoetes* larva, 23
   in *Amphioxus*, 10, 12
   in mammals, 249, 250
   in *Molgula*, 5
   in *Necturus*, 244, *245*
   in *Squalus*, 235, 236, *236*
Phrenic arteries, 306
Phrenic nerves, 253
Phrenic veins, 293
Pia mater, 207, 220
Pia-arachnoid mater, 203
Pig embryo, pectoral limb bud of, *264*
Pigment spot, 9
Pineal body, 210, 217
Pineal eye, 164
   in *Ammocoetes* larva, 23
   in lamprey, 15, *164*
Pinna, 38, 174, 178
Piriform area, 210
Piriformis, 145, 150
Pisiform bone, in cat, 100
   in *Chelydra*, 94
Pit organs, 162
Pituitary gland, 200
   in sheep, 210
Placenta, 326, 336, 337
   deciduous, 336
   discoidal, 336
   endotheliochorial, 337
   hemoendothelial, 337
   nondeciduous, 336
   zonary, 336
Placodermi, 26
Placoid scales, 33, *33*
Planes of body, *339*, 340
Plantigrade locomotion, 37, 96
Plastron, 92, *92*
Platysma, in mammals, 129, 160
   in primitive tetrapods, 120
Pleura, 252
Pleural cavity, 246, 252
Pleurapophysis, 52
   in cat, 68
Pleurocentra, 51
Pleuroperitoneal cavity, 224
   in lamprey, 20
   in *Necturus*, 240-243
   in *Squalus*, 229-235
Pleuroperitoneal membranes, 226
Plexus, 221
   posterior chorioid. See *Posterior chorioid plexus*.

Plica, 243
Plica semilunaris, 169
Plica vocalis, 251
Polypterus, 28
Pons, 206, 213
Popliteal fossa, 142, 146
Popliteus, 126
Portal system, renal. See *Renal portal system*.
Portal veins, 263, 267
Postaxial border, 35
Postcleithrum, 87
Postcranial axial skeleton, 41-44, 51-56, 66-72
Posterior (defined), 339
Posterior branchiomeric muscles, 160
Posterior cardinal sinus, 269
Posterior cardinal veins, 264
   in lamprey, 21
   in *Necturus*, 284, 285
   in *Squalus*, 269
Posterior cervical arteries, *297*
Posterior cervical veins, *297*
Posterior chamber, 167
Posterior chorioid plexus, in *Necturus*, 205
   in *Squalus*, 186
Posterior commissure, 217
Posterior coracoid, 89
Posterior cranial fossa, 79
Posterior facial vein, 298, 300
Posterior horns, 83
Posterior intestinal vein, 267
Posterior lacerate foramen, 80
Posterior lienogastric vein, 267
Posterior palatine canal, 81
Posterior pancreaticoduodenal vein, 292
Posterior peduncle, 213
Posterior perforated substance, 212
Posterior systemic veins, in mammals, 292
   in *Necturus*, 284
Posterior trapezius muscles, 137
Posterior trunk muscles, 130
Posterior utriculus, 177
Posterior vena cava, in mammals, 252, 291, 292, 308
   in *Necturus*, 284
   in primitive tetrapods, 281
Posterior ventrolateral artery, 278
Posterior zygapophyses, in cat, 68
   in *Necturus*, 53
Posteromedial thigh muscles, 144, 149
Postfrontal base, 64
Posthepatic arteries, *294*
Posthepatic veins, *294*
Posthyoid muscles, in mammals, 156
   in *Squalus*, 115
Postorbital bone, 63
Postorbital process, in cat, 74
   in *Squalus*, 46
Posttemporal arch, 87
Posttemporal fenestrae, in *Chelydra*, 62
   in primitive tetrapods, 56
Prearticular bone, 66
Preaxial border, 35
Precerebral cavity, 47
Precerebral fenestra, 47
Prefrontal bone, 63
Preganglionic neurons, 182
Prehyoid muscles, in mammals, 157
   in *Squalus*, 115
Premaxilla, in cat, 76

Premaxilla (*Continued*)
  in *Chelydra*, 63
  in *Necturus*, 58
Premolars, 82
Preorbitalis, 118
Preparation of specimens, 341-344
Prepuce, 330
Preserving of specimens, 341
Presphenoid bone, 78
Primary lamellae, 237
Primary palate, 64
Primates, 29
Primitive tetrapods, appendicular muscles of, 123-127
  appendicular skeleton of, 88-96
  circulatory system of, 280-290
  coelom of, 239-245
  digestive system of, 239-245
  excretory system of, 322-325
  external features of, 34-36
  head skeleton of, 56-61
  integumentary derivatives of, 36
  muscular system of, 119-128
  nervous system of, 203-206
  postcranial axial skeleton of, 51-56
  reproductive system of, 322-325
Processus vaginalis, 329
Procoracohumeralis, 124
Procoracoid process, in *Necturus*, 90
  in primitive tetrapod, 88
Proctodeum, 226
Profundus nerve, 190
Pronation, 112
Pronephric duct, 312
Pronephric kidney, 24
Pronephros, 312
Prootic bones, in *Chelydra*, 64
  in *Necturus*, 59
Prootic foramen, 64
Prootic ossification, 60
Proprioceptors, 161
Propterygium, 87, 88
Prosencephalon, 184
  in *Ammocoetes* larva, 23
Prostate, 330
Prostatic utricle, 336
Prototheria, 29
Protraction, 111
Proximal (defined), 339
Pseudobranch, 32, 238
Psoas major, 151
Psoas minor, 132, 151
Pterygiophore cartilages, 44
  in *Squalus*, 87
Pterygoid bones, in *Chelydra*, 64
  in *Necturus*, 59
Pterygoid canal, 81
Pterygoid fossa, 75
Pterygoid process, 75
Pterygoquadrate cartilage, 49
Ptyalin, 246
Pubic cartilage, 90
Pubis, in cat, 102
  in *Chelydra*, 94
  in primitive tetrapod, 89
Puboischiadic bar, 88
Puboischiadic fenestra, 95
Puboischiadic plate, 90
Puboischiofemoralis externus, 126

Puboischiofemoralis internus, 126
Puboischiotibialis, 126
Pubotibialis, 125
Pudendum femininium, 333
Pulmo, 252
Pulmonary. See also *Lung*.
Pulmonary arteries, in mammals, 302
  in *Necturus*, 288
Pulmonary circulation, 280
Pulmonary ligament, in mammals, 252
  in *Necturus*, 242
Pulmonary pleura, 252
Pulmonary trunk, 291, 302
Pulmonary valve, 310
Pulmonary veins, in mammals, 292, 309
  in *Necturus*, 288
Pulp cavity, 33
Pulvinar, 211
Pupil, in mammals, 169
  in *Squalus*, 166, 167
Pyloric region, in mammals, 255
  in *Necturus*, 242
  in *Squalus*, 233
Pyloric sphincter, in mammals, 255
  in *Necturus*, 243
Pyramids, 215
  renal, 327

Quadrate bone, in *Amia*, 50
  in *Chelydra*, 64
  in *Necturus*, 59
  in primitive tetrapods, 56
Quadratojugal bone, 63
Quadratus lumborum, 132, 152
Quadriceps femoris, 144, 146, 148, 149, 151

Rabbit. See also *Mammals*.
  abdominal viscera of, 254, *260*
  anterior systemic veins of, 298-302
  brachial plexus of, 221-223
  digestive organs of, 247, 259-261
  ear of, 178-179
  excretory system of, 326-328
  female reproductive system of, 331-336
  injection of, 343
  internal structure of heart in, 308-310
  male reproductive system of, 328-331
  pectoral muscles of, 134, *136*
  pelvic muscles of, 146-152, *147*
  peritoneal cavity of, 254-261
  spinal cord of, 219-223
  thorax of, 252-254
  venous system in, 292-295
Raccoon, carotid circulation in, *304*
Radial cartilage, 44
  in *Squalus*, 87, 88
Radial notch, 99
Radial tuberosity, 99
Radiale, 93
Radices of aorta, 287
Radius, in cat, 99
  in *Chelydra*, 93
  in *Necturus*, 90
  in primitive tetrapod, 88
Raja, valvular intestine of, *234*
Ramus, communicating, 181
  dorsal, 181, 276
  of dentary, 82
  ventral. See *Ventral ramus*.

Raphe, 111
Rat, deeper lymphatic vessels of, *311*
Receptors, 161
Rectal gland, 231
Rectovesical pouch, 258
Rectum, 246, 258
Rectus abdominis, in mammals, 132, 153
  in *Necturus*, 121
Rectus cervicis, in *Necturus*, 122
  in *Squalus*, 115
Rectus femoris, 144
  deep head of, 149
  superficial head of, 149
Rectus muscles, in mammals, 170
  in *Squalus*, 165, 188, 190
Red nucleus, 206
Renal. See also *Kidney*.
Renal arteries, in mammals, 307, 326
  in *Necturus*, 289
  in *Squalus*, 278
Renal collar, 300
Renal papilla, 327
Renal pelvis, 327
Renal portal system, in *Necturus*, 283
  in *Squalus*, 267
Renal portal vein, 283
Renal pyramid, 327
Renal sinus, 326
Renal tubule, 312
Renal veins, 267, 293, 326
Renal vesicle, 6
Renes. See *Kidney*.
Reproduction, in mammals, 336
  in *Squalus*, 320
Reproductive system, 312-337
  female, in cat, 331-336
    in rabbit, 331-336
  in primitive tetrapods, 322-325
  male, in cat, 328-331
    in rabbit, 328-331
  study of, 315
Reptile, head skeleton of, 62-66
  skull of, *62*
Reptilia, 28
Respiratory system, development of, 226-229
  in fishes, 229-239
  in lamprey, 20
  in *Necturus*, 244
  in primitive tetrapods, 239-245
Respiratory tube, 20
Restiform body, 213
Rete cords, 315
Rete testis, 315
Retina, 167
Retraction, 111
Retractor bulbi, in mammals, 170
  in *Necturus*, 205
Retroperitoneal (defined), 316
Rhachitomous vertebrae, 51
Rhinal fissure, 210
Rhombencephalon, 184
  in *Ammocoetes* larva, 23
Rhomboideus, 140, 152
Rhomboideus capitis, 152
Rhomboideus major, 140
Rhomboideus minor, 140
Rhynchocephalia, 28
Ribs, dorsal, 44
  false, 70

Ribs (*Continued*)
  in *Necturus*, 53
  true, 70
  ventral, 44
  vertebral, 70
  vertebrocostal, 70
Rostral fenestrae, 47
Rostrum, 46, 47
Rotunda, 81
Round ligament, 255, 332
Round window, 174
  in *Chelydra*, 64
  in mammals, 178
  in primitive tetrapod, 178
Rugae of stomach, 233
  in cat, 255
  in *Necturus*, 242

Saccoglossus, 1, *2*
Sacculus rotundus, 259
Saccus vasculosus, 200
  in *Necturus*, 206
Sacral artery, median, 308
Sacral vertebrae, 52
  in cat, 67, 69
Sacrospinalis, 132, 155
Sacrum, 67
Sagittal crest, 75
Sagittal section, 340
Salamandra, *243*
Salivary glands, 246, 247
Sarcopterygii, 28
Sartorius muscle, 142, 149
Scalenus muscle, 153
Scapholunar bone, 99
Scapula, in cat, 98
  in *Chelydra*, 92
  in *Necturus*, 90
  in primitive tetrapod, 88
Scapular artery, transverse, 303
Scapular deltoid, 124
Scapular process, 86
Scapular vein, transverse. See *Suprascapular vein*.
Scapulocoracoid cartilage, in primitive tetrapod, 88
  in *Squalus*, 85
Sciatic nerve, 145, 150
Sclera, 167
Sclerotic bones, 167
Scrotum, 38, 328
Sebaceous glands, 38
Secondary palate, in mammals, 73, 75, 249
  in *Chelydra*, 64
Selachii, 26
  visceral arch of, *118*
Sella turcica, 200
  in cat, 79
Semicircular canals, 46, 176
Semilunar ganglion, 191
Semilunar notch, 99
Semilunar valves, in mammals, 310
  in *Squalus*, 279
Semimembranosus, 144
Semimembranosus accessorius, 150
Semimembranosus proprius, 150
Seminal vesicle, in rabbit, 330
  in *Squalus*, 317
Semispinalis capitis, 155
Semispinalis cervicis, 155

Semitendinosus, 143, 150
Sense organs, 161-179
  in lamprey, 18
Sensory column, 182, 187
Sensory neurons, 182
Sensory organs, 161
Septum, dorsal skeletogenous, 41
  horizontal skeletogenous, 41, 113
  interatrial. See *Interatrial septum.*
  interbranchial, 237
  interventricular, 310
  mediastinal, 253
  nasal, 173
  pellucidum, 215
  transverse. See *Transverse septum.*
Serratus, 124
Serratus dorsalis muscle, 153
Serratus ventralis muscle, 140, 152
Sesamoid bone, 94
Sex glands, accessory, 325
Sexually indifferent stage, 313
Shank, in primitive tetrapod, 88
  muscles of, 146, 152
Sheep brain, *209*
  dissection of, 217-219, *218, 219*
  external features of, 209-215
  meninges of, 207-209
  sagittal section of, 215-217, *216*
Shoulder, deeper muscles of, 138-140
  superficial muscles of, 137
Shoulder girdle, 84
Sinuatrial aperture, 278
Sinuatrial valve, 278
Sinus, air, 79
  cardinal, anterior, 193
  coronary, 291
  frontal, 79, 173
  hepatic. See *Hepatic sinus.*
  hyoidean, 270
  orbital, 269
  renal, 326
  urinary, 320
  urogenital. See *Urogenital sinus.*
Sinus venosus, 266
  in lamprey, 21
  in *Necturus,* 281
  in *Squalus,* 278
Siphon, excurrent, 3
  in *Squalus,* 319
  incurrent, 3
Skeletogenous septum, horizontal, 113
Skeleton, appendicular. See *Appendicular skeleton.*
  axial. See *Axial skeleton.*
  dermal, 40
  of head. See *Head skeleton.*
  preparation of, 344
  visceral, 40-83
Skinning, 129
Skull, composition of, in cat, 76-78
    in *Chelydra,* 63
  foramina of, 80-81
  general features of, in cat, 74-75
    in *Chelydra,* 62
  of *Amia, 50*
  of Necturus, 58, *58*
  of primitive land vertebrate, *57*
  of reptile, *62*
  sagittal section of, 79
Soft palate, 75, 249

Somatic motor column, 182
  in *Squalus,* 187
Somatic motor neurons, 182
Somatic muscles, 103
Somatic nervous system, 182
Somatic sensory column, 182
  in *Squalus,* 187
Somatic sensory neurons, 182
Somatic sensory organs, 161
Sperm sac, 317
Spermatheca, 325
Spermatic cord, 329
Spermatophores, 324
Sphenodon, 28
Sphenoid bone, 78
Sphenoid sinus, 79, 173
Sphenopalatine foramen, 81
Sphincter colli, 127
Spinal cord, 180
  in *Ammocoetes* larva, 23
  in lamprey, 18
  in mammals, 219-223
  in *Squalus,* 203
Spinal ganglion, 220
Spinal nerves, 181
  dorsal roots of, 181, 203
    in *Amphioxus,* 11
  in mammals, 219-223
  ventral roots of, 181
    in *Amphioxus,* 11
    in fishes, 203
Spinalis, 155
Spine, neural. See *Neural spine.*
Spinodeltoid, 138
Spinotrapezius, 137
Spinous process, 68
Spiracle, 227
  in *Squalus,* 32, 237
Spiracular artery, 273
Spiracular valve, 32, 238
Spiracularis, 118
Spiral valve, in lamprey, 20
  in rabbit, 260
  in *Squalus,* 233
Splanchnic nerves, 306
Spleen, 262. See also *Lien.*
  in mammals, 256
  in *Necturus,* 240
  in *Squalus,* 230
Splenial bone, 61
Splenic artery, 289
Splenic veins, in mammals, 292
  in *Necturus,* 282
Splenium, 215
Splenius, 155
Squalus, 48. See also *Dogfish* and *Fishes.*
  appendicular muscles of, 116-117
  arterial system in, 270
  caudal vertebrae of, 42-43
  chondrocranium of, 45, 47
  cranial nerves of, 187-200
  digestive organs of, 232-235
  dorsal surface of brain of, 184-187
  ear of, 175-177
  external features, 31-33
  eye of, 165-168
  female urogenital system in, 319
  fins of, 85-87, *86*
  head skeleton of, 45

Squalus (*Continued*)
  hepatic portal system in, 266
  male urogenital system in, 316-319
  myomeres of, 113-115
  nose of, 171-173
  pectoral girdle of, 85-87
  pelvic fin muscles in, 117
  pleuroperitoneal cavity of, 229-235
  respiratory organs of, 235
  venous system of, 266-270
  ventral surface of brain in, 200-201
  ventricles of brain in, 201-202
  vertebral regions of, 42-43
  visceral organs of, 230
Squamata, 28
Squamous bone, in cat, 77
  in *Chelydra*, 64
  in *Necturus*, 59
Stapedial artery, 273
Stapedius, 179
Stapes, 174
  in *Chelydra*, 64
  in mammals, 77, 179
  in *Necturus*, 59
  in primitive tetrapod, 57
Statoacoustic nerve, 192
  in *Necturus*, 205
  in sheep, 214
  in *Squalus*, 177
Sternocleidomastoid muscles, 135, 160
Sternohyoid, 156
Sternomastoid, 136
Sternothyroid, 156
Sternum, 52
  in cat, 70
  in *Necturus*, 53
Stomach, greater curvature of, 233
  in mammals, 255
  in *Molgula*, 5
  in *Necturus*, 240
  in *Squalus*, 230, *231*
  lesser curvature of, 233
Stomochord, 3
Stomodeum, 226
Stratum basale, 30
Stratum corneum, 30
Styloglossus, 158
Stylohyal process, 83
Stylohyoid, 160
Styloid process, lateral, 99
Stylomastoid foramen, 80
Subarachnoid space, 208
Subarcuals, 128
Subcardinal veins, 268
Subclavian artery, in mammals, 302
  in *Necturus*, 289
  in *Squalus*, 276
Subclavian veins, in mammals, 296, 298
  in *Necturus*, 285
  in *Squalus*, 270
Subcoracoscapularis, 124
Subdural space, 220
Subendostylar coelom, 12
Sublingual gland, 248
Submandibular duct, 247
Submandibular gland, 247
Submaxillary gland, 247
Subpharyngeal gland, 25
Subscapular fossa, 98

Subscapular vein, 270
Subscapularis, 140
Substantia alba, 217
Substantia grisea, 217
Subsupracardinal anastomoses, 300
Subtemporal fenestrae, in *Chelydra*, 62
  in primitive tetrapods, 56
Subvertebralis, 122
Sudoriferous glands, 38
Sulcus (sulci), collateral, 210
  coronary, 308
  dorsal intermediate, 215
  dorsal median, 214, 220
  dorsal lateral, 215, 220
Sulcus limitans, 187
Superficial constrictor muscle, 117, 119, 238
Superficial fascia, 111
Superficial head of rectus femoris, 149
Superficial ophthalmic branch of facial nerve, 191
Superficial ophthalmic branch of trigeminal nerve, 190
Superficial ophthalmic foramina, 47
Superficial ophthalmic trunk, 184
Superior (defined), 339
Superior colliculi, 212
Superior concha, 173
Superior epigastric artery, 303
Superior epigastric vein, 296, 298
Superior lobe of brain, 200
Superior mesenteric artery, in mammals, 306
  in *Squalus*, 278
Superior mesenteric vein, 292
Superior oblique muscles, in mammals, 170
  in *Squalus*, 165, 190
Superior rectus muscle, 165, 188
Superior thyroid artery, 304
Supination, 112
Supracardinals, 300
Suprachoroidea, 167
Supracleithrum, 87
Supracoracoideus, 124
Supraoccipital bone, 64
Supraorbital canal, 164
Supraorbital crest, 46
Suprarenal glands. See *Adrenal glands*.
Suprarenolumbar arteries, 307
Suprarenolumbar veins, 293
Suprascapular artery, 303
Suprascapular cartilage, in *Necturus*, 90
  in *Squalus*, 85
Suprascapular nerve, 222
Suprascapular vein, 297, 300
Supraspinatus, 138
Supraspinous fossa, 98
Supratemporal canal, 164
Surangular, 66
Sweat glands, 38
Swim bladder, 228
Sylvius, cerebral aqueduct of. See *Cerebral aqueduct of Sylvius*.
Sympathetic ganglia, 306
Sympathetic subdivision of autonomic nervous system, 182
Synapsid condition, in cat, 72
  in reptiles, 29
Synotic tectum, 45, 60
Systemic veins, 263
  in *Squalus*, 268
  posterior. See *Posterior systemic veins*.

Tachyglossus, 29
Taeniae coli, 260
Tail, in mammals, 37
  in *Necturus*, 34
  in *Squalus*, 31, 113-115
Talus, 102
Tapetum lucidum, 167
Tarsal bones, in cat, 102
  in *Chelydra*, 96
  in *Necturus*, 91
  in primitive tetrapods, 88
Tarsus, in cat, 102
  in *Necturus*, 34
Teats, 38
Tectum of brain, in sheep, 212
  in *Squalus*, 186
Teeth, in cat, 82
  in lamprey, 15
  in *Necturus*, 61
  in primitive tetrapod, 58
  lateral, 61
  medial, 61
Tela chorioidea, in sheep, 214
  in *Squalus*, 186
Telencephalon, 184
  in sheep, 209
Teleostei, 28
Temporal fenestra, 72, 74
Temporal fossa, 74
Temporal line, 75
Temporalis, 158, 159
Tendon, 111
Tensor fasciae latae, 142, 147
Tentacles, in *Ammocoetes* larva, 23
  in *Amphioxus*, 10
Tentorium, in cat, 79
  in sheep, 207
Tenuissimus, 143, 148
Teres major, 139, 222
Teres minor, 139
Terrestrial vertebrate, preparation of, 341
Testes, in lamprey, 22
  in *Necturus*, 242, 323
  in *Squalus*, 231, 316
Testicular arteries, in mammals, 307
  in *Squalus*, 278
Testicular veins, in mammals, 293
  in *Necturus*, 284
  in *Squalus*, 269
Tetrapoda, 26
Tetrapods, primitive. See *Primitive tetrapods*.
Thalamus, in sheep, 211
  in *Squalus*, 186
Thecodont, 82
Therapida, 29
Theria, 29
Thigh, in mammals, 37
  in primitive tetrapod, 88
  muscles of, 142, *145*, 147, *149*
Third ventricle of brain, in sheep, 210, 216
  in *Squalus*, 186
Thoracic artery, 303
  in cat, *297*
Thoracic duct, 310
Thoracic nerves, 222
Thoracic veins, 296
Thoracic vertebrae, 66, 67
Thoracics, ventral, 222
Thoraciscapularis, 124

Thoracolumbar fascia, 130
Thorax, 37, 252-254, *253*
Thymus, 227, 254
Thyreohyoid, 157
Thyrocervical trunk, 303
Thyrohyal bone, 83
Thyroid artery, 304
Thyroid cartilage, 251
Thyroid gland, 228
  in mammals, 251
  in *Necturus*, 286
  in *Squalus*, 115, 271
Tibia, crest of, 102
  in cat, 102
  in *Chelydra*, 96
  in *Necturus*, 91
  in primitive tetrapod, 88
Tissue fluid, 262
Tongue, in lamprey, 15
  in mammals, 249
  in primitive tetrapods, 239
  primary, in *Squalus*, 236
Tonsillar fossa, 227, 250
Tonsils, palatine, 250
Trabeculae, 45
  in *Necturus*, 60
Trabeculae carneae, 310
Trabecular horns, 60
Trachea, 246, 250, 251
Transverse colon, 259
Transverse fibers of pons, 213
Transverse foramen, 69
Transverse jugular vein, 300
Transverse process, in cat, 68
  in *Necturus*, 53
Transverse scapular artery, 303
Transverse scapular vein. See *Suprascapular vein*.
Transverse section, 340
Transverse septum, 224
  in lamprey, 20
  in *Necturus*, 244
  in *Squalus*, 235
Transverse sheet of muscle, 12
Transversi ventrales, 128
Transversus, 122
Transversus abdominis, 131
Transversus costarum, 153
Transversus thoracis, 153
Trapezium, 100
Trapezius muscles, 135, 137, 160
Trapezoid body, 213
Trapezoid process, 100
Triceps, 125
Triceps brachii, 140
Tricuspid valve, 309
Trigeminal nerve, in *Necturus*, 205
  in sheep, 213
  lingual branch of, 247
  mandibular branch of, 191
  maxillary branch of, 191
  superficial ophthalmic branch of, 190
Trigemino-facial foramen, 47
Trilaminar yolk sac, 320
Triquetrum, 99
Trochanter, 102
Trochanteric fossa, 102
Trochlea, 98, 170
Trochlear nerve, in *Necturus*, 205
  in sheep, 212

Trochlear nerve (Continued)
  in Squalus, 185, 190
Trochlear notch, 99
True ribs, 70
Trunk, in lamprey, 14
  in mammals, 37
  in Necturus, 34
  in Squalus, 31
  muscles of. See Trunk muscles.
Trunk muscles, in Necturus, 121
  in Squalus, 114
  posterior, 130
Trunk myomeres, 113-115
Trunk vertebrae, 52, 53
Tuber cinereum, 211
Tubercle of cuneate nucleus, 215
Tubercle of humerus, 138, 139
Tubercle of nucleus gracilis, 215
Tuberculum, 51
  in Necturus, 53
Tuberculum impar, 246
Tuberosity, of ischium, 102
  of tibia, 102
Tunic, 3
Tunica vaginalis, 329
Turbinal bones, 72, 79
Turtle. See also Chelydra.
  appendicular skeleton of, 91-96
Tympanic bone, 73, 77
Tympanic bulla, 74
Tympanic cavity, 174, 227
  in mammals, 178
  in Chelydra, 63
  in primitive tetrapod, 177
Tympanohyal process, 83
Tympanum, 174
  in mammals, 38, 178
  in primitive tetrapods, 177

Ulna, in cat, 99
  in Chelydra, 93
  in Necturus, 90
  in primitive tetrapod, 88
Ulnar nerve, 223
Ulnare, 93
Ultimobranchial bodies, 227
Umbilical arteries, 264
Umbilical cord, 337
Uncinate process, 100
Ungual phalanx, 100
Unguligrade locomotion, 37, 96
Ungues, 39
Ureter, 312, 326
Urethra, 328
Urinary bladder, 229
  in mammals, 258, 328
  in Necturus, 242, 322
  in primitive tetrapods, 240
  in Squalus, 318
Urinary duct, 317
Urinary papilla, 31, 320
Urinary sinus, 320
Urochordata, 1, 3-6
Urodela, 28
Urodeum, in mammals, 335
  in Squalus, 320
Urogenital papilla, in lamprey, 17
  in Squalus, 317

Urogenital sinus, in lamprey, 22
  in mammals, 333
  in Squalus, 317
Urogenital system, female, in cat, 332
  in Necturus, 324
  in Squalus, 319
  male, in Necturus, 323
  in rabbit, 327
  in Squalus, 316-319
Urogenital union, cords of, 315
Urorectal fold, 335
Uterine tube, 331
Uterine villi, 321
Uterus, 258, 332, 333
  cervix of, 333, 334
  duplex, 334
  in Squalus, 319
Utriculi, 177

Vagina, 333
Vagosympathetic trunk, 303
Vagus foramen, 46
Vagus nerve, in mammals, 214, 303
  in Necturus, 205
  in Squalus, 192
Valvular intestine, in Raja, 234
  in Squalus, 230
Vas deferens, 315
Vasa efferentia. See Ductuli efferentes.
Vascular tunic, 167
Vastus intermedius, 144, 148
Vastus lateralis, 144, 148
Vastus medialis, 144, 149
Veins, 262. See also names of veins.
Velum, in Amphioxus, 9
  in lamprey, 21
Velum transversum, 186
Vena cava, 291, 308
  posterior. See Posterior vena cava.
Venae lienalis, 292
Venous system, in mammals, 292-295
  in Necturus, 282-285, 283
  in Squalus, 266-270
Ventral (defined), 339
Ventral abdominal vein, 282
Ventral aorta, 263
  in lamprey, 21
  in Squalus, 270
Ventral fin, 7, 9, 13
Ventral hyoid constrictor, 118
Ventral intercalary plates, 44
Ventral median fissure, 215, 220
Ventral mesentery, 224
Ventral pancreas, 228
Ventral ramus, 181
  in fish, 278
  in mammals, 220
Ventral ribs, 44
Ventral skeletogenous septum, 41
Ventral thoracics, 222
Ventral visceral arteries, 276
Ventricle(s) of brain, 180
  in Squalus, 201-202
  fifth, 216
  fourth. See Fourth ventricle of brain.
  third. See Third ventricle of brain.
  of heart, 266
  in lamprey, 21

Ventricle of heart (*Continued*)
   in mammals, 291
   in *Necturus*, 281
Ventriculus, 255
Ventrolateral artery, posterior, 278
Vermiform appendix, 260
Vermis, 212
Vertebrae, caudal, 69
   cervical, 69
   lumbar, 68
   sacral, 69
   thoracic, 67
Vertebral arch, 68
Vertebral artery, in mammals, 303
   in *Necturus*, 287
Vertebral canal, in cat, 68
   in *Necturus*, 53
   in *Squalus*, 42
Vertebral regions, 52
Vertebral ribs, 70
Vertebral skeleton, 66
Vertebral veins, 296, 298, 300
Vertebrata, 1, 26
Vertebrates, evolution of, 26-30
   external anatomy of, 30-39
   phylogenetic tree of, 27
   primitive, 14-25
Vertebrocostal ribs, 70
Vesica fellea, 255
Vesica urinaria, 328
Vesical artery, 308
Vesical vein, in mammals, 295
   in *Necturus*, 282
Vesicouterine pouch, 258
Vestibule, 179, 249, 333
Vestibulocochlear nerve, in sheep, 214
   in *Squalus*, 177
Vibrissae, 39
Villi, 246, 258
Viscera, abdominal, 254, *260*
Visceral arch, in selachian, *118*
   in *Squalus*, 48, 238
Visceral arteries, lateral, 276
Visceral motor column, 182
   in *Squalus*, 187
Visceral motor neurons, 182
Visceral nervous system, 182
Visceral organs, coelom and, *225*

Visceral organs (*Continued*)
   in *Necturus*, 240
   in *Squalus*, 230
Visceral pericardium, in mammals, 254
   in *Necturus*, 244
Visceral peritoneum, in mammals, 254
   in *Necturus*, 240
   in *Squalus*, 230
Visceral pleura, 252
Visceral sensory column, 182
   in *Squalus*, 187
Visceral sensory neurons, 182
Visceral sensory organs, 161
Visceral skeleton, 40-83
Vitelline veins, 263
Viteodentine, 33
Vitreous body, chamber of, 167
Viviparous, 321
Vocal cords, 251
Vomer bones, in cat, 78, 173
   in *Chelydra*, 64
   in *Necturus*, 59
Vomeronasal organs of Jacobson, 173
Vulva, 38, 333

Wheel organ, 9
White matter, 180, 217, 220
Winslow, epiploic foramen of, 256
Wolffian duct, 312
Wrist. See *Carpus.*

Xiphihumeralis, 134
Xiphiplastra, 92
Xiphisternum, 70

Yolk sac, 229
   in mammals, 325
   in *Squalus*, 320, *321*
Yolk stalk, 229

Zonary placenta, 336
Zonule, 167
Zygapophyses, posterior. See *Posterior zygapo-physes.*
Zygomatic arch, 73, 74, 77